Calvin Cutter

New Analytic Anatomy, Physiology and Hygiene

Calvin Cutter

New Analytic Anatomy, Physiology and Hygiene

ISBN/EAN: 9783337365349

Printed in Europe, USA, Canada, Australia, Japan

Cover: Foto ©berggeist007 / pixelio.de

More available books at **www.hansebooks.com**

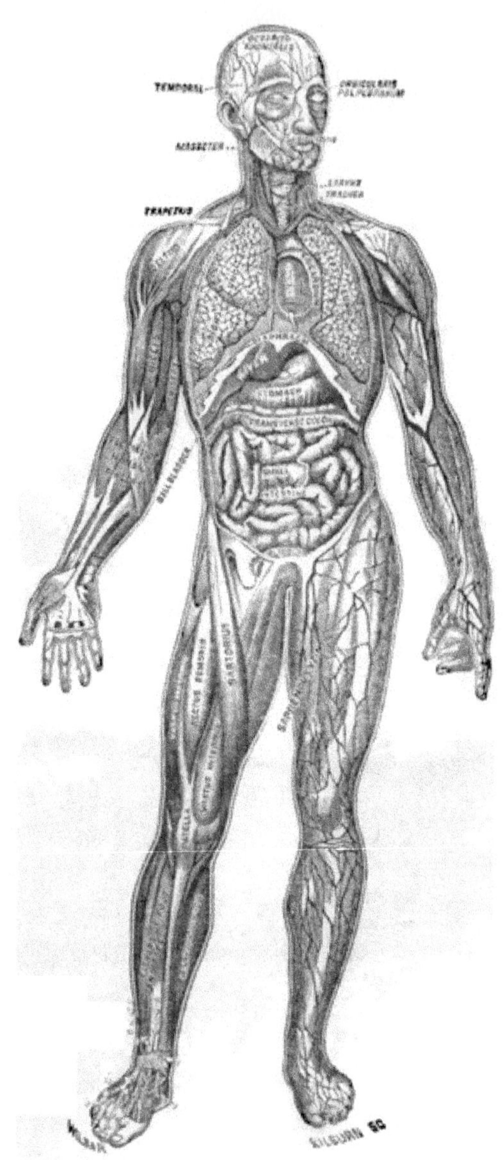

New Analytic Anatomy, Physiology and Hygiene.

NEW ANALYTIC

ANATOMY, PHYSIOLOGY AND HYGIENE

HUMAN AND COMPARATIVE.

FOR COLLEGES, ACADEMIES AND FAMILIES.

WITH QUESTIONS.

BY CALVIN CUTTER, M.D.

WITH NUMEROUS ENGRAVINGS.

PHILADELPHIA
J. B. LIPPINCOTT & CO.
1873.

LIPPINCOTT'S PRESS
PHILADELPHIA

TO MY AFFECTIONATE WIFE,

EUNICE P. CUTTER,

WHO UNTIRINGLY AIDED ME IN THE PREPARATION OF MY FORMER WORK IN 1849, AND IN ITS REVISION IN 1852,

AND TO OUR TWO SONS,

JOHN CLARENCE CUTTER AND WALTER POWERS CUTTER,

TRUSTING THAT THEY WILL SEE THAT ALL TIMELY REVISION AND EMENDATION ARE MADE IN FUTURE, TO MEET THE DEMANDS OF INCREASING INTELLIGENCE,

THIS TREATISE IS DEDICATED BY

HUSBAND, FATHER AND AUTHOR.

PREFACE.

The solicitation of my publishers, and the request of many teachers, have induced me to review and remodel my school-book on Outline Anatomy, Physiology and Hygiene, adapting it to the advanced position of teachers, schools and the community. My former work was published in 1849, and thoroughly revised in 1852. Several hundred thousand copies of the revised edition have been published. It has been translated and published in five different languages, by the missionaries in Asia and Europe.

In general arrangement, the present Treatise is modeled after the former. The aim has been to improve the analysis; to bring the Chemistry and Histology to the present advanced state of these sciences; to make the Anatomy and Physiology concise and definite, the Hygiene plain and practical; to introduce some Comparative Anatomy; and to furnish illustrating cuts, both apposite and artistic.

I am under great obligation to Joseph Leidy, M.D., of Philadelphia, who kindly permitted the use of his original illustrating cuts from his very valuable work upon Human Anatomy; also to the works of Marshall and Owen, and other scientific men, whose writings have been quoted and opinions adopted.

As my physical frame is much enfeebled from wounds received while surgeon in the volunteer army, I am under special obligations to Miss Ada L. Howard, Principal of Ivy Hall, Bridgeton, N. J., not only for the ready pen, but for much detailed investigation, for simplifying the abstruse and erudite statements of our strictly scientific works, and for bringing into close contact, relationship and harmony—in a word, *unifying*—what, without skillful combination, would be isolated and fragmentary.

To the educational men and women, to all desiring knowledge of themselves, physically, intellectually and morally, this small volume is respectfully submitted.

CALVIN CUTTER.

WARREN, MASS., July, 1870.

TO TEACHERS.

ALLOW me to suggest that the method of study and instruction of this work should be Analytical, with Synthetical Reviews; that the Headings of the several chapters may be used as TOPICS; that each subject should be thoroughly considered, viewed in its relations to other subjects, and, if possible, investigated beyond the limits of this elementary work; that the Chemistry and Histology should receive due attention, as the underlying basis of the Anatomy, Physiology and Hygiene; that, as far as possible, the subject should be made an *object* study—the Chemistry, by simple experiments, the Anatomy, by examinations of parts of domestic animals; also, that Outline Anatomical Diagrams or Charts are as desirable as a map in History or Geography; that, in case of limited time or other necessity, the Comparative Anatomy, Histology and Chemistry, one or all, may be omitted (though with great loss to the pupil), and the remaining sections will be well adapted to each other.

ANALYSIS OF CONTENTS.

DIVISION I.

CHAPTER I.—GENERAL REMARKS.

CHAPTER II.—GENERAL HISTOLOGY.

CHAPTER III.—GENERAL CHEMISTRY.

DIVISION II.

MOTORY APPARATUS.

CHAPTER IV.—THE BONES.

CHAPTER V.—THE MUSCLES.

DIVISION III.

NUTRITIVE APPARATUS.

CHAPTER VI.—THE DIGESTIVE ORGANS.

CHAPTER VII.—ABSORPTION.

CHAPTER VIII.—THE CIRCULATION.

CHAPTER IX.—ASSIMILATION.

CHAPTER X.—THE RESPIRATORY AND VOCAL ORGANS.

DIVISION IV.

SENSORIAL APPARATUS.

CHAPTER XI.—NERVOUS SYSTEM.

CHAPTER XII.—THE ORGANS OF SPECIAL SENSE.

APPENDIX.

CHAPTER XIII.

For Treatment of Wounds, see ¶ 363. For Recovery of Drowned Persons, see ¶ 430. For Treatment of Burns, see ¶ 610. For Treatment of Frost-Bite, see ¶ 612.

ANATOMY, PHYSIOLOGY AND HYGIENE.

DIVISION I.

CHAPTER I.

GENERAL REMARKS.

§ 1. THE THREE KINGDOMS OF NATURE COMPARED.—*Essential distinctions between the Mineral, Vegetable and Animal Kingdoms.—Nature of the Life-force.—Vitalized and Non-Vitalized Bodies compared.—Plants and Animals compared.*

1. "LAPIDES CRESCUNT; VEGETABILIA CRESCUNT ET VIVUNT; ANIMALIA CRESCUNT, VIVUNT ET SENTIUNT,"* was the Linnæan distinction between the three great kingdoms of Nature. Though imperfect, it is still suggestive of the boundaries of each division. The *Mineral* kingdom includes all things naturally destitute of life; the *Vegetable* kingdom, all organizations having a *certain type of life*, but no power to feel or to will; the *Animal* kingdom, those possessing a *higher* type of life and the powers of *sensation* and *voluntary* motion.

2. INORGANIC, or MINERAL bodies are made up of atoms combined and arranged according to certain mechanical and chemical laws. ORGANIC, or VEGETABLE and ANIMAL bodies are combinations of like atoms, according to the same laws controlled by Vitality or the Life-force. Plants have a

* "Stones grow; Plants grow and live; Animals grow, live and feel."

vegetable vitality—animals an *animal vitality*. Of the real character of this life-force we know nothing. Nature works in her inner laboratory with "No admittance" upon her door. We are at liberty to examine her products, but the mighty principle upon which they are wrought she holds fast as a secret unrevealable to us with our present limitations.

3. Among the *Distinctions* between ORGANIZED, or VITALIZED, and UNORGANIZED, or NON-VITALIZED bodies, are the following: An *Organized* body consists of an assemblage of parts called organs, having a mutual relation to, and dependence upon, each other; these taken together constitute an *individual*, a *being;* therefore the parts when separated are incomplete, as is seen in a divided plant. Not so with the *Unorganized* body: each fragment of a rock possesses all the essential characteristics of the original mass. *Organized* bodies, being subject to constant waste from vitalized activities, demand nourishment; *Unorganized* bodies, being permanent in their nature, require no food. *Organized* bodies grow by means of particles of matter conveyed to their interior and there assimilated; *Unorganized* bodies increase in size by simple layers upon the exterior: the former have a limit in size; the latter have no natural limit. *Organized* bodies have their period of duration: decay and death await every living animal and vegetable; but, from the nature of the *Inorganic* world, we speak of the mountains as everlasting. *Organized* bodies have their particles arranged in lines generally more or less curved, with varying angles, as in animals and plants; *Unorganized* bodies have their lines straight, with angles mathematically exact, as in the crystal of common salt. *Organized* bodies *reproduce* themselves, each species after its own kind; *Unorganized* bodies have no such power of reproduction.

4. The *Distinctions* between ANIMALS and PLANTS are important. *Animals* take in oxygen and give out carbonic acid gas; *Plants* take in carbonic acid gas and give out oxygen. *Animals* subsist upon the products of the animal and vegetable kingdoms; *Plants*, upon those of the mineral

kingdom. *Animals*, possess the power of sensation and voluntary motion; *Plants*, neither.

5. These distinctions are obvious and definite in the higher grades; but in the descending scale we recognize a gradual approach of plants and animals to each other, and likewise to the mineral kingdom; so that, in the lower forms of life, all perceptible traces of organization disappear, and, like converging radii, the three kingdoms of Nature blend in one common centre.

§ 2. Definition of Terms.

6. An organized body consists of parts called *Or'gans*. A collection of organs so arranged that their combined actions shall produce a given result is called an *Appara'tus*. The definite, peculiar use of an organ or apparatus is called its *Function: Example.*—The digestive apparatus consists of the organs—teeth, stomach, liver, etc.—whose combined functions result in the digestion of food.

The description of the form and position of these organs is called ANAT'OMY;* the description of their functions, PHYSIOL'OGY;† the examination of the conditions most favorable to their health, HY'GIENE.‡

7. The organs are composed of a variety of structures, called *Tissues*, which are themselves composed of *Cells*. The description of the form, color, constituents and origin of these tissues and cells, or their minute anatomy, is called HISTOL'OGY;§ the science which treats of their ultimate elements is called CHEM'ISTRY.‖

* Gr., *ana*, through, and *tomē*, a cutting.

† Gr., *phusis*, nature, and *logos*, a discourse.

‡ Gr., *hugieinon*, health.

§ Gr., *histos*, a web, and *logos*, a discourse.

‖ Ar., *kimia*, hidden art.

CHAPTER II.

GENERAL HISTOLOGY.

§ 3. CELLS.—*Unity of Plan exhibited in Plants and Animals.—Simple Cells.—Adaptation to Different Offices.—Modes of Multiplication.*

8. WHEREVER we find the work of the INFINITE, there we find *Unity of Plan.* Whatever the extent of the applications of this plan, whatever its modifications, there is still more or less apparent the distinct *central idea.* Amid the seemingly great diversity of substances in plants and animals, there appears a beautiful and remarkable exhibit of this *Unity.*

9. PROTOPLASM* is the formal basis of all living bodies. *Animal Protoplasm,* or *Blastema,*† as it is often called, is an albuminous fluid, generally regarded as identical with the *liquor sanguinis,* or fluid portion of the blood, in which the red corpuscles are suspended. Floating in this protoplasm are numerous minute spheroidal cells, and an infinitude of smaller bodies having the appearance of dots called granules. From this *organizable fluid* every part of living beings is formed; here is *Unity of Substance.*

10. The simple *Nucleated cell* is the earliest *organic form* of every living thing, and increase of size is but an increase of the number of cells. There are sundry very low animals, each of which is structurally a nucleated cell, a colorless blood-corpuscle leading an independent life; a step higher come those which are little more than aggregations of similar cells; and at length, as the vital functions become more and more differentiated, appear those with cells variously modified, forming increasingly well-defined and complicated organs, till they seem to reach perfection in man.

---

* Gr., *protos,* first, and *plasma,* formed. † Gr., *blastos,* a germ.

11. In the plant-world we find the same plan pursued; under the microscope, the vegetable and the animal cell *appear* essentially the same, but they are by no means identical. In examining the *nucle'oli* of animal cells, *little circular bodies* dart across the field of view. These seem to possess the power of voluntary movement; and, had we the requisite refinement of sight, we should doubtless be able to classify even these minute bodies as accurately as we now do the fully-developed animal. In the vegetable cell these are never seen.

FIG. 1.

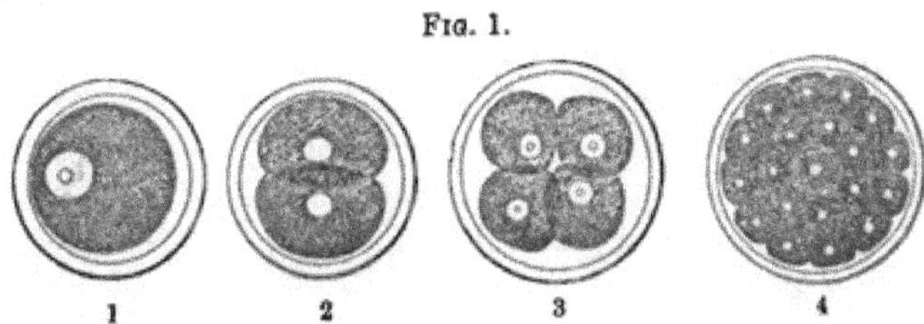

FIG. 1 (*Leidy*). AN IDEAL CELL.—1, Cell with its wall, protoplasm, nucleus and its nucleolus. 2, The same divided into two. 3, The same divided into four cells. 4, The same divided into many cells. The dark portion, the protoplasm; the white spot, the nucleus; the inner small circle, the nucleolus. Magnified.

12. It appears, then, that the lowest and the highest organism—the fungus and man—have, in their earliest development, a *unity of form* of which the type is the simple cell.

13. A SIMPLE CELL consists of a delicate sac containing protoplasm, in which is another very minute sac, called the *nu'cleus*, which contains yet another sac—the *nucleolus*, or little nucleus. Very minute particles, or granules, are also seen. A good example of a simple animal cell, on a large scale, is an egg: the lining of the shell is the cell-wall or sac; the white is the contained *protoplasm;* the yolk is the *nucleus;* and its germ-spot is the *nucleolus.*

14. Cells in the course of their development are subject to numberless modifications—the animal cell, to subserve various purposes in the animal economy; the vegetable cell, in the vegetable economy. As if under the immediate control of intelligence, they select each its own appropriate substance,

rejecting all else. One set of cells has for its office the production of motion; another set is for the purpose of secretion; another, for assimilation; another, for absorption; still another, for reproduction; and so on, through all the dissimilar offices of the animal economy.

15. Cells vary in size and shape; the normal form is probably spheroidal, as in cells of fat; but they often become many-sided, sometimes flattened, as in the cuticle, and sometimes elongated into a simple filament, as in fibrous tissue or muscular fibre.

FIG. 2.

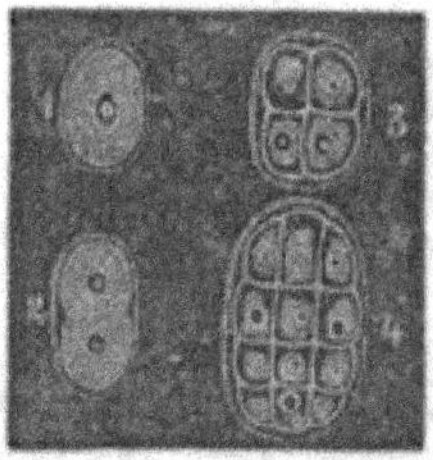

FIG. 3.

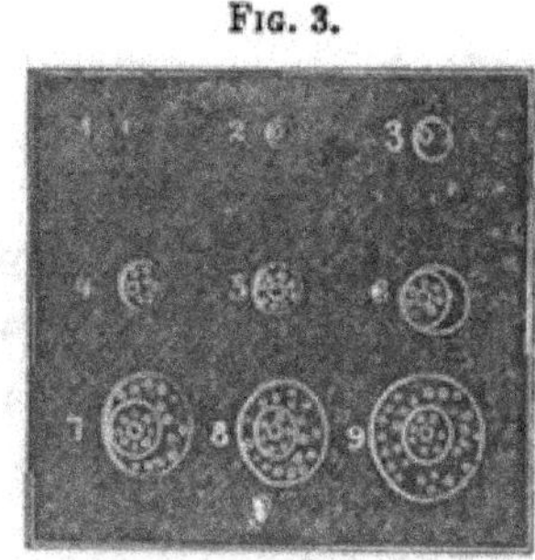

FIG. 2 (*Leidy*). PROCESS OF MULTIPLICATION, OF CARTILAGE CELLS.—1, Simple cartilage cell from the embryo. 2, Increase of cartilage cells by division of the primary cell. 3, 4, Groups of cartilage cells, from an adult articular cartilage. Magnified.

FIG. 3 (*Leidy*). PROCESS OF DEVELOPMENT OF AN ORGANIC CELL FROM A GRANULE.—1, A granule. 2, A vesicle developed upon the granule; the two constituting the nucleus and contained nucleolus. 3, The same, increased in size. 4, 5, Granules developed in the contained liquid of the nucleus. 6, The cell-wall developed on the nucleus. 7, 8, 9, Successive increase of the cell, and development of granular contents. Magnified.

16. Cells multiply in three ways: 1st, A cell may elongate, contracting in the middle like an hour-glass or dumb-bell, by the infolding of the cell-wall, till a complete division is made and two cells are formed, each with its own share of the original nucleus; the new cells divide in a similar manner, and like divisions are repeated indefinitely; 2d, Another form of multiplication is by the division of the nucleus *within the cell;* each part appropriates a portion of the fluid, and at length vesicles are formed, the old cell-wall breaks, and the vesicles develop into perfect cells; and 3d, Cells are sometimes developed *de novo* from the protoplasm, which contains nuclei and granules.

17. Cells have their period of growth, of perfection and of decay. While the vital force directs and controls the chemical and mechanical agencies, they tend to preserve and build up the system; but when the vital powers yield, they tend to its decay, and, "as if they were the grave-diggers of Nature, fulfill the old motto—'Earth to earth and dust to dust.'"

§ 4. PRIMARY TISSUES.—*Fibrous Tissue.—Areolar.—Cartilaginous.—Adipose.—Sclerous.—Muscular.—Tubular.—Nervous.*

18. By the various aggregations and transformations of cells the different tissues of the body are formed, and their individual characters depend upon the peculiar selecting power of these cells.

19. The PRIMARY TISSUES are reducible to the following: the *Fi'brous*, the *Are'olar* and the *Cartilag'inous*, which, collectively, form the *Connect'ive* tissues; and the *Ad'ipose*, the *Scle'rous*, the *Mus'cular*, the *Tu'bular* and the *Ner'vous* tissues.

20. The object of the CONNECTIVE tissues seems to be, mainly, that of binding together organs and their parts. It has few nerves and blood-vessels, and is, therefore, except when inflamed, nearly insensible, and attended with little hemorrhage under surgical operation.

21. The FIBROUS form of connective tissue is composed of minute filaments arranged in parallel and somewhat wavy bundles, marked with faint cross-waves. It is strong, unyielding and glistening. The fibrous tissue has two distinct forms—the *White Fibrous* and the *Yellow Fibrous.*

22. The WHITE FIBROUS tissue is formed of white, glistening, inelastic bands, having longitudinal creasings, but not admitting of separation into filaments of determinate size. This tissue, by long boiling, is entirely resolved into *Gel'atin.* The white fibrous tissue is found under three forms: *Mem'brane, Lig'ament* and *Ten'don.*

23. The YELLOW FIBROUS tissue is composed of yellow elastic bands separable into their component filaments. It is called the *Elas'tic* tissue, elasticity being its chief charac-

teristic. It does not gelatinize by boiling. It is found in the middle coat of the arteries, in the vocal cords, between the vertebræ, and in many other places where elasticity is needed. The proportion between the white and the yellow fibrous tissues, when found together, varies—the greater the elasticity required, the greater the proportion of yellow elastic fibres.

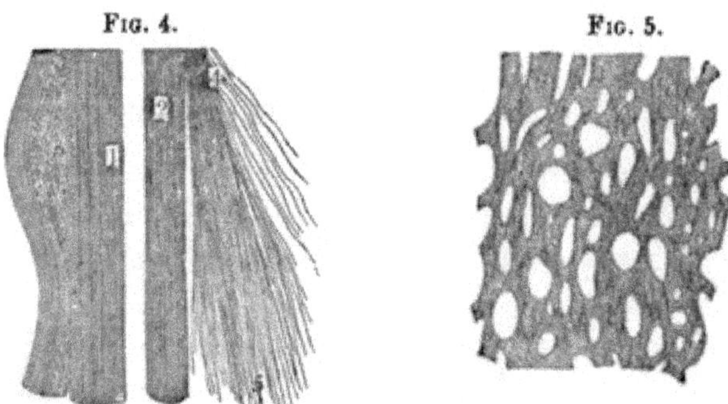

FIG. 4 (*Leidy*). FIBROUS TISSUE.—1, Portion of tendon, exhibiting its composition of prismatic bundles of fibrous tissue, the filaments all parallel to one another. 2, A few bundles drawn from the others, exhibiting their union by delicate crossing filaments of connective tissue. 3, One of the varieties of fibrous tissue. 4, A single bundle, more highly magnified, with a portion (5) of the filaments fretted out.

FIG. 5 (*Leidy*). ELASTIC TISSUE. Highly magnified.

Observation.—In rheumatism the connective white fibrous tissue is the part chiefly affected; hence, the large joints and the loins, where this tissue is most abundant, suffer most. Where there is predisposition to rheumatism, the tendency to it may be lessened and attacks relieved by increasing the amount of clothing over the part affected.

24. The AREOLAR form of connective tissue consists of bands of the fibrous, both of the white and yellow, which interweave in every direction, leaving open spaces between, called cells; hence this tissue is sometimes named *Cel'lular.* These spaces communicate through the body, and contain a fluid resembling the serum of the blood. Although the connective areolar tissue enters into the composition of all organs, it never loses its individuality. In the nerves and muscles it shares neither the sensibility of the one nor the contractility of the other.

Observation.—The swelling of the feet so often seen in feeble persons shows the peculiarity of this tissue, which allows the fluid to pass from part to part and accumulate in the lowest portion of the body, while a recumbent position restores the original shape. Great excess of the fluid produces general dropsy. The free communication between all parts of this tissue is still more remarkable in regard to air. Sometimes, when an accidental opening has been made from the air-cells of the lungs into the adjacent tissue, the air in respiration penetrates every part of the surrounding tissue, and even of the entire body, till inflation endangers life from suffocation. Butchers often avail themselves of this fact, inflating their meat to give it a fat appearance.

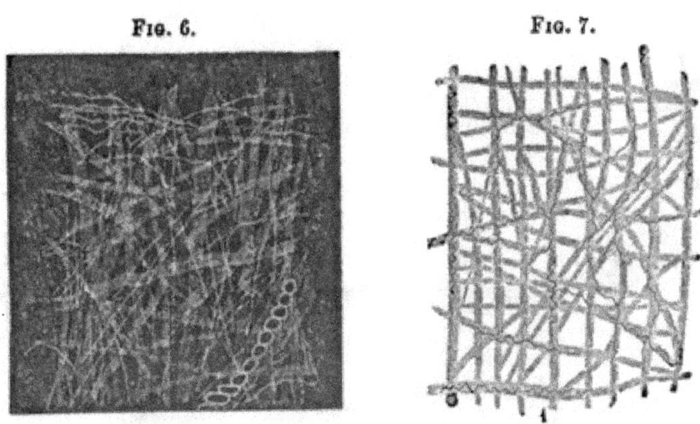

FIG. 6. FIG. 7.

FIG. 6 (*Leidy*). PORTION OF CONNECTIVE TISSUE, from the axilla, exhibiting its composition of bundles and filaments of fibrous tissue crossing in every direction. The rounded bodies represent a single row and a portion of small groups of fat cells. Magnified.

FIG. 7 (*Leidy*). 1, PORTION OF CONNECTIVE TISSUE, from that which envelops the flexor tendons of the fingers as they pass beneath the annular ligament, treated with acetic acid. The pale, dotted portion is intended to represent the fibrous element fading away; the blacker, tortuous lines and nets represent the mixture of elastic tissue. 2, 3, Simple tortuous fibres and a net of elastic tissue. Magnified.

25. CARTILAGINOUS tissue consists of a solid *mat'rice*, apparently homogeneous in structure, resembling ground glass. In this are imbedded nucleated cells, sometimes arranged simply, but usually in groups. It has no perceptible nerves nor blood-vessels. Cartilage is elastic and flexible, but inextensible—qualities admirably essential to its use in the

formation of the joints and in giving to other organs form and strength, without too much rigidity. This tissue constitutes the articular cartilages, the cartilage of the ribs, of the larynx (except the epiglottis), of the trachea and its divisions, and of the nose. The bones usually originate in cartilage, which disappears as bony matter is deposited; such cartilage is called *temporary*, while that which continues till later years is called *permanent*.

FIG. 8.

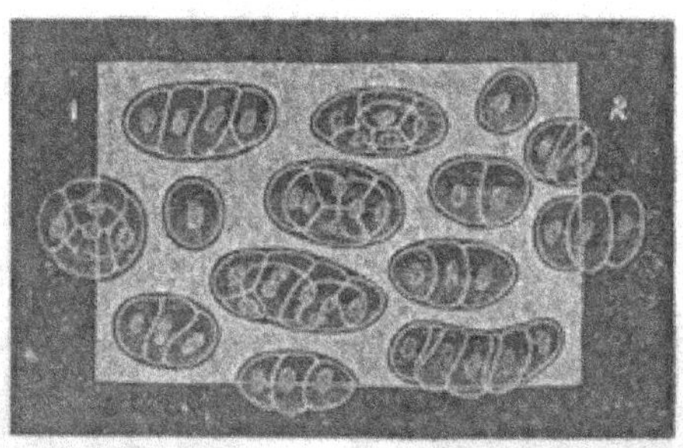

FIG. 8 (*Leidy*). CARTILAGE—section through the thickness of the oval cartilage of the nose. 1, Toward the exterior. 2, Toward the interior surface; highly magnified. It exhibits groups of cartilage cells imbedded in a homogeneous matrice.

FIG. 9.

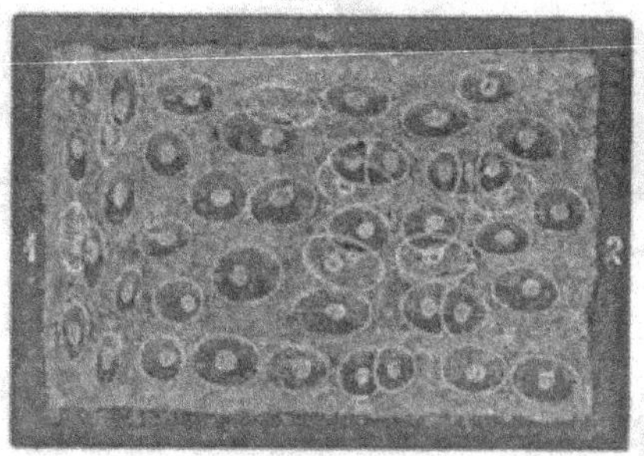

FIG. 9 (*Leidy*). SECTION OF FIBRO-CARTILAGE FROM THE AURICLE OF THE EAR.—The cells are seen imbedded in a fibrous matrice. 1, Exterior surface, where the cells are parallel to it. 2, Toward the middle. Highly magnified.

26. When the matrice assumes a fibrous condition, *Fibro-cartilages* are formed, as in the intervertebral disks, the inter-articular cartilages, the epiglottis, the cartilages of the ear and Eustachian tube, and those of the eyelids. Between pure fibrous tissue and pure cartilage there are various degrees of

intermixture. Fibro-cartilage unites the elasticity of cartilage with the toughness of fibrous tissue, and is therefore well adapted to the firmest union of bones accompanied with moderate flexibility.

27. Adipose tissue has the peculiarity of not being *essential* to the constitution of any organ. It is composed of delicate aggregated cells, of nearly spheroidal form, containing a substance called fat. It is found in the interspaces of areolar tissue beneath the skin and around the heart and kidneys; while none is ever found within the skull, the lungs and the eyelids, where its presence would interfere with their several

Fig. 10. Fig. 11.

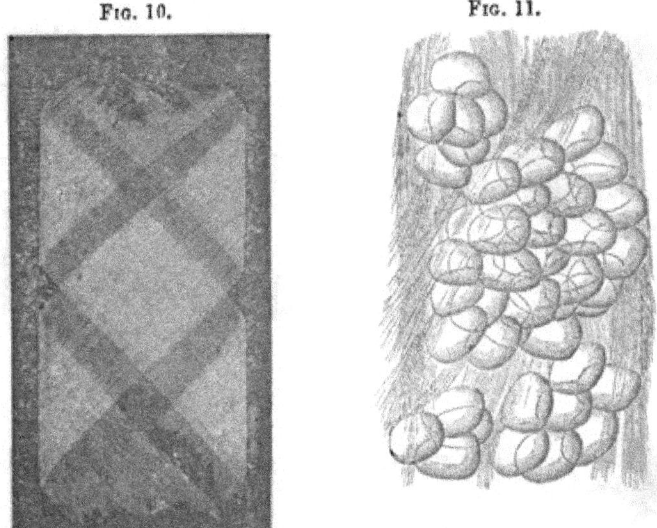

Fig. 10 (*Leidy*). Crossing Bands of fibrous connective tissue.

Fig. 11 (*Leidy*). Adipose Tissue, with Connective Tissue, from the superficial fascia of the abdomen; highly magnified. The groups of fat vesicles are observed contained in the meshes of connective tissue.

functions. Fat accumulates more readily than other matter, and is the earliest removed in disease. It is a storehouse of nutriment, always ready for use, and a non-conductor of heat; it also gives roundness and beauty to the form.

28. Sclerous tissue is found in the bones and teeth. Its composition and arrangement vary at different periods of life.

29. MUSCULAR tissue is composed of fibres, which are themselves composed of minute fibres, called *fi'brillæ*, or filaments. The fibres of this tissue are of two kinds—*non-striated* and *striated*. The former are soft, pale, smooth, either roundish or flattened, and indistinctly granulated, having no markings, or striæ; the latter are soft, yellowish, prismatic, and composed of quadrangular particles so arranged as to give transverse striæ. This tissue has for its peculiar charac-

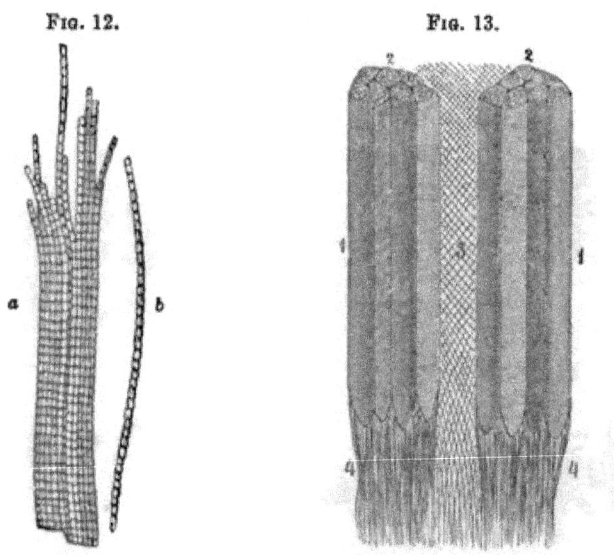

FIG. 12 (*Leidy*). FIBRILS FROM A MUSCULAR FIBRE OF THE AXOLOTL, A BATRACHIAN REPTILE; highly magnified. *a*, Bundle of fibrils. *b*, An isolated fibre.

FIG. 13 (*Leidy*). TWO PORTIONS OF A MUSCULAR FASCICULUS, from the trapezius muscle; highly magnified. 1, Two portions of a muscular fasciculus, composed of prismatic striated fibres terminating below, in rounded extremities, among the fibrous tissue of the commencing tendon. 2, Cut extremities of the fibres, showing their prismatic form. 3, Delicate sheath, composed of obliquely-crossing filaments of fibrous tissue. 4, The fibres of the commencing tendons. Partly a diagram.

teristic, *contractility*, and is the instrument upon which the sensible motions of the body depend. It is a good conductor of electricity, and very sensitive to that agent. It has within itself constant electrical currents, called, collectively, the *muscular current.*

30. TUBULAR tissue consists of a network of minute tubes,

called *cap'illary** vessels. These vessels connect the terminal extremities of the arteries with the commencement of the veins, but are otherwise closed, and never communicate except by imbibition with the structures through which they pass. Their walls are composed of exceedingly thin, transparent, structureless membrane containing scattered nuclei. They vary in size, being largest in the bones, and smallest in the brain and in the lungs. This tissue is found in all parts of the body, excepting the substance of the teeth, the cartilage of the joints, the transparent part of the eye, the *epithe'lial* tissue, the hair and the nails.

FIG. 14. FIG. 15.

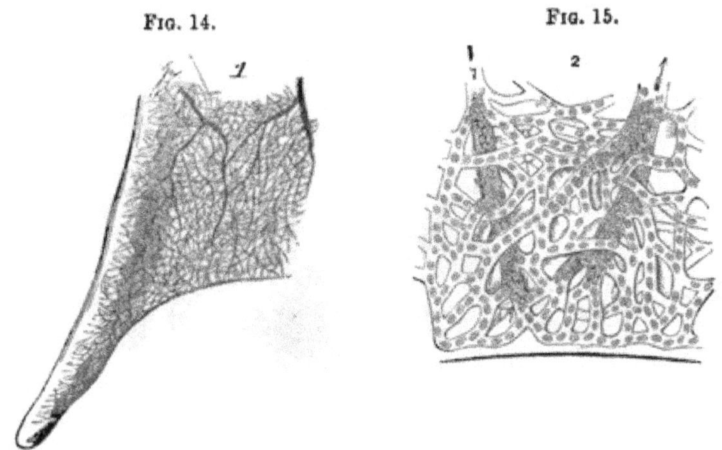

FIG. 14 (*After Wagner.*) A PIECE OF THE WEB OF A FROG'S FOOT, slightly enlarged, showing the fine capillary network connecting the terminations of the arteries with the commencement of the veins.

FIG. 15 (*Allen Thomson*). MINUTE PIECE OF THE MARGIN OF THE FROG'S WEB, showing the ultimate capillaries, connecting the end of a small artery with the beginning of a minute vein. The oval blood-corpuscles are seen in these vessels, and the arrows entering and passing out of the artery and vein indicate the course of the blood-current; magnified about thirty diameters.

31. The NERVOUS tissue is distinguished from all other tissues by its *sensibility*. Like the muscular tissue, it has constant electrical currents. It forms the essential substance of the brain, spinal cord and nerves. This tissue contains

---

* Lat., *capillus*, a hair.

three distinct microscopical elements—*Nerve-Cells*, or *Ganglionic Corpuscles; Gray* or *Gelatinous* fibres; and *White* or *Tubular* fibres.

Fig. 16.

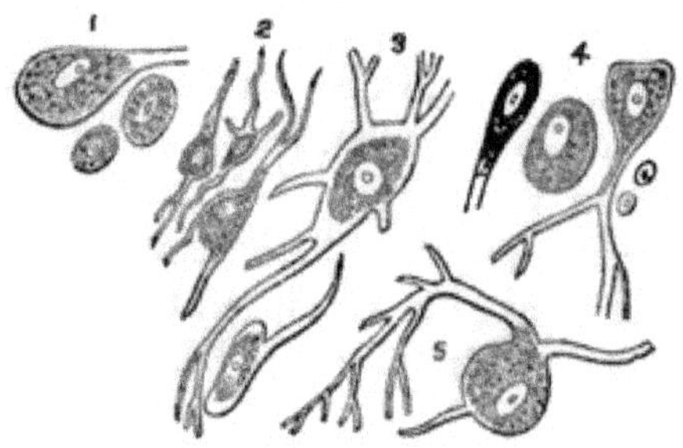

Fig. 16 (*Kolliker and Hannover*). 1, Nucleated cells from a sympathetic ganglion. 2, Branched or stellate cells from the gray substance of the spinal cord. 3, Branched cells from the medulla oblongata. 4, Simple and branched cells from the convolutions of the brain. 5, A large cell from the gray substance of the brain; magnified one hundred diameters.

Fig. 17.

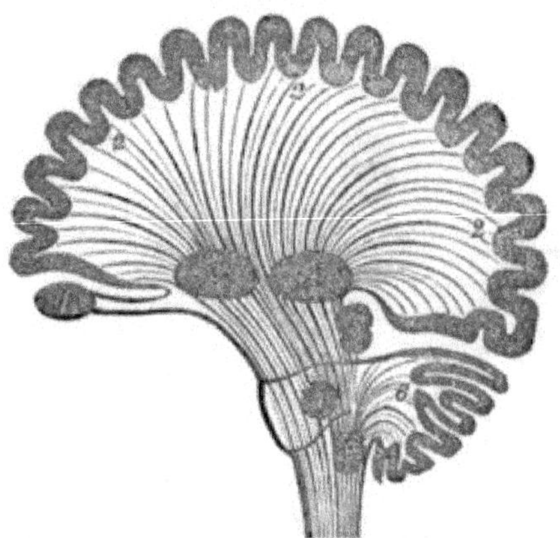

Fig. 17. Diagram of Human Brain, in Vertical Section, showing the situation of the different ganglia and the course of the fibres. 1, Olfactory ganglion. 2, Hemisphere. 3, Corpus striatum. 4, Optic thalamus. 5, Tubercula quadrigemina. 6, Cerebellum. 7, Ganglion of tuber annulare. 8, Ganglion of medulla oblongata.

32. The Ganglionic Corpuscles are cell-bodies containing pulpy matter, with one or more nuclei surrounding colored granules. These cells vary in shape, being roundish, pear-shaped, or branched in a caudate or stellate manner, these offsets being continuous with the cell-wall and its contents, and

often entering another cell and connecting the two. These nerve-cells are found in the brain, spinal cord and ganglia, and at the extremities of the nerves of sight and hearing.

33. The GRAY or GELATINOUS fibres are soft and granular, with no distinct medullary sheath. They contain many dark nuclei, and are most abundant in the sympathetic ganglia and its branches.

34. The WHITE or TUBULAR fibres are microscopic tubes. The walls are structureless membrane enclosing a layer of medullated matter resembling fluid fat, which acts as a sheath; within this is a firmer part, or core, called the *band-axis*, or axis cylinder; this is albuminous.

35. The gray substance is most abundant in the outer part of the brain, and the white in the inner; but the two intermix more or less in every part of the nervous system.

§ 5. MEMBRANES.—*Basement Membrane.—Epithelium.—Serous Membrane.—Synovial Membrane.*

36. BASEMENT MEMBRANE is an exceedingly thin, delicate, structureless layer of protoplasm or blastema, resembling, under the microscope, a film of transparent gelatine. Upon it, in various parts of the body, are imbedded minute *epithelial** cells. The membrane formed by these cells is called *epithe'lium.* The relation of this structureless membrane to the epithelium gives it the name of *Basement Membrane.*

FIG. 18.

FIG. 18 (*Leidy*). DIAGRAM EXHIBITING THE RELATIVE POSITION OF THE COMMON ANATOMICAL ELEMENTS OF SEROUS AND MUCOUS MEMBRANES, THE GLANDS, THE LUNGS AND THE SKIN.—1, Epithelium, secreting cells or epidermis, composed of nucleated cells, and occupying the free surface of the structure mentioned. 2, Basement layer, represented much thicker than natural, in comparison with the other layers. 3, Fibrous layer, in which the arteries and veins (4) terminate in a capillary network. Magnified.

---

* Gr., *epi*, upon, and *tithemi*, I cover or place.

37. From difference in form and other peculiarities, the EPITHELIUM is divided into several varieties—as the *Squamous* Epithelium, consisting of several layers of thin scales, which are flattened cells having a nucleus and a few scattered granules, as in the mucous membrane of the mouth; the *Pavement* Epithelium, consisting of from one to four layers of nucleated cells, six-sided and regularly arranged like the blocks of a pavement (whence the name), as in the serous membranes; the *Columnar* Epithelium, consisting of a

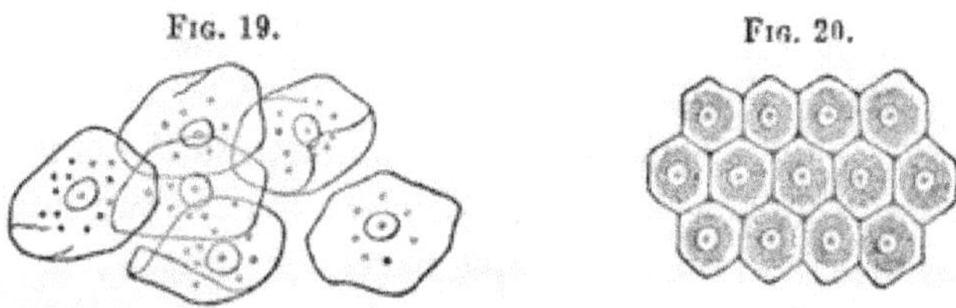

FIG. 19. FIG. 20.

FIG. 19 (*Leidy*). SQUAMOUS EPITHELIUM, consisting of nucleated cells transformed into broad scales, from the mucous membrane of the mouth; highly magnified.

FIG. 20 (*Leidy*). PAVEMENT EPITHELIUM, from a serous membrane, highly magnified, and seen to consist of flat, six-sided nucleated cells.

single layer of six-sided columnar cells, with a conical prolongation terminating in a progeny of developing cells, as in the mucous membrane of the stomach and intestines; the *Ciliated* Epithelium, having cells possessing at their free extremity fine filamentary processes of the cell-wall, resembling the eye-lashes (whence the name). During life these cilia are endowed with a power of moving rapidly backward and forward in a wave-like manner, reminding one of the movement of a field of grain swept by a gentle breeze. Currents are thus produced in liquids, conveying them from one part to another. This kind of epithelium is found on the mucous membrane of the upper part of the nose and pharynx, the Eustachian tube and all the respiratory organs.

38. Beneath the basement membrane, and in contact with it, is a very dense and vascular layer of areolar and elastic tissue. This triple arrangement of epithelium, basement membrane and fibro-areolar tissue, constitutes the serous, the

synovial and the mucous membranes, the skin, the ducts of all glands, and the inner coat of the blood-vessels and the lymphatics.

39. The Serous Membrane is that portion which lines the walls of certain closed cavities or sacs. It is smooth, shining and moistened by a fluid called *se'rum*, which the membrane secretes; as the *pleu'ra, peritone'um, pericar'dium, arach'noid*, etc.

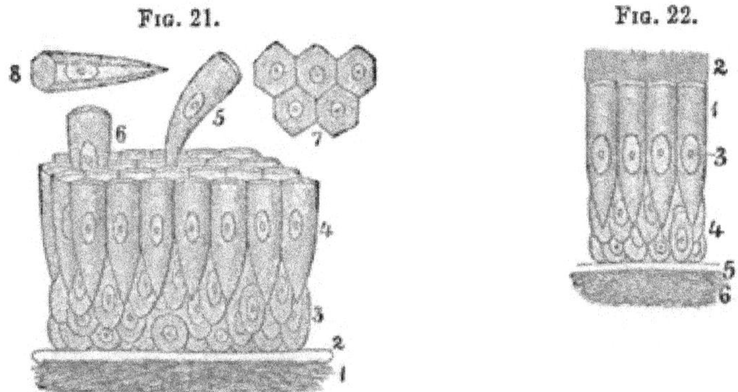

Fig. 21 (*Leidy*). Diagram of a Vertical Section of the Mucous Membrane of the Small Intestines; highly magnified. 1, Fibrous layer, in which the blood-vessels are distributed. 2, Basement membrane. 3, Young nucleated cells. 4, Layer of columnar cells. 5, 6, Cells in the act of being shed or thrown off. 7, Free ends of the columnar cells, exhibiting their six-sided form. 8, A single columnar cell, exhibiting its actual form at all parts.

Fig. 22. Diagram of a Vertical Section of the Bronchial Mucous Membrane.—1, Columnar ciliated epithelial cells. 2, Cilia. 3, Nuclei. 4, Young cells. 5, Basement membrane. 6, Fibrous layer.

40. The Synovial Membrane resembles the serous very closely as regards structure and the closed sacs. It also secretes a fluid, called *syno'via*, which is more viscid than that of the serous membrane. It has fringe-like processes hanging loosely in the joints, having large epithelial cells, which probably secrete the synovial fluid. This membrane covers the cartilages, and lines the ligaments which enter into the composition of the joints.

Observation.—When the synovial membrane is ruptured, the synovia escapes into the surrounding areolar tissue, and what is popularly known as the "weeping sinew" is formed.

Similar tumors in the joints of lower animals are called "windgalls."

41. There are two MUCOUS MEMBRANES—the *Gastro-Pulmonary* and the *Urinary*. These do not form closed sacs, like the serous and synovial membranes, but both open to the surface. The mucous membranes secrete a viscid fluid, called *mu'cus*, and in their glandular recesses are formed various secretions, as *sali'va, bile, tears,* etc. These membranes vary in different parts both in thickness and appearance. In the nasal and air passages, the membrane is smooth, rugose or ridgy in the stomach, papillous in the tongue and villous in the intestines.

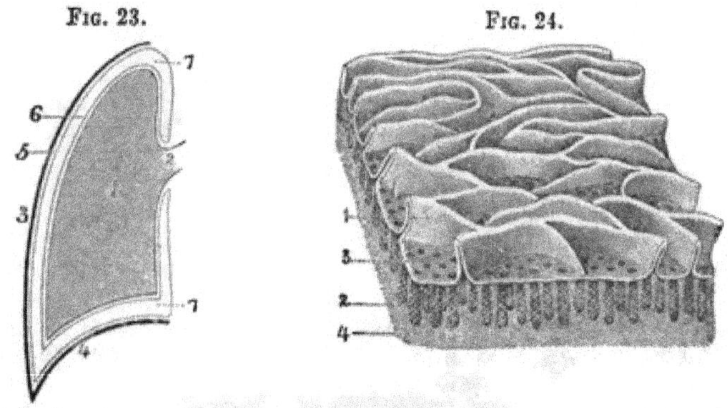

FIG. 23 (*Leidy*). DIAGRAM EXHIBITING THE RELATION OF A SEROUS MEMBRANE (*the pleura*) TO THE ORGAN IT INVESTS AND THE CAVITY IT LINES.—1, Lung. 2, Root of the lung, which is the only attached portion of the organ, all others being free. 3, Side of the thorax. 4, Diaphragm. 5, Parietal pleura. 6, Pulmonary or reflected pleura. 7, Cavity of the pleura. Magnified.

FIG. 24 (*Leidy*). MUCOUS MEMBRANE FROM THE JEJUNUM.—1, Villi resembling valvulæ conniventes in miniature. 2, Tubular glands: their orifices. 3, Opening on the free surface of the mucous membrane. 4, Fibrous tissue. Magnified.

42. The GASTRO-PULMONARY MUCOUS membrane commences at the mouth, enters the nostrils, passes between the eyelids, dips into the deep parts of the ear, lines the trachea and the air-tubes of the lungs, and the alimentary canal from one extremity to the other.

43. The URINARY MUCOUS membrane lines the ducts connecting the kidneys and the bladder, of which it forms the interior coat; also the passages to the skin.

44. The skin is continuous with the mucous membranes, and will be described hereafter.

Observation.—Like tissues readily assume similar conditions; hence diseases of the skin or of the serous or mucous membranes are often transferred from one to the other. In diseases of the skin—as measles, scarlet fever, etc.—if the surface becomes damp or chilled, there is danger of their being transferred to the mucous membrane of the air-passages, stomach or intestines. In chronic diseases of the mucous membrane—as coughs, catarrh, diarrhœa or dysentery—the skin is usually cold, dry and inactive. By improving its condition much relief will be afforded.

CHAPTER III.

GENERAL CHEMISTRY.

§ 6. *Solids and Fluids.—Proximate Constituents.—Inorganic.—Organic. —Nitrogenous.—Non-Nitrogenous.—Ultimate Chemical Elements.*

45. The human body is composed of solids and fluids, reducible, by chemical analysis, to the same constituents and elements. In different periods of life the proportion of fluids and solids varies; the former being more abundant in youth than in old age. This is one reason why the limbs in childhood are soft and smooth, but in later years hard and wrinkled.

46. If the tissues of the body are subjected to chemical analysis, they yield about ninety substances, called *Proximate Constituents*, these being the first chemical compounds into which the tissues resolve themselves. In living beings vitality is, as it were, "the architect who plans the building and sees that the requisite materials are procured by the chemical processes and worked up according to his will." Hereupon arise many new substances which cannot be artificially imitated; these are called *Organic proximate constituents.* Those substances found in the inorganic kingdom also, and capable of artificial imitation, are called *Inorganic proximate constituents.*

47. Of the Inorganic Proximate Constituents, water is the most abundant: it exists in all the tissues; next to this, in relative quantities, are Phosphates of Lime, of Magne'sia, of Soda and of Potas'sa; Carbonates of Lime, of Soda and of Potassa; Chloride of Sodium (common Salt) and of Potassium; and Fluoride of Cal'cium. Some compounds contain Iron, Sil'ica, Manganese', and perhaps some accidental substances, as Lead, Copper and Alu'minum. Am-

mo'nia, in combination, is found in the urine. Ox'ygen, Ni'trogen and Carbon'ic Acid gas exist in a free state.

48. The ORGANIC PROXIMATE CONSTITUENTS are of two classes. One class contains the chemical element *Azote'*,* or *nitrogen;* hence its compounds are called *az'otized* or *nitrog'enous;* the other has no azote, and its compounds are named *non-azotized* or *non-nitrogenous.*

49. The NITROGENOUS class contains Albu'men and its allied substances, called albuminoids. Some of the most important are—*Albu'minose, Fi'brin, Mus'culin, Glob'ulin, Hæm'atin, Ca'sein, Cartila'gin, Sal'ivin, Pep'sin, Pancrea'tin, Mu'cin, Neu'rin, Ker'atin, Elas'tin, Mela'nin* and *Biliverd'in;* also some acids, as the *Cer'ebric, Chol'ic* and *U'ric.*

50. ALBUMEN and the albuminoids, together with fatty matter (non-nitrogenous), are the great nutritive substances of the animal economy. Albumen† is well known in the white of an egg; whence its name. It is found in the substance of the brain and nerves; in the fluid part of the blood; in the moisture that pervades the muscles and other tissues; in the lymph and chyle; and in the mucous, serous and synovial secretions. It coagulates by the action of heat and alcohol, and is dissolved by weak acids and alkalies.

51. ALBUMINOSE is found in the chyle and blood in a liquid condition, and is a result of the digestion of albuminous, fibrinous, musculinous and caseous matter of food; unlike albumen, it is not coagulated by heat.

52. FIBRIN is a soft, white, stringy substance, obtained from freshly-drawn blood by whipping it with fine sticks or wires. It coagulates spontaneously, assuming the form of minute threads, or *fi'brils;*‡ whence its name. Fibrin is also found in the chyle, lymph and serous secretions. It is precipitated and hardened by alcohol, and redissolved by weak acid.

53. MUSCULIN is a peculiar form of fibrin that exists in

---

* Gr., *a*, not, and *zoē*, life.

† Lat., *albus*, white.

‡ Lat., *fibra*, a thread.

the muscles, or flesh. Its characteristic property is *contractility.* Boiling hardens it, while weak acids render it more soluble.

54. GLOBULIN and HÆMATIN form the contents of the red globules of the blood. Hæmatin contains about seven per cent. of iron; but the color of the blood is now supposed not to depend on the iron, but a peculiar substance named *cru'orin.*

55. CASEIN resembles albumen in its general properties, but, unlike albumen, when in solution it is not coagulated by heat, but by acids. It exists in solution in milk with *lac'tin* (milk sugar) and salts. It forms the curd in soured milk, the casein being coagulated by the *lactic acid* formed from decomposed lactin.

56. CARTILAGIN is the principal constituent of the connective tissues, as the so-called bone cartilage, true cartilage, ligaments, tendons, fibrous membranes, dermis and the areolar tissue. The basis of bone cartilage is *os'teine,* with which are blended salts of lime. The basis of true cartilage is called *chon'drigen.* Unlike albumen, cartilagin is insoluble in water and does not coagulate by heat, but is liquified by boiling and changed into *gel'atin,* or glue.

57. SALIVIN is found in the saliva. It has the peculiar property of changing starch into a kind of gum called *dex'trine,* the dextrine into *glu'cose,* or grape sugar, and this into *lactic acid.*

58. PEPSIN is a remarkable and potent substance secreted by the glands of the mucous membrane of the stomach. This secretion is a peculiar principle of the gastric juice, and, when slightly acidulated, has the property of quickly dissolving coagulated albumen, blood, meats, fish, cheese and many other substances.

59. PANCREATIN is the active principle of the secretion of the pancreas. It has three distinct actions—1st, on starch; 2d, on fat; and 3d, on albuminous matter.

60. MUCIN is a substance found in the different varieties of mucus, imparting to them their viscid character. It is usually mixed with other fluids.

61. NEURIN is also an albuminoid substance connected with the brain and nerves, upon which the peculiar characteristics of the nervous system are supposed to depend.

62. KERATIN is the peculiar albuminoid principle giving the horn-like character to the hair, nails and cuticle.

63. ELASTIN is the substance peculiar to the elastic tissue. It is insoluble in all common fluids.

64. MELANIN is a blackish-brown coloring matter found in the choroid coat and the iris of the eye, in the hair and in the epidermis. It is most abundant in the black and brown races, but it also exists in the yellow and white races.

65. BILIVERDIN is the coloring matter of the bile. It is yellow in transmitted light, and greenish in reflected light. On exposure to the air in its natural fluid condition, it absorbs oxygen and assumes a bright grass-green color.

66. Beside the before-mentioned constituents, none of which are acid but mucin, there are several acids, among which may be named the *Cerebric* acid found in the gray substance of the brain; *Cholic* acid in the bile; and *Uric* acid in the urine.

67. The groups of NON-NITROGENIZED or non-azotized substances are—the fats, sugars and starch. The fats are most abundant. These are insoluble in water, but are dissolved by heat, alcohol and ether. They are found in the brain, muscles, blood and chyle.

68. The FATS of the human body are composed mostly of *o'lein* (liquid fat), and *ste'arin* and *mar'garin* (solid fats), margarin being most abundant, and stearin least. The fats are derived from the fatty components of food, and also from transformed saccharine compounds. When boiled with an alkali, as in the manufacture of soap, they decompose into fatty acids, margaric, stearic and oleic acids, and a sweet, viscid substance called *glyc'erine.*

69. SUGARS are of different kinds, as *Glu'cose* (grape sugar), in the blood and chyle; Liver sugar, in the liver; *Lac'tin* (milk sugar), in milk; *In'osit* (muscle sugar), in muscles. LACTIN, in contact with azotized matter, or a *ferment*, easily

decomposes, forming lactic acid. All these saccharine and acid substances are soluble in both water and alcohol. STARCH granules are found in the brain, and are named *Corpora Amyla'cea.* These exhibit the chemical reactions of vegetable starch, and are colorless and transparent, resembling in appearance the granules of Indian corn.

70. The ULTIMATE CHEMICAL ELEMENTS enter into the composition of the body in about the following percentage proportions:

GASES.	Oxygen	72.
	Hydrogen	9.1
	Nitrogen	2.5
	Chlorine	.085
	Fluorine	.08
SOLIDS.	Carbon	13.5
	Phosphorus	1.15
	Calcium	1.3
	Sulphur	.1476
	Sodium	.1
	Potassium	.026
	Iron	.01
	Magnesium	.0012
	Silicon	.0002
		100.0000

71. The greater part of the oxygen and hydrogen exist in a state of water, but the dried residue still contains some gaseous as well as solid elements.

72. Carbon is the most abundant element. In the inevitable decomposition of the body, while its hydrogen and nitrogen, with part of its carbon and oxygen, are restored to the inorganic world in the shape of *water*, *carbonic acid* and *ammonia*, the rest of its carbon and oxygen, its chlorine and fluorine, its phosphorus and sulphur, and its metallic bases, calcium, sodium, potassium, magnesium and iron, with a trace of silicon and manganese, revert to the condition of *inorganic salts* and *earths*—viz., carbonates, sulphates and phosphates, chlorides and fluorides of the above-named saline and earthy bases.

Fig. 25.

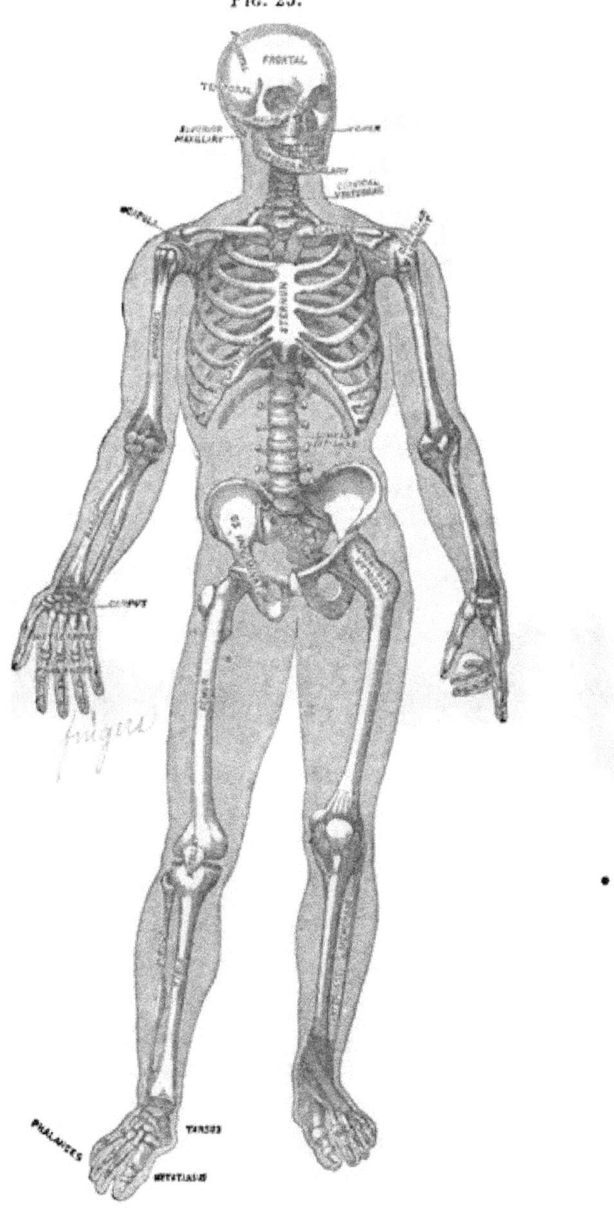

DIVISION II.

MOTORY APPARATUS.

73. In all the movements connected with the merriments of childhood, with the ceaseless industry of the toiling millions, with the hymning of the praises of the great I Am,—in a word, in every movement of the body, certain organs are brought into action, which, taken collectively, constitute the Motory Apparatus. The organs of this apparatus are the *Bones* and *Joints*, the *Muscles* and *Motor Nerves*.

CHAPTER IV.

THE BONES.

§ 7. Anatomy of the Bones.—*The Skeleton and its Uses.—Number and Classification of the Bones.—Bones of the Head.—Of the Trunk.—Of the Upper Extremities.—Of the Lower Extremities.—The Joints.—Definition and Classification.—Immovable Joints.—Mixed.—Movable.—Peculiar Forms of Movable.*

74. The Internal Framework of the human body consists of Bones, which, united by strong ligaments,* constitute the *Skeleton.*

75. These bones number two hundred and eight, besides the teeth. For convenience they are classed as the bones of the *Head*, the *Trunk* and the *Extremities.*

76. The Bones of the Head are divided into those of the *Skull*, the *Face* and the *Ear.*

77. The Skull is composed of eight bones—the *Front'al*, occupying the portion called the forehead; the two *Tem'poral*.

---

* Lat., *ligo*, I bind.

covering the part commonly known as the temples; the two *Pari'etal,* forming the essential part of the projection on the upper and lateral parts of the head and uniting in the median line upon the top of the skull; the *Occip'ital,* at the posterior part of the skull, resting upon the atlas vertebra and having a large orifice for the passage of the spinal marrow; the *Sphe'noid,* situated across the base of the skull, extending from side to side, having many depressions and processes, articulating with all the bones of the cranium and five of those in the face, and serving as a point of attachment for twelve pairs of muscles; and the *Eth'moid* bone, between the sockets of the eyes and behind the base of the nose.

Fig. 26.

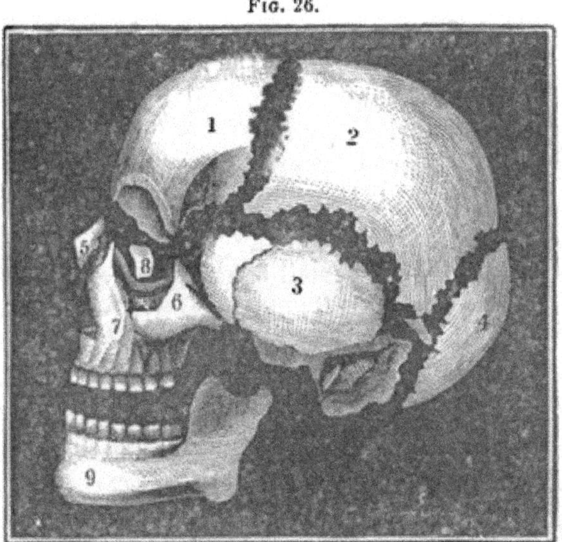

Fig. 26. Bones of the Head.—1, Frontal bone. 2, Parietal bone. 3, Temporal bone. 4, Occipital bone. 5, Nasal bone. 6, Malar bone. 7, Upper jaw. 8, Os unguis. 9, Lower jaw.

78. The skull-bones are formed of two plates united by porous bone-substance. The *external* plate is fibrous and tough; the *internal,* dense and hard, hence called the *vitreous* or glassy plate. These bones are united by *sut'ures;* the external plate having notched edges fitted together as in the dovetailing of carpentry; the internal, plane edges in simple

apposition. From infancy to the twelfth year the sutures are imperfect; from that time to forty, distinctly marked; and in old age, nearly obliterated.

Observation.—We find no less diversity in the form and texture of the skull than in the expression of the face. The head of the New Hollander is small, that of the African is compressed, while that of the Caucasian is distinguished for its beautiful oval form. In texture the Greek skull is close and fine, while the Swiss is softer and more open.

79. The FACE has fourteen bones—the two *Na'sal*, forming the bridge or base of the nose; the two *Ma'lar* (cheek-bones); the two *Lach'rymal;* the two *Superior Max'illary*, articulating with two bones of the skull and all the bones of the face excepting the lower jaw; the two *Palate* bones, forming the orbits of the eyes, the outside of the nose, and the most of the roof of the mouth known as the hard palate; the two *Tur'binated*, in the nostrils; the *Inferior Maxillary*, or mandible, the only movable bone of the face, articulating with the temporal bones; and the *Vo'mer*, which separates the nostrils from each other.

80. The EAR has three small bones, which aid in hearing.

81. The BONES OF THE TRUNK number fifty-four—twenty-four *Ribs;* twenty-four bones in the *Spinal Column;* four in the *Pel'vis;* the *Ster'num* (breast-bone); and the *Os Hyoi'des* (at the base of the tongue). These bones, with the soft parts attached, are so arranged as to form two cavities, called the *Tho'rax* (chest) and the *Ab'domen.*

82. The THORAX is formed by the sternum in front, the ribs at the sides and the twelve dorsal vertebræ at the back. The natural form of the chest is conical, with the apex above; but fashion, in a multitude of instances, has inverted the order. The thorax contains the heart, the lungs and the large blood-vessels.

83. The STERNUM is situated in the middle line of the front of the chest, and is held in place chiefly by the ribs. Each side is marked by seven pits for receiving the cartilages of the corresponding true ribs. In childhood the sternum

consists of several cartilaginous pieces, which ossify and unite in later years.

84. The RIBS are connected with the spinal column, twelve on each side. The first seven, called *True ribs*, are connected with the sternum by means of cartilage; of the remaining five, called *False ribs*, three are connected by cartilage with each other, while the two lower are free at their anterior extremity, hence called *floating* ribs. In length, the ribs increase from the first to the eighth, then again diminish to the twelfth; in breadth, they gradually diminish from the first to the last; in direction, the first rib is horizontal, all the others are oblique and downward.

FIG. 27.

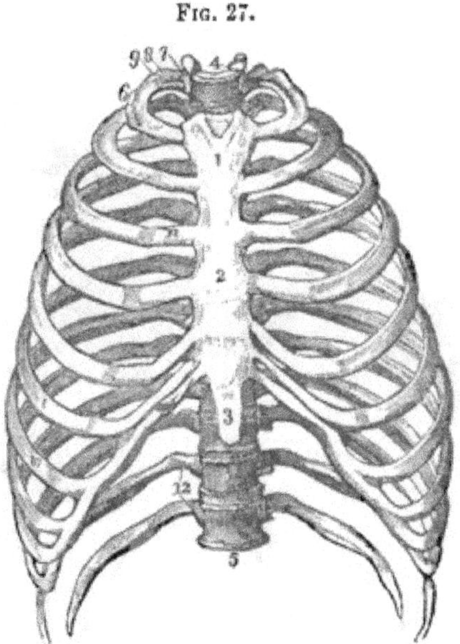

FIG. 27. THE FRONT VIEW OF THE THORAX.—1, 2, 3, The sternum. 4, 5, The spinal column. 6, 7, 8, 9, The first ribs. 10, The seventh rib. 11, Cartilage of the third rib. 12, The floating rib.

85. The SPINAL COLUMN is composed of twenty-four bones, called *Vert'ebræ.* Each vertebra consists of a main part, called the *body*, and seven projections, called *processes;* four of these, employed in binding the bones together, are called *articulatory;* two of the remaining, *transverse;* and the other,

spinous. The last three give attachment to the muscles of the back. The projections are so arranged that immediately behind the bodies of the vertebræ a canal is formed for the *Medul'la Spina'lis* (spinal cord), sometimes called the pith of the back-bone.

86. The VERTEBRÆ are arranged in three classes, according to their situation; the seven of the neck are called *Cer'vical;* the twelve of the back, *Dor'sal;* and the five of the loins, *Lum'bar* vertebræ.

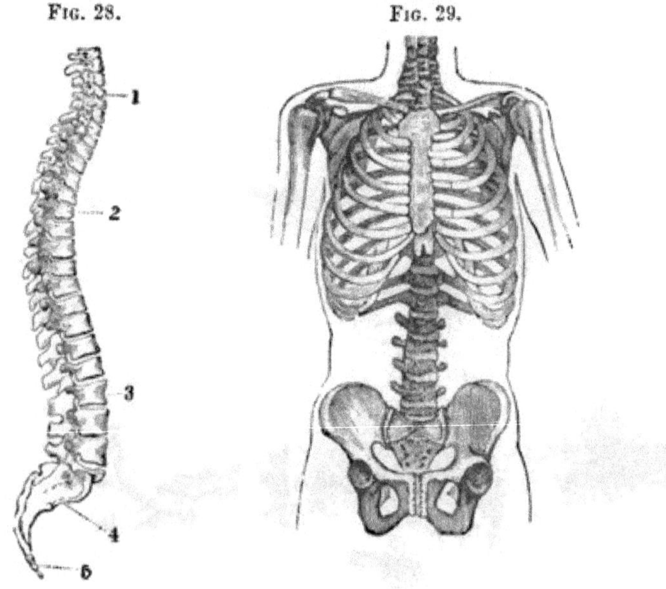

FIG. 28. THE SPINAL COLUMN, Lateral view. 1, 2, 3, The vertebræ. 4, Sacrum. 5, Coccyx.

FIG. 29. THE CHEST AND PELVIS, Front view.

87. The CERVICAL VERTEBRÆ are smaller than those of other regions; they are concave above and convex below; hence, when articulated, they lock one into the other. The processes are short, bifid and horizontal, permitting a partial rotary movement. The upper vertebra articulating with the occipital bone is called the *At'las;* the second, the *Ax'is.*

88. The DORSAL VERTEBRÆ furnish support for the ribs,

which are so connected with the transverse processes as to impede rotation. The spinous processes are very long and extend obliquely downward. The bodies are larger than those of the cervical vertebræ, and increase in size according to the weight to be sustained.

89. The Lumbar Vertebræ have broad bases, with large, strong and horizontal processes. These vertebræ show those transitional changes which are calculated, by an easy gradation, to unite separate vertebræ into solid bone.

Fig. 30. Fig. 31.

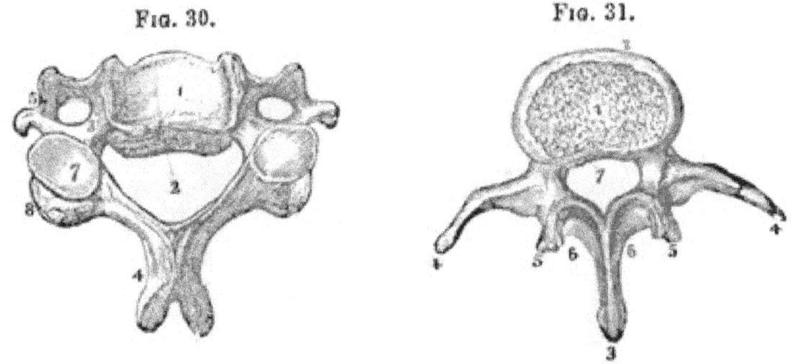

Fig. 30. A Vertebra of the Neck.—1, The body of the vertebra. 2, The spinal canal. 4, The spinous process, cleft at its extremity. 5, The transverse process. 7, The inferior articulating process. 8, The superior articulating process.

Fig. 31. A Cervical Vertebra.—1, The cartilaginous substance that connects the bodies of the vertebræ. 2, The body of the vertebra. 3, The spinous process. 4, 4, The transverse processes. 5, 5, The articulating processes. 6, 6, A portion of the bony bridge or arch that assists in forming the spinal canal (7).

90. Covering the front of the bodies of the vertebræ is the *Anterior Vertebral Ligament*, consisting of a broad range of fibres closely blended and of variable length. This ligament, besides joining the vertebræ firmly together, gives attachment to the pharynx, the œsophagus, the thoracic duct, the aorta and other large blood-vessels. On the posterior part of the bodies of the vertebræ, within the spinal canal, is the *Posterior Vertebral Ligament*, remarkably smooth and shining. Both the anterior and posterior ligaments adhere very closely to the intervertebral substance and to the bodies of the vertebræ.

91. Between the *arches* of the vertebræ behind, are the

Yellow Ligaments (*Ligamenta Subflava*), filling up the intervals and completing the spinal canal from the axis to the sacrum; they are two in number on each side, making twenty-three pairs. They are attached to the anterior surface of the lower part of the arches above and the posterior surface of the upper part of those below, between the spinous and the transverse processes, and are separated by a vertical fissure. Their greatest length is in the neck, but their greatest thickness is in the lumbar series. They combine great strength with remarkable elasticity, thus differing, in an essential particular, from ordinary ligaments.

Fig. 32.

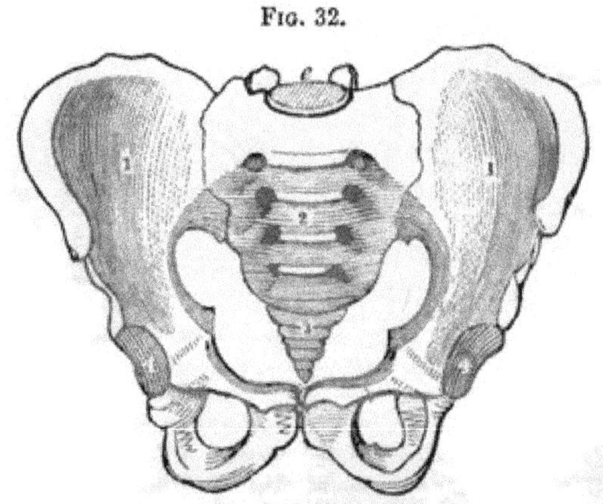

Fig. 32. Front View of the Pelvis.—1, 1, The innominata. 2, The sacrum. 3, The coccyx. 4, 4, Socket. *e*, The junction of the sacrum and lower lumbar vertebra.

92. Between every pair of true vertebræ are discoidal plates of white fibrous tissue, called *Intervertebral Ligaments.* These ligaments are inelastic in structure, but very elastic by arrangement, for the decussation of the concentric laminæ enables them to yield to pressure and to resume their original position when the pressure is removed; and this elasticity is greatly increased in the whole disk by the presence of a central pulp that occupies the hollow circle within the ligamentous structure.

93. The Pelvis is composed of the two *Innomina'ta* (nameless bones), the *Sa'crum* and the *Coc'cyx.*

94. The INNOMINATUM is the hip-bone; in it is a deep socket for the head of the thigh-bone. In the centre of this cavity is a depression to which the round ligament of the thigh-bone is fixed.

95. The SACRUM (so called because offered by the ancients in sacrifice) is a wedge-shaped bone, between the innominata. In early life it is composed of five vertebræ, which become united in later years. It is the basis of the vertebral column. The texture of the sacrum is very light and spongy.

96. The COCCYX,* at the lower extremity of the spinal column, varies at different ages: in infancy it is cartilaginous; in adult age, formed of four pieces of bone or vertebræ; in after life it becomes a continuous, blended structure.

97. The UPPER EXTREMITIES contain sixty-four bones: the *Scap'ula* (shoulder-blade); the *Clav'icle* (collar-bone); the *Hu'merus* (arm-bone); the *Ra'dius* and *Ul'na* (fore-arm); the *Car'pus* (wrist); the *Meta-car'pus* (palm of the hand); and the *Phalan'ges* (fingers and thumb.)

98. The SCAPULA, a flat, thin, triagular bone, is situated upon the upper and back part of the chest. It lies upon muscles by which it is held in place and moved in different directions.

99. The CLAVICLE,† shaped like the italic *f*, is attached at one extremity to the sternum, and at the other to the scapula.

100. The HUMERUS is a long, cylindrical bone, joined at the elbow with the ulna of the fore-arm; and at the scapular extremity lodged in the *glé'noid* cavity.

101. The ULNA‡ is the small bone of the fore-arm, and occupies the inner side. It articulates with the humerus at the elbow, forming a perfect hinge-joint.

102. The RADIUS§ is placed on the outside (the thumb side) of the fore-arm, and nearly parallel to the ulna. It is larger than the ulna, and articulates with it, both at the

---

* From *cuckoo*, on account of resemblance to the cuckoo's bill.

† Lat., *clav'is*, a key. ‡ It., an *ell*, a *measure*. § Lat., a *spoke*.

elbow and at the wrist. The radius also articulates with the first row of bones at the wrist forming the wrist-joint.

103. The CARPUS has eight bones, arranged in two rows, and so firmly bound together, as to permit little movement. One row articulates with the fore-arm, the other with the metacarpus.

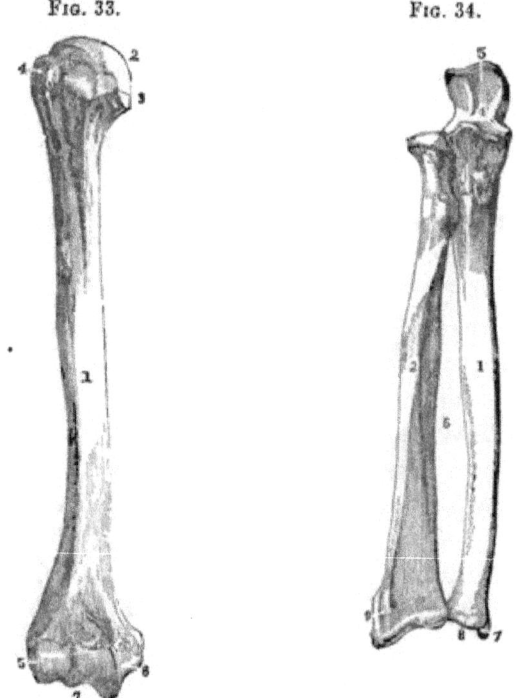

FIG. 33. THE HUMERUS.—1, The shaft. 2, The large, round head that is placed in the glenoid cavity. 3, 4, Processes for attachment of muscles. 5, A process, called the external elbow. 6, A process, called the internal elbow. 7, The articulating surface upon which the ulna rolls.

FIG. 34. THE FORE-ARM.—1, The ulna. 2, The radius. 3, The upper articulation of the radius and ulna. 4, The articulating cavity, in which the lower extremity of the humerus is placed. 5, The upper extremity of the ulna, called the olecranon process, which forms the point of the elbow. 6, Spaces between the radius and ulna, filled by the intervening ligament. 7, The styloid process of the ulna. 8, The surface of the radius and ulna, where they articulate with the bones of the wrist. 9, The styloid process of the radius.

104. The METACARPUS* has five bones; upon four of

* Gr., *meta*, after or beyond, and *karpos*, wrist.

which, are placed the first range of finger-bones, and upon the other, the first thumb-bone. The metacarpal bone of the thumb is the shortest, and it is also disconnected with and divergent from the others.

105. The PHALANGES* of the fingers have three bones, while the thumb has but two. The fingers are named in succession, the thumb, the index, the middle, the ring, and the little finger.

FIG. 35.

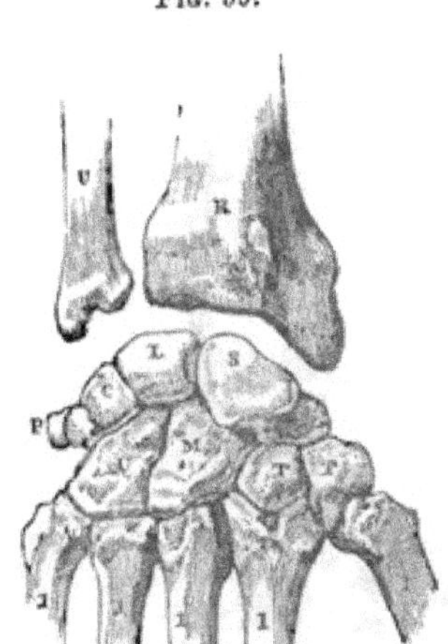

FIG. 36.

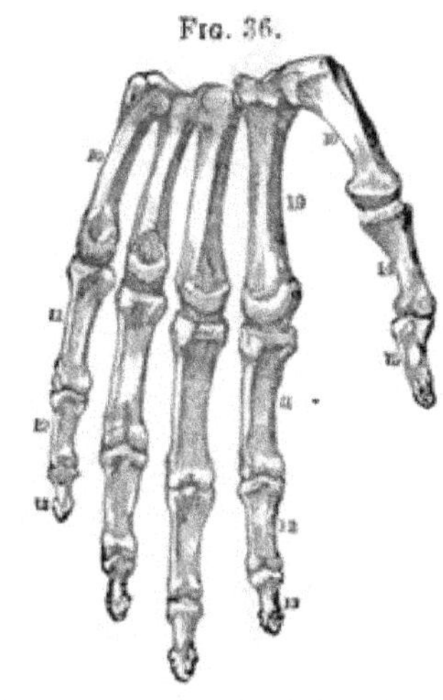

FIG. 35. THE WRIST.—U, The ulna. R, The radius. S, The scaphoid. L, The semilunar. C, The cuneiform. P, The pisiform. The last four form the first row of carpal bones. T, T, The trapezium and trapezoid. M, Magnum. U, Unciform. The last four form the second row of carpal bones. 1, 1, 1, 1, Metacarpal bones.

FIG. 36. THE HAND.—10, 10, 10, The metacarpal bones of the hand. 11, 11, First row of finger-bones. 12, 12, Second row of finger-bones. 13, 13, Third row of finger-bones. 14, 15, The bones of the thumb.

106. The LOWER EXTREMITIES contain sixty bones: the *Fe'mur* (thigh-bones); the *Patel'la* (knee-pan); the *Tib'ia* (shin-bone); the *Fib'ula* (small bone of the leg); the *Tar'sus* (instep); the *Metatar'sus* (middle of the foot); and the *Phalan'ges* (toes).

107. The FEMUR† is the strongest and longest bone of the skeleton. It supports the weight of the head, trunk and upper extremities.

---

* Gr., *row.* † Lat., *thigh.*

108. The PATELLA* is a small chestnut-shaped bone, placed on the anterior part of the lower extremity of the femur, and connected with the tibia by a strong ligament.

109. The TIBIA† is situated at the fore and inner part of the leg. It is triangular in shape.

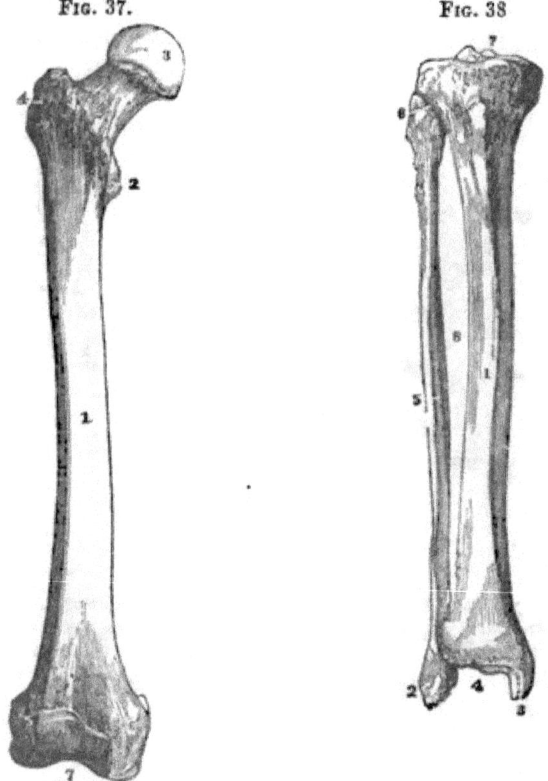

FIG. 37. THE FEMUR.—1, The shaft. 2, A projection (trochanter minor), to which are attached strong muscles. 4, The trochanter major, to which are attached the large muscles of the hip. 3, The head of the femur. 5, The external projection or condyle of the femur. 6, The internal projection or condyle. 7, The surface of the lower extremity of the femur, that articulates with the tibia, and upon which the patella slides.

FIG. 38. THE BONES OF THE LEG.—1, The tibia. 5, The fibula. 8, The space between the two, filled with the interosseous ligament. 6, The articulation of the tibia and fibula at their upper extremity. 2, The malleolar process, or external ankle. 3, The inter-malleolar process, or internal ankle. 4, The surface of the lower extremity of the tibia, that unites with a tarsal bone to form the ankle-joint. 7, The upper extremity of the tibia, upon which the lower extremity of the femur rests.

---

* Lat., *little dish.*

† Lat., a *flute.*

110. The FIBULA* is smaller than the tibia and of similar shape. It is firmly bound to the tibia at each extremity.

111. The TARSUS is formed of seven irregular bones, firmly bound together by a few large and strong ligaments, and by a great number of short fibres that extend between the contiguous bones, both on the back and sole of the foot.

112. The METATARSUS consists of five bones, they bear a close resemblance to the metacarpus of the hand. The tarsal and metatarsal bones are so united as to give the foot the form of a double arch.

113. The PHALANGES of the toes have fourteen bones, each of the small toes having three, and the great toe, two rows.

FIG. 39.

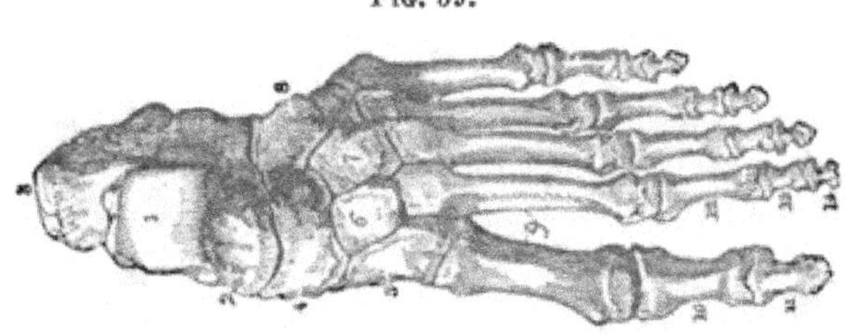

FIG. 39. THE UPPER SURFACE OF THE BONES OF THE FOOT.—1, The surface of the astragulus, or ankle-bone, where it unites with the tibia. 2, The body of the astragulus. 3, Calcis, or heel-bone. 4, The scaphoid. 5, 6, 7, The cuneiform. 8, The cuboid. 9, 9, 9, The metatarsal bones. 10, 11, The phalanges of the great toe. 12, 13, 14, The phalanges of the other toes.

114. The JOINTS are formed by the ends of bones, usually enlarged and variously united according to the purposes to be subserved. Generally, one surface is somewhat convex and the other correspondingly concave, the two parts being beautifully fitted to each other; associated with these, are the Cartilages, Ligaments, and the Synovial Membrane. With slight modifications, the classification of joints now adopted was proposed by Galen nearly two thousand years ago; all the articulations being distributed into three groups—the *Immovable* (Synarthrosis); the *Mixed* (Amphiarthrosis); and the *Movable* (Diarthrosis).

* Lat., a *clasp*.

115. The IMMOVABLE JOINTS include the several kinds of suture. A suture is called *Serrated* when the zigzag edges are united as in the external plate of the skull; *Squa'mose*, when the edges are beveled so that one overlaps the other as in the union of the temporal and parietal bones; *Lim'bous* when the borders of the adjacent bones are elevated, as in the union of the parietal and occipital bones. Sometimes a *false* suture occurs called *Harmonia*, where the opposed edges are smooth and even, as in the internal plate of the skull, the upper jaw-bones, the palate, and other bones. The fitting of the teeth into their sockets, as a nail is driven into a board, is called *Gomphosis:** these are improperly classed with the joints.

116. The MIXED JOINTS are those in which the opposed surfaces of the bones are joined directly together by some intermediate soft substance, which is fibrous externally, and more or less cartilaginous toward its central part; as between the bodies of the vertebræ, and the two upper parts of the sternum.

117. The MOVABLE JOINTS are the most perfect articulations, being *freely movable*, for which purpose, they are covered with cartilage where the surfaces are in contact, and provided with synovial membrane, and connecting ligaments. They are of three kinds—the *Planiform*, the *Hinge*, and the *Ball and Socket* joints. The *Planiform* is found where the surfaces of the bones are more or less plane, and the movements gliding, as in the articulations of the tarsus and metatarsus, the carpus and metacarpus; in those of the collar-bone; of the lower jaw; of the articular processes of the cervical and dorsal vertebræ; of the ribs with the vertebræ, and their costal cartilages with the sternum; and of the tibia with the fibula: The *Hinge joint* (Ginglyform), where there is motion in two directions only, backward and forward, as at the knee and the elbow; the ankle and the wrist; and the joints of the phalanges of the fingers, and the toes. In this joint, the end

* Gr., *gomphos*, a nail.

of one bone is so modeled as to present a median groove, and two lateral projections; while the end of the other has a median projection, and two lateral grooves: The *Ball and Socket* joint (Enarthrosis), also called *Rotary*, where there is free movement in all directions; it consists of a cup-like cavity in one bone, and a rounded extremity to fit it, in the other bone, as seen in the hip and shoulder joints; the socket at the hip is called the *Acetab'ulum;** at the shoulder, the *Glenoid* cavity.

118. There are certain forms of movable joints which require special description. The articulation between the upper ends of the radius and ulna, may be called a *ring* or *collar* joint; for the convex head of the radius plays into a very shallow cavity of the ulna, and is bound to it by a ring of ligament passing around the head of the radius. At the wrist, by a similar arrangement, the head of the ulna plays into the radius. The joint permitting the rotary motion of the head, is somewhat similar; a tooth-like projection of the axis playing into the anterior part of the ring of the atlas is held in position by transverse ligaments crossing behind it. The articulation allowing the nodding motion of the head, really consists of two separate joints, one on each side of the median plane; these joints are between the lower skull-bone and the atlas.

§ 8. HISTOLOGY OF THE BONES.—*Formation of Temporary Cartilage. Intra-Cartilaginous Mode of Bone-Formation.—Intra-Membranous Mode. Periosteum. Endosteum. Cartilages of the Joints. Synovial Membrane. Ligaments.*

119. The primitive basis, or *plasma* of the bone, is a subtransparent, glairy matter containing numberless minute corpuscles. It gradually acquires firmness, and nucleated cells appear, indicating the change into cartilage. As these cells increase in number and size, they become aggregated in rows, or columns, with intercellular tracts where ossification

* LAT., *acetum*, vinegar, from resemblance to the vinegar-cup of the ancients.

is about to begin. In the cartilaginous basis of long bones, these rows are vertical to the *ends;* in that of flat bones, to the *margin.*

120. The first appearance of *bone* is that of minute granules in the intercellular tracts. Afterward, the cartilage corpuscles become filled with these granules in all parts excepting their nuclei, which remain isolated in the bony substance. From these proceed minute canals, which become enlarged, forming the cavities called *Lacu'næ.* These everywhere connect with each other by minute tubes, called *Canalic'uli.* One layer

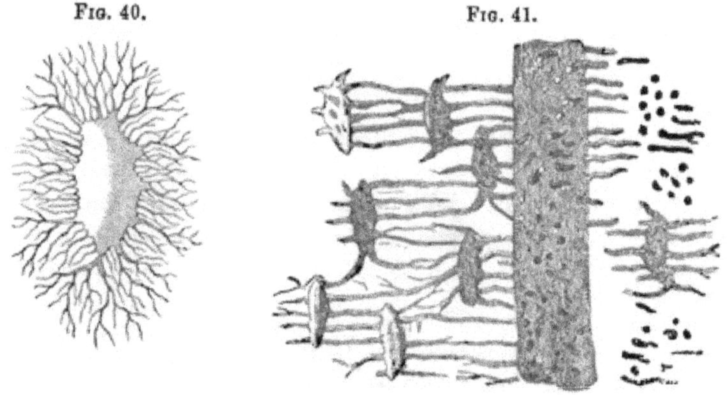

Fig. 40. Fig. 41.

Fig. 40 (*Leidy*). An Osseous Lacuna, exhibiting its numerous diverging canaliculi; highly magnified.
Fig. 41 (*Lessing*). Haversian Canal, lacunæ and connecting canaliculi.

of cells after another is thus converted into bony plates, till the whole column is filled excepting a fine central tube called the *Canal of Havers.* This microscopic, osseous cylinder is called an *os'sicle,* and is a true miniature of any one of the long bones. The compact portion of all bones is made up of these ossicles, which, under the microscope, resemble bundles of pipe-stems, placed side by side; the interspaces being filled with lamellated bone-substance. This mode of bone-formation, is called *Intra-cartilaginous.*

121. There is another mode of bone-development widely distributed through the body, called *Intra-membranous;* as seen in the flat bones. The membrane in which this ossifica-

tion takes place seems to consist of soft, amorphous matter, containing granular corpuscles. From certain points, called centres of ossification, the growing bone shoots into the soft substance in the form of transparent fibres which gradually become charged with earthy salts. The number of these centres varies in different bones—the parietal having one; the frontal, two; the occipital, four; and others, still more. It is supposed that the corpuscles become the lacunæ of the bone. By the fibres radiating from the centres of ossification, little grooves are formed which become the canals of Havers. The lacunæ connect with each other and with the Haversian canals; thus establishing a free communication between all parts of the bone.

FIG. 42.

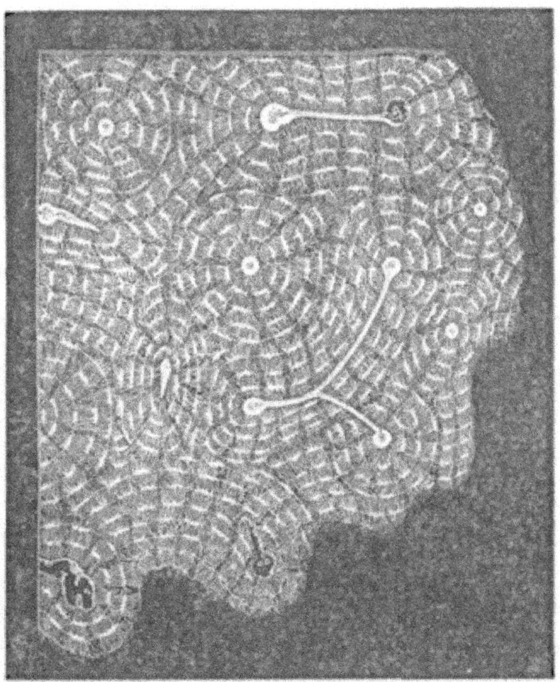

FIG. 42 (*Leidy*). TRANSVERSE SECTION OF BONE FROM THE SHAFT OF THE FEMUR; highly magnified. The large circular orifices are transverse sections of the Havers canals, surrounded by concentric layers of osseous substance. Between the latter are seen the lenticular excavations or lacunæ intercommunicating by means of canaliculi.

122. The long bones are hollow cylinders compact upon the exterior, and cancellated, or spongy within. This open texture increases toward the ends, which it entirely fills excepting the very thin, hard wall. The cylindrical cavity is filled with a yellowish fat called *Medul'la*, consisting of soft, delicate adipose cells.

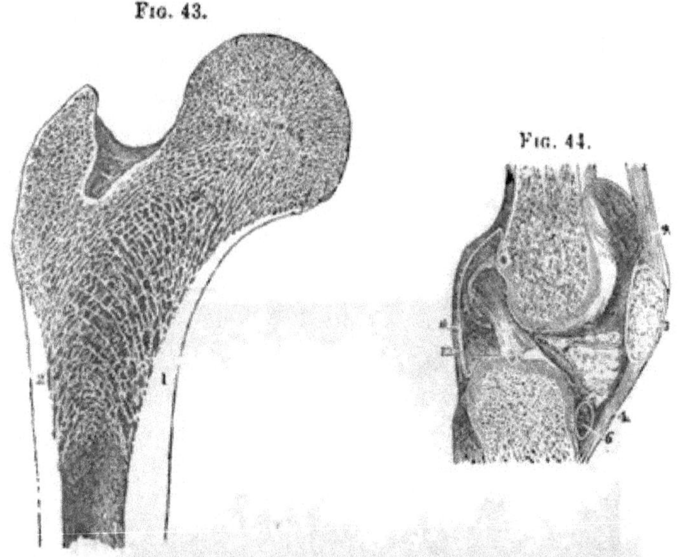

Fig. 43 (*Leidy*). Longitudinal Section of the Proximal Extremity of the Femur, exhibiting the arrangement of the spongy substance. 1, 2, Positions in which the compact substance appears to resolve itself into a series of arches.

Fig. 44. A Vertical Section of the Knee-Joint.—1, The femur. 3, The patella. 5, The tibia. 2, 4, Ligaments of the patella. 6, Cartilage of the tibia. 12, The cartilage of the femur. * * * *, The synovial membrane.

123. With the exception of the cartilage-tipped extremities, the bones are invested with a dense, white-fibrous membrane, called *Peri'osteum;** and even at the joints it may be traced over the capsular ligaments, thus realizing the opinion of the ancients that the periosteum formed a complete sac for the whole skeleton. Nor is this true of the external only; for, continuous with the periosteum, and lining the medullary cavity and various openings of the bone, there is a web-like

* Lat., *peri*, around, and *os*, a bone.

and very vascular membrane of extreme tenuity, called *Endos'teum,** or *Internal Periosteum.*

124. In order to facilitate the movements of bones upon each other, they are covered at the joints, with a thinnish layer of CARTILAGE or gristle—a tough, elastic, pearly-white substance, very smooth on the free surface (25). Upon convex surfaces, it is thickest in the centre, while upon concave surfaces, it is thickest toward the circumference. This cartilage is sometimes interposed as a ring, forming a movable socket, which, like the friction-wheels of machinery, aids the motion of the joint. This arrangement is seen in the lower jaw, the cartilage being attached to the synovial membrane, but perfectly movable and following the movements of the jaw, thus preventing dislocation.

125. The SYNOVIAL MEMBRANE (40) secretes a viscid fluid, called *Syn'ovia,* which lubricates the movable joints. This membrane is of three kinds;—the *Articular Capsules,* the *Bursæ Mucosæ,* and the *Sub-cutaneous* Synovial Capsules. The *Articular Capsule* forms a complete sac, which covers the articular surface of one bone, and is thence reflected to the other, adhering closely to the borders of each of the cartilaginous surfaces. The *Bursæ Mucosæ* are pouches of synovial membrane interposed between bones and the tendons that play upon them like cords upon pulleys; they also occur where tendons or muscles move upon ligaments, fibro-cartilages, or upon each other. Tendons moving through grooves of bone are enclosed in a synovial tube which is reflected upon itself so as to line the groove within which the motion takes place. The *Sub-cutaneous Capsule,* or membrane, is found wherever the skin is frequently moved over a resisting part, as between the skin and patella at the knee. Wherever a number of tendons move upon one another, this membrane is folded around and among them; it appears to have the same function as the bursæ mucosæ.

126. Outside the synovial membrane, and more or less

* Gr., *within,* and *os,* bone.

connected with it, are the special *ties* of the joints, called LIGAMENTS. These ligaments are composed of white-fibrous tissue and are named *Cap'sular*, *Band-like*, and *Funic'ular.* The *Capsular Ligaments* are cylindrical sacs, extending completely around the joints, and blending with the periosteum. This form is found with the ball-and-socket joint. The hip and the shoulder joints furnish perfect examples of the ligamentous capsule. The *Band-like Ligaments* are broad bands of parallel fibres, found with the hinge-joint; and sometimes, where great strength is needed, as accessory to the capsular ligament. The *Funicular Ligaments* are cords round or flat, which extend from one bone to another, sometimes within and sometimes without the joint. An example is seen in the two ligaments crossing each other within the knee-joint, also in the single ligament within and connecting the ball-and-socket joint of the hip.

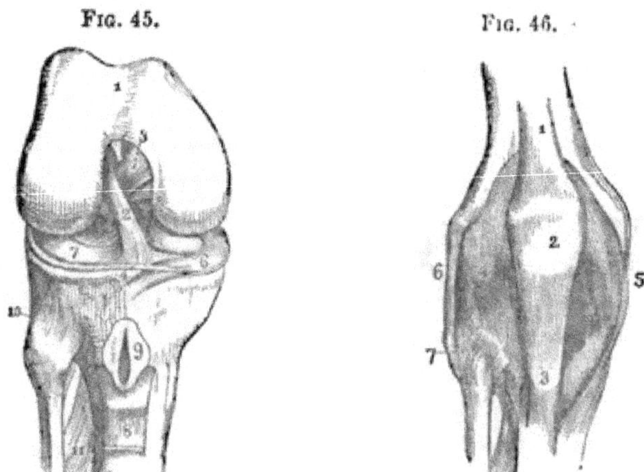

FIG. 45 (*Leidy*). THE RIGHT KNEE-JOINT, laid open from the front. 1, Articular surface of the femur. 2, 3, Crucial ligaments. 4, Insertion of one of these ligaments into the tibia. 6, 7, Internal and external semi-lunar fibro-cartilages. 8, Ligament of the patella turned down, so as to exhibit the synovial bursa (9) beneath. 10, Superior tibio-fibular articulation. 11, Interosseous membrane.

FIG. 46 (*Leidy*). FRONT VIEW OF THE RIGHT KNEE-JOINT.—1, Tendon of the quadriceps extensor muscle. 2, Patella. 3, Ligament of the patella, or tendinous insertion of the muscle just mentioned. 4, 4, Capsular ligament. 5, 6, Internal and external lateral ligaments. 7, Superior tibio-fibular articulation.

§ 9. Chemistry of the Bones.—*Chemical Composition of the Bones.—Experiments showing Earthy and Animal Matter.*

127. Bones are composed of both animal and mineral matter; the animal matter being in excess in early life, and the mineral, in old age. The average proportion is about thirty-three per cent. animal matter (*cartilage* and *blood-vessels*); and sixty-seven per cent. mineral, of which fifty-one parts are bone-earth (*phosphate of lime*); eleven parts chalk (*carbonate of lime*); the remaining parts are fluor spar (*fluoride of calcium*); *phosphate of magnesia;* and common salt (*chloride of sodium*).

Observation.—To show the earthy without the animal matter, burn a bone in a clear fire, and it becomes white and brittle, the animal part having been consumed. To show the animal without the earthy matter, immerse a slender bone, for a few days, in weak acid (one part muriatic acid and six parts water) and it becomes flexible, the earthy matter having been removed.

§ 10. Physiology of the Bones.—*General Uses of the Bones.—Adaptation of their Structure to their Uses. Skill as shown in the Skull.—In the Spinal Column.—In the Ribs.—In the Pelvis.—In the Upper Extremities.—In the Lower Extremities.—In the Long Bones. The uses of the Joints. Classification of the Joints.—Of Movable Joints. Function of the Synovia. Of the Cartilages. Of the Ligaments. Of the Periosteum. Perfection of this part of the Animal Fabric.*

128. The Bones serve as the framework of the system; as bases for the attachment of muscles; as levers for the organs of locomotion; as pulleys for the passage of tendons; and as protection for the delicate internal organs. In their adaptation to their several offices, they exhibit a perfection of mechanism worthy the infinite mind of the Divine Architect.

129. In the minutest structure of the bones as revealed by the microscope, we find the delicate tissues so disposed as to give the greatest amount of strength and lightness, and a certain degree of elasticity—qualities essential to the per-

formance of their several offices. In their more general structure, we see regard to the same qualities, the exterior being dense and compact, the interior, spongy, or cancellated. Take any bone, or series of bones, and note their peculiar configurations and the purposes to be subserved, and there appears the same marked evidence of special care and skillful mechanism.

130. In the arrangement of the SKULL for the protection of the brain, the oval form (the form best adapted to resist pressure equally applied on all sides); the thickened base where the most important part of the brain lies; the strong and narrow prominences, both in front and back, where most exposed to violence; the tough and hard plates to resist the penetration of sharp substances; the intervening spongy layer to diminish vibrations; the separate bones, and the serrated unions of the external plates, also to lessen shocks; the simple contact of plane edges in the internal vitreous plate, where zigzag edges would be easily broken; the projections, depressions, and apertures for the safe passage of nerves and blood-vessels—all combine to accomplish the one object, *protection.*

131. To construct the SPINAL COLUMN was no easy mechanical problem. These offices were to be taken into the account: 1. It must support the head; 2, furnish an axis of support for the other parts of the body; 3, allow a bending and somewhat rotary movement; 4, furnish a basis for the attachment of muscles; 5, provide passages and protection for the spinal cord and nerves; 6, the whole must be arranged with reference to the importance and delicacy of the brain.

132. To furnish proper support, it must possess firmness and strength; these are secured by the broad bases of the vertebræ, which increase in size according to the increase of weight to be supported, by the powerful ligaments extending the length of the column, and by the interlocking of the projections of the vertebræ: To secure the necessary rotary movement, there are short horizontal processes in the cervical and lumbar regions, while those of the dorsal vertebræ, where

movement is not required, are long and extend downward; flexibility and firmness are two qualities difficult to unite, but here the union is effected by making the vertebræ of short bones and increasing their number, and by the mechanical arrangement of the fibres of the disks between the vertebræ: To furnish an attachment for muscles, are the dense *processes*, which vary in size, form, and direction, according to the requirements of the muscles to be attached: To provide for the spinal cord, each vertebra has a perforation, and these openings, coming directly over each other, form a bony canal, while at the sides of this canal, are apertures for the passage of nerves; the spinal cord is still farther protected by the elastic ligaments between the arches of the vertebræ, allowing no openings in the bending of the body; and by the number of the vertebræ, for were there only three or four bones, the cord would be injured at every angle: The number of vertebræ, the cartilage cushions, and the four curves of the column, all tend to secure the brain from shocks it would otherwise receive, from walking, leaping, and running.

133. The RIBS serve to protect the delicate organs of the chest. These slender bones should be elastic and movable: The first quality is secured by the cartilaginous union to the sternum; the second, by their cartilages, their articulations with the spine, and their oblique position.

134. The PELVIS not only furnishes support for the upper part of the body, and the articulations of the lower extremities, but also serves as a base for the attachment of the powerful erector muscles of the spine, the muscles for moving the lower limbs, and the muscles which shut in the abdominal and pelvic cavities.

135. The form and proportion of the UPPER EXTREMITIES relate to the *hand*, which belongs exclusively to man, and gives the power of execution to the human mind: Thus, the arm is longer than the fore-arm, and this, longer than the hand, securing greater mobility, flexibility, and power of adaptation as we approach this delicate organ of prehension. It is the

relative position of the four fingers to the thumb, however, which principally stamps the character of the hand, as this construction permits its adaptation to every shape, and gives that complete dominion which it possesses over the various forms of matter.

136. The LOWER EXTREMITIES have a strong analogy to the upper, the differences being only such as are necessary to constitute them organs of locomotion rather than of prehension; hence, their *solidity*, at the expense of their *mobility*. The tarsal and metatarsal bones form a strong arch toward the inner and lower surface of the foot, protecting the vessels, nerves, and tendons, passing from the leg to the foot, and the opposite.

FIG. 47.

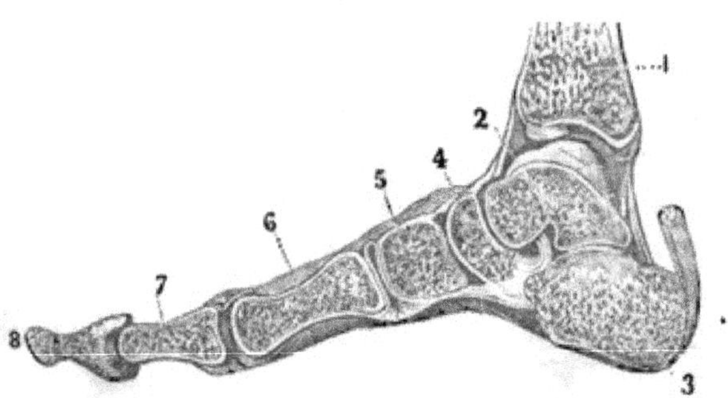

FIG. 47. A SIDE VIEW OF THE BONES OF THE FOOT, SHOWING ITS ARCHED FORM.—The arch rests upon the *heel* behind, and the *ball* of the toes in front. 1, The lower part of the tibia. 2, 3, 4, 5, Bones of the tarsus. 6, The metatarsal bone. 7, 8, The bones of the great toe.

137. The *shafts of long bones* are made hollow, giving not only lightness but strength, according to the well-known principle in mechanics, that, with a given amount of material, a hollow cylinder will sustain more weight than a solid one, both being of the same height: (The same principle is illustrated in the culms of grasses.) We find the walls most compact, and the cavity broadest at the middle of the cylinder, the part subjected to the greatest strain; we find the extremities enlarged, to give a broader surface for the attachment of muscles, and often to change the direction in the

action of muscular power; also more spongy in texture, to increase the size without corresponding increase of weight.

138. THE JOINTS. The uses of the *joints* are to enable the body to sustain greater weight (as several short pillars will support more weight than a single pillar of the same height and thickness); to diminish the force of blows or shocks; to afford freedom of movement; to provide fulcrums for the various levers; to modify the direction in the action of muscular power, and to determine the plane of action.

139. For simple union without movement, we find the Immovable joint; for great strength and little movement, the Mixed joint; and for full freedom of movement, the Movable joint. Of the movable joints for motion in one plane and two directions, we find the Hinge-joint; for the gliding movement, the Planiform joint; and for free rotary motion, the Ball-and-Socket joint. At the hip, the socket is very deep, giving strength; at the shoulder, it is shallow, giving unrestricted motion; everywhere the configuration and the use correspond.

140. The use of the SYNOVIA is to enable the surfaces of the bones to move more easily upon each other, preventing friction and consequent wear. No machine of human invention manufactures for itself the necessary lubricating fluid, but, in the animal mechanism, it is supplied in proper quantities, applied in the proper places, and at the proper time.

141. CARTILAGE tips the articular extremities of bones, facilitating the sliding motion, and deadening shocks; and in various parts of the body it serves as an elastic cushion, yielding on compression and regaining its form when the pressure is removed.

142. The function of the LIGAMENTS is to bind together the bones of the system. By them the lower jaw is bound to the temporal bones, and the head to the neck; they extend the length of the spinal column, between the vertebræ, and from one spinous process to another; they bind the ribs to the vertebræ, and the sternum; the sternum to the clavicle; the clavicle to the first rib and the scapula; the scapula to

the humerus; the bones of the fore-arm at the elbow-joint and also at the wrist; the bones of the wrist to each other, and to those of the hand; and these to each other, and to those of the fingers and thumb: In the same manner they bind the bones of the pelvis together; and these to the femur, or thigh-bone; and this to the two bones of the leg and the patella, or knee-pan; and so on to the ankle, foot and toes, as in the upper extremities. The bones of the wrist and those of the foot are as firmly fastened as if bound by clasps of steel.

143. The PERIOSTEUM serves to transmit blood-vessels into the bone, thus furnishing nutriment; it gives insertion to muscles, tendons and ligaments; obviates the effects of friction; strengthens the whole skeleton as an investing membrane, and possesses some agency in the process of ossification.

144. We have noticed but a few of the many wonderful examples of skill, wisdom and benevolence exhibited in the internal framework of the animal fabric. Each bone, however small, illustrates some profound principle of science; each is perfect in its adaptation to a specific use. The whole structure is a faultless piece of mechanism, in which every known principle of architecture and dynamics has been brought into service.

§ **11.** HYGIENE OF THE BONES.—*Effect of Exercise upon the Bones of Children.—Effect of Compression.—Of Stooping. Treatment of Fractures.—Of Sprains.—Of Felons.*

145. *The health of the bones is promoted by regular exercise.* The kind and amount of labor, should be adapted to the age, health, and development of the bones; neither the cartilaginous bones of the child, nor the brittle bones of the aged man, are adapted to long-continued and severe exercise. While protracted exercise in childhood is injurious, moderate and regular labor favors a healthy development and consolidation of the bones. In middle age, the proportions of animal and earthy matter are usually such as to give the proper degree

of flexibility, firmness and strength, with little liability to injury.

146. *The lower extremities of the very young, are not adapted to sustaining much weight;* hence, to induce a child to walk, or to stand by chairs, while the bones of the lower limbs are imperfectly developed, is ill-advised and productive of serious injury. The "bandy" or "bow" legs are thus produced. The benches or chairs for children in a school-room, should permit the feet to rest upon the floor, otherwise the weight of the limbs below the knee, may cause the flexible bone of the thigh to become curved; the chairs should also have suitable backs, and the child be allowed frequent change of position.

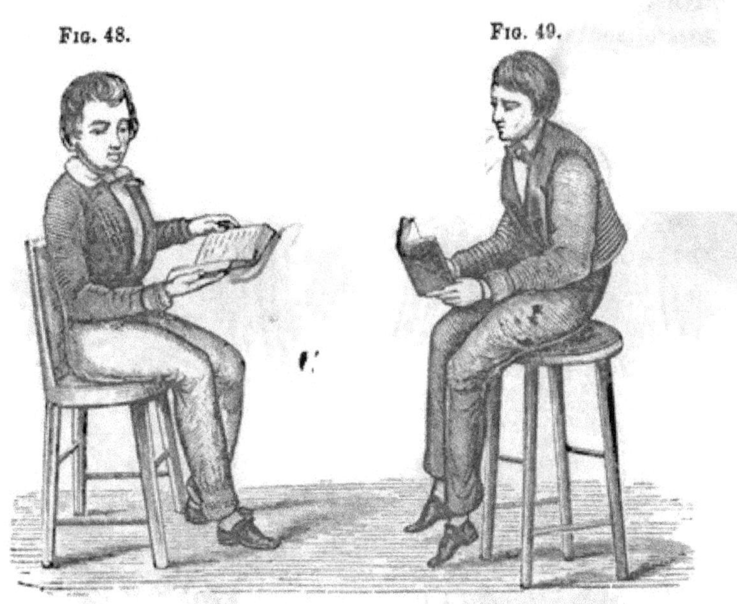

Fig. 48. Natural Position when the feet are supported.
Fig. 49. Unnatural Position when a seat is too high.

147. *Compression of the chest should be avoided.* In youth, the ribs are very flexible, and a small amount of pressure will increase their curvature, particularly at the lower part of the waist. By tight or "snug" clothing the ribs are drawn down, and the space between them lessened, so that in some

instances, the anterior extremities of the lower ribs are brought quite together; hence, the apparel should be loose and supported by the shoulders, both for children and adults.

148. *An erect position both in sitting and standing should be carefully maintained.* The spinal column naturally curves from front to back, but not from side to side. The admirable arrangement of the bones and cartilages permits a great variety of motions and positions, the elasticity of the cartilages always tending to restore the spine to its natural position; but if a stooping, or a lateral curved posture be continued for a long time, the compressed edges of the cartilages lose their power of reaction, and finally, one side becomes thinned, while the other is thickened. These wedge-shaped cartilages produce permanent curvature of the spine, which is often attended with disease of the spinal cord.

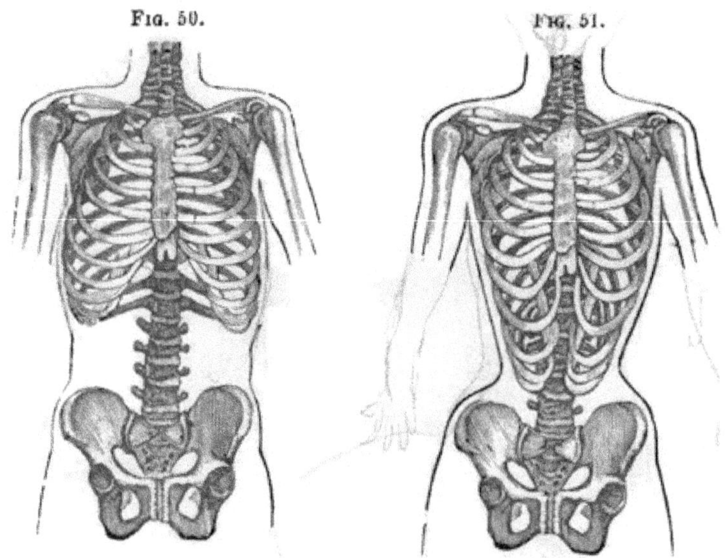

Fig. 50. Fig. 51.

Fig. 50. A Chest well proportioned.
Fig. 51. A Chest fashionably deformed.

149. The student, seamstress, and artisan, frequently acquire a stooping position, by inclining forward to bring their books or work nearer the eyes. The desk of the pupil is

often higher than the elbow as it hangs from the shoulder at rest, consequently, in drawing, writing, and often in studying, one shoulder is elevated and the other depressed, distorting the spine. In the daily employments of life, children should early be taught to use the left hand and shoulder more freely. Distortions of the chest necessarily accompany deformity of the spine, and disease of the heart and lungs follows, compared to which, the loss of symmetry is a minor consideration.

Fig. 52.

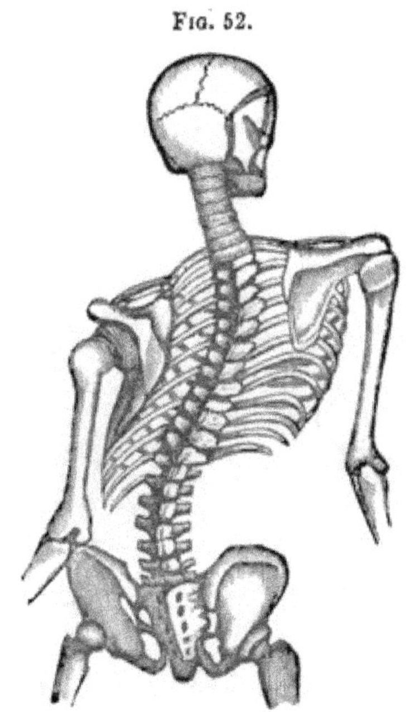

Fig. 52. A Deformed Thorax and Spinal Column.

150. Eminent physicians, both in this country and Europe, state that, among the fashionably educated, not one female in ten escapes deformities of the shoulders and spinal column. The student, to prevent as well as to cure slight curvatures of the spine, should walk with a book, or a heavier weight upon the head. Porters and laborers of some countries, bear very great burdens upon their heads, and walk at a rapid pace with comparative ease. Itinerant toymen carry securely

their trays of fragile merchandise, because the head and spinal column are erect.

Fig. 53.

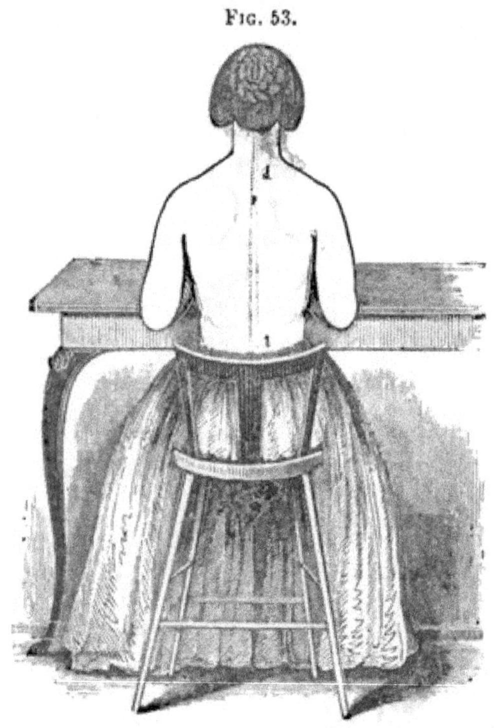

Fig. 53. Correct Position in Sitting.—1, 1, The spinal column straight; the shoulders of equal height.

151. *Fractured or diseased bones and ligaments should receive special attention.* In *fractured bones*, a surgeon's care is not only needed to adjust the parts, but for several weeks, to watch the reunion, that the limb may not be crooked or shortened. In *sprains*, the ligaments are not usually lacerated, but strained and twisted, causing much pain, and afterward inflammation and weakness of the joints. To effect a cure, there should be absolute *rest* for days, and perhaps weeks. More persons are crippled from ill-cared-for sprains, than fractured bones. Persons enfeebled by disease, particularly scrofula, cannot be too assiduous, in adopting an early and

proper treatment of injured joints, to prevent the affection, called "white swelling."

Observation.—The disease called "*Felon*" is an inflammation that commences in or beneath the periosteum. It is attended with severe, throbbing pain, and the unyielding structure of the parts prevents much swelling. The only *successful* treatment of this painful affection, is an *early, free* opening through the periosteum to the surface of the bone. The earlier the incision is made, the less the risk and the suffering.

Fig. 54.

Fig. 54. Incorrect Position in Sitting.—1, 1, Three curves of the spine; the shoulders of unequal height.

Fig. 55.

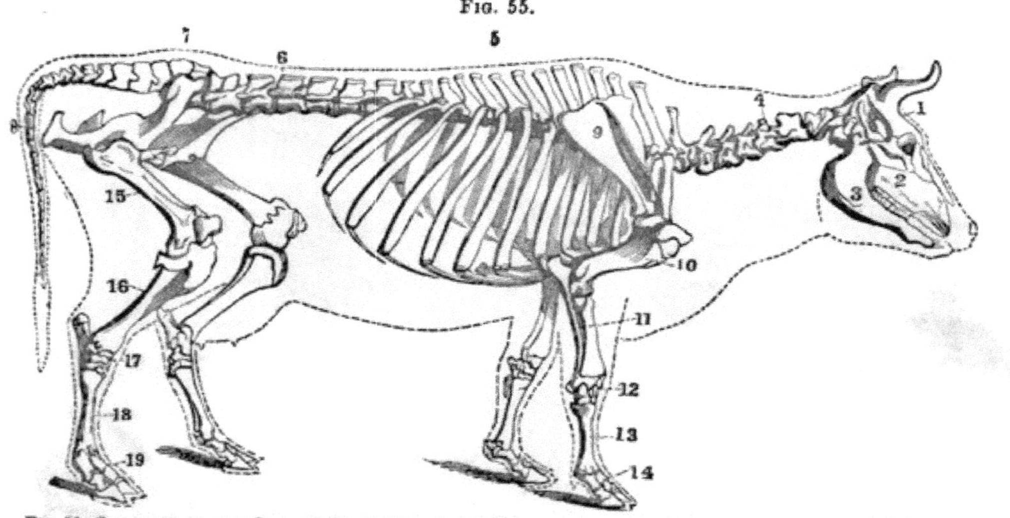

Fig. 55. Skeleton of the Cow.—1, Frontal bone of the head. 2, Upper jaw (superior maxillary). 3, Lower jaw (inferior maxillary). 4, Cervical vertebræ. 5, Dorsal vertebræ. 6, Lumbar vertebræ. 7, Sacral vertebræ. 8, Caudal vertebræ. 9, Scapula. 10, Humerus. 11, Radius and ulna. 12, Carpus. 13, Metacarpus. 14, Phalanges (toes). 15, Femur. 16, Tibia. 17, Tarsus. 18, Metatarsus. 19, Phalanges. In this fig. the same terms are used as for the corresponding bones in man (see fig. 25). The common names vary.

§ 12. Comparative Osteology.—*Classification of Animals according to their Plan of Structure. Classification of Vertebrates. Compare Spinal Columns of Vertebrates.—Bones of the Head.—Of the Thorax.—Of the Extremities.*

152. The tissues, cells, and chemical composition of all animals are essentially the same, but their different appointments in the plan of creation require special conformations. Animals have therefore been arranged, according to their *plan of structure*, into four sub-kingdoms:—1. Vertebrata, including man and other animals having an *internal* skeleton with a *back-bone* as its basis, the bones being composed chiefly of phosphate of lime: 2. Articulata, comprising animals having an *external* skeleton made up of similar segments, or rings, consisting mostly of the carbonate of lime; as insects, lobsters and worms: 3. Mollusca, including soft-bodied animals covered with a hard shell consisting of one or two pieces, also composed of carbonate of lime; as cuttlefish, oysters, clams, and snails: 4. Radiata, having no proper skeleton or shell-covering, but parts more or less symmetrically arranged about a vertical axis; as star-fishes, sea-anemones and coral animals.

153. The Vertebrata are classified as Mammals (including man, monkeys, bats, quadrupeds, etc.), Birds, Reptiles and Fishes.

154. The Vertebral Column of all *Mammals* is similar to that of man. The cervical vertebræ, with two exceptions, number seven: the dorsal, average thirteen; the lumbar, from three to seven; the sacral, usually four; the caudal, from four (the number of the coccyx in man) to forty-six. The length of any part of the column seems to depend not so much upon the *number* of the vertebræ as upon their *length;* thus we find seven cervical vertebræ in the long-necked giraffe and in the short-necked mole. In *Birds*, the flexibility of the neck enables any part of the body to be reached by the beak. This is owing to the ball-and-socket articulations, and to the great number of cervical vertebræ, which in the swan are twenty-four; in the ostrich, eighteen; and in the domestic

cock, thirteen. The dorsal vertebræ vary from seven to eleven, and are generally consolidated into one; but in birds that do not fly, they remain distinct and movable. The lumbar and sacral vertebræ are united into one. The last caudal vertebra has a large, strong process for the support of the large feathers. In *Reptiles*, the vertebræ vary in number from some twenty-four, as in the frog, to four hundred, as in some snakes, as the Python. Perhaps about one hundred is the average number. In *Fishes* there are but two kinds of vertebræ, the dorsal and the caudal, and these vary in number from twenty to two hundred. The vertebral bodies present a conical, cup-like depression on each side, which contains a gelatinous fluid having the same use as the elastic intervertebral substances in mammals.

FIG. 56.

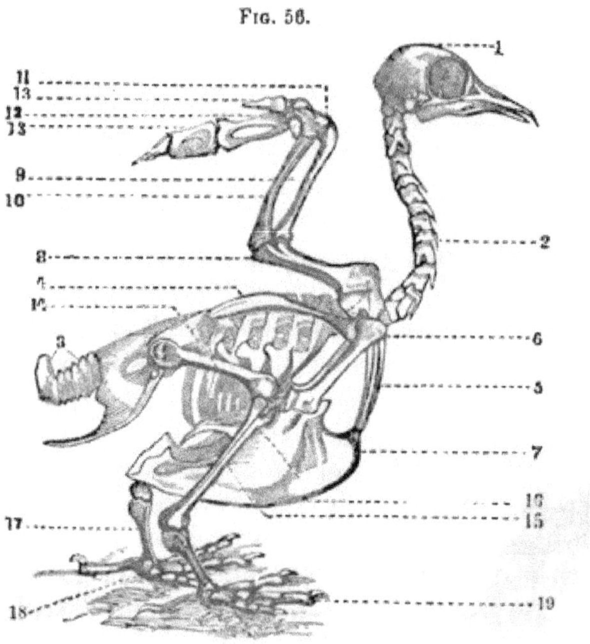

FIG. 56. SKELETON OF A BIRD.—1, The head. 2, Cervical vertebræ. 3, Dorsal and lumbar vertebræ. 4, Scapula. 5, Clavicle. 6, Coracoid bone. 7, Sternum. 8, Humerus. 9, Radius. 10, Ulna. 11, Carpus. 12, Metacarpus. 13, 13, Phalanges (fingers). 14, Femur. 15, Tibia. 16, Fibula. 17, Tarsus. 18, Metatarsus. 19, Phalanges (toes).

155. The Bones of the Head of all the *Mammals* resemble, in many points, those of man. In some quadrupeds, as the horse and the cow, the frontal bone is in two parts; in others, the two parietal bones are united: Between the two upper maxillary bones, are two small bones called *inter-maxillary;* the lower jaw consists of two pieces. In *Birds*, the bones of the head, in number and position, resemble mammals, but they are early united, leaving no trace of the sutures. The superior mandible, or upper jaw, of the bird is so articulated with the cranium as to admit of motion independent of the lower jaw (which never occurs in mammals), and the inferior mandible, instead of being articulated directly with the cranium, is connected through the intermedium of a distinct bone called the *Os Quadratum.* In *Reptiles* the head-bones are irregular in form, and greatly vary in number. In *Fishes* the bones of the head are numerous and irregular, and their study is a matter of much interest in acquiring a full knowledge of Natural History.

156. The obvious use of the Clavicle is to maintain the shoulders apart; hence in quadrupeds, where its presence would be a defect, it is wanting, as in the horse and cow. The clavicles of *Birds* are peculiar; they unite at their anterior extremity, forming a forked bone called *furcula*, or wish-bone. In birds of powerful flight, as the eagle, the clavicles are very strong; in others, as the domestic turkey, they are weak. Connecting the scapula to the sternum is the *cor'acoid* bone, which is placed side by side with the furcula, and is the main source of support to the wings in flight. In some *Reptiles*, as the tortoise, both the clavicle and the coracoid bone are found, while in others, as serpents, both are wanting. In *Fishes* the true clavicle is wanting, but in some species there is a modified form of the coracoid bone, free at its lower extremities, which may, perhaps, be considered as homologous with the coracoid bone or clavicle of the higher animals.

157. The Scapula is present in all *Mammals* and *Birds*, and most *Reptiles* and *Fishes*.

Fig. 57.

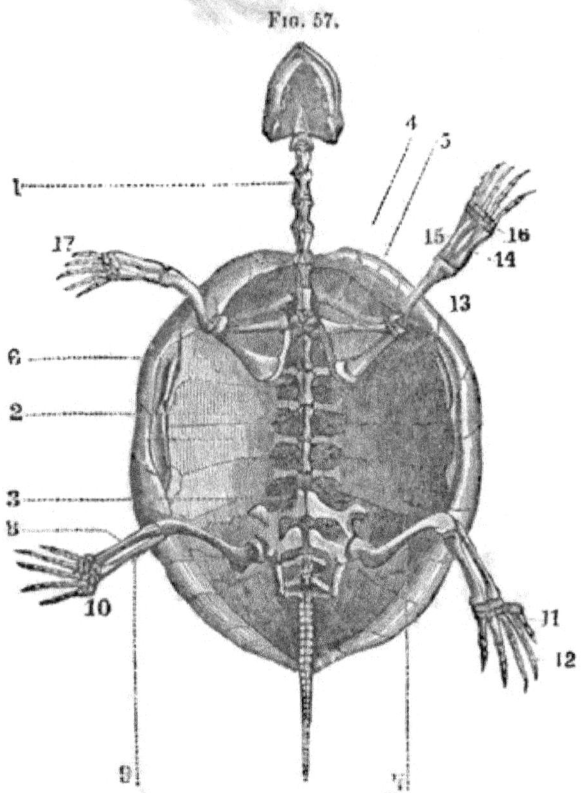

Fig. 57. Skeleton of a Tortoise.—1, Cervical, 2, Dorsal, 3, Lumbar vertebræ. 4, Scapula. 5, Clavicle. 6, Coracoid bone. 13, Humerus. 14, Ulna. 15, Radius. 16, Carpus. 17, Phalanges (fingers). 7, Femur. 8, Tibia. 9, Fibula. 10, Tarsus. 11, Metatarsus. 12, Phalanges (toes).

158. The Sternum of birds is the largest bone in their bodies. It has upon its anterior surface a *ridge* resembling the keel of a ship, for the support of the pectoral muscles used in flying. The size is proportioned to the powers of flight; hence in the little humming-bird, which is on the wing most of the day, it reaches the maximum of development. Of the *Reptiles*, serpents have no sternum; but in turtles, it has an extraordinary development, and extends from the base of the neck to the commencement of the tail, forming the ventral part of the shell-covering. In some *Fishes*, the sternum is represented by a *chain* of bones.

159. The RIBS are much alike in all *Mammals*. In *Birds*, the cartilage that unites the rib to the sternum is osseous, giving solidity to the chest. In some *Reptiles*, as lizards, crocodiles, and other reptiles formed in the same way, the ribs are more numerous than in mammals and birds, and protect the abdomen as well as the chest. In the turtle, the ribs are expanded, forming the dorsal part of its shell, or the roof of its portable dwelling-house. In serpents, the lower or anterior extremities of the ribs have no cartilage: they aid in progressive movement, or crawling, as, under the skin, their ends can be placed on the ground like feet. In some *Fishes*, the ribs are wanting; in others, they are very complete and surround the trunk; in still others, they are connected with a chain of bones representing the sternum.

FIG. 58.

FIG. 58. THE SKELETON OF A HADDOCK.

160. The HUMERUS is usually a long, hollow bone, with a rounded head at the upper extremity, but in animals that swim or burrow, it is short and flattened at the ends for the attachment of muscles, thus enabling the fore-limbs to be used with much force. In *Birds*, the humerus is larger than the femur, contrary to the relative proportion in man.

161. The RADIUS and ULNA are present in most *Mammals*, but they are incapable of moving upon each other. In *Birds*, they are longer in proportion than in Mammals, especially in birds of flight.

162. The CARPUS of *Mammals* is made up of two rows, but the number of bones varies from five to eleven. In *Birds*, it is represented by two short bones. The METACARPUS usually consists of five elongated bones, but, in the horse, only one bone with two rudimentary ones are found. Most mammals have five fingers, of which the thumb is generally rudimentary. In some *Birds*, the thumb is entirely wanting, also the little finger. The middle finger is longest, consisting of two and even three bones.

163. The POSTERIOR EXTREMITIES of quadrupeds are usually less modified than the anterior. In *Birds*, the Femur is short; the Tibia is the chief or longest bone of the hind limb; the Fibula is a small bone united at various distances down the tibia in different birds. A single bone represents the Tarsus and Metatarsus; this supports or carries the toes, which in birds never exceed four in number. In some *Reptiles*, as the tortoise, lizard and others, the anterior and posterior limbs are composed of bones which, in number, form, position and functions, much resemble the corresponding ones in mammals and birds: in the serpent tribe, the limbs are wanting. In *Fishes* the extremities are rudimentary, being represented by fins.

Suggestion.—The osteology of the three *lower* sub-kingdoms of animals is replete with interest and instruction, but the necessarily limited space of this elementary school-book entirely precludes their consideration: allow us to advise all who can command the leisure, to extend this study to the beautiful and wonderful works of creation as seen in these parts of the garden of the Lord.

CHAPTER V.

THE MUSCLES.

§ 13. ANATOMY OF THE MUSCLES.—*Law of Muscular Contraction.—Consequent Forms of Muscles. Modes of Attachment of Muscles.—Number and General Arrangement. Muscles of the Head and Neck.—Of the Upper Extremities.—Of the Trunk.—Of the Lower Extremities.*

164. THE CHARACTERISTIC PROPERTY of muscles is *contractility*, and the law is, that they shall *contract toward the centre.* To accomplish this, there must be diversity of form, adapting them to different positions; hence, muscular fibres are longitudinal, terminating at each extremity in a tendon, forming a *spindle-shaped* or *fusiform muscle;* disposed like the rays of a fan, converging to a tendinous point, a *radiate muscle;* converging to one side of a tendon running the whole length of a muscle, as one side of the plume of a feather to its shaft, a *penniform muscle;* converging to both sides of the tendon like an entire feather, a *bi-penniform muscle;* or running in a circular direction, an *orbicular*, or *sphincter muscle.*

FIG. 59.

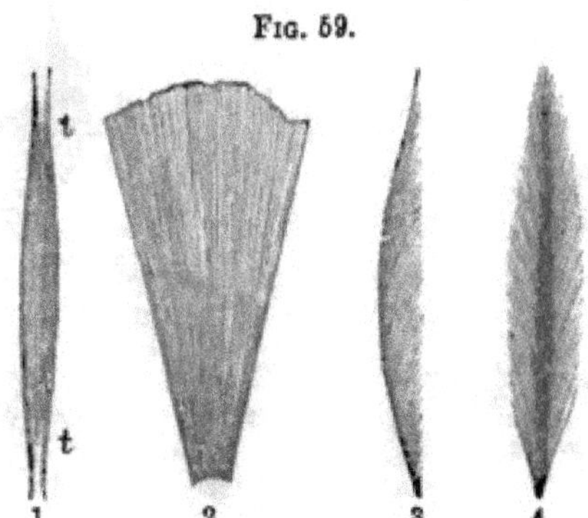

FIG. 59. 1, A REPRESENTATION OF THE DIRECTION AND ARRANGEMENT OF THE FIBRES in a fusiform or spindle-shaped muscle. 2, In a radiated muscle. 3, In a penniform muscle. 4, In a bi-penniform muscle. *t*, *t*, The tendons of a muscle.

165. Muscles are usually attached by both their extremities to the bones, either directly, or indirectly by means of the inelastic but flexible tendons, which may be cord-like (either round or flattened) or flat and broad, supporting the organs which they surround, and named *Aponeuroses*, or *Fasciæ*. Sometimes the muscle is attached to bone by one extremity only, the other being fixed to the skin or other soft part, as certain of the muscles of the face: sometimes there is no connection with bone, as in the orbicular muscle of the mouth. When a muscle is attached to bone by one extremity only, that attachment is called its *origin*, the other being termed its *insertion;* when attached at both ends to the bones, the attachment nearer the centre of the body, and which is usually the more fixed point, is called the origin, while that more distant and movable is named the insertion. Muscles may have one, two, three, or many points of origin, and some muscles have more than one point of insertion.

FIG. 60.

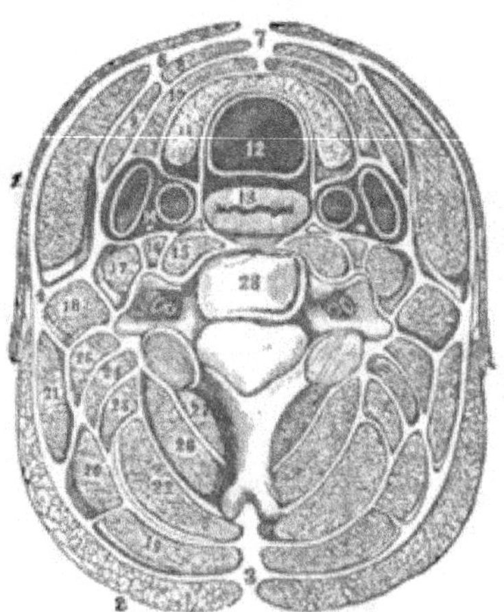

FIG. 60. A TRANSVERSE SECTION OF THE NECK.—The separate muscles, as they are arranged in layers, with their investing fasciæ, are well represented. 12, The trachea. 13, The œsophagus. 14, Carotid artery and jugular vein. 28, One of the bones of the spinal column. (The figures in the white spaces represent fascia; other figures, muscles; as the system is symmetrical, figures are placed only on one side.)

166. The NUMBER of muscles in the human body is more

than five hundred. In general, they form about the skeleton two layers, distinguished as *superficial*, and *deep-seated* muscles; yet in some parts there are three, four, five, and even six layers.

167. With the exception of twelve *single* muscles, they are arranged in pairs. Each muscle has its antagonist; when one contracts, the other relaxes. The muscles passing over the back of a joint are usually called *Extensors*, because they serve to extend the part beyond the joint; while those lying in front of the joint are, for the opposite reason, called *Flexors.*

Examples.—1st, Clasp the arm midway between the shoulder and elbow, with the thumb and fingers of the opposite hand; when the arm is bent, the inside muscle is hard and prominent, and the tendon near the elbow, rigid, while that upon the opposite side is relaxed: straighten the arm, and the outside muscle swells and becomes firm, while the inside muscle and its tendon are relaxed.

2d, Clasp the fore-arm about three inches below the elbow, then open and shut the fingers rapidly, and the alternate contraction and relaxation of the muscles on the opposite sides of the arm are felt, the movements corresponding to those of the fingers: when the fingers bend, the inside muscles contract, and the outside ones relax; when the fingers open, the inside muscles relax, and the outside contract. This action of antagonist muscles may be felt in all the different movements of the limbs.

MUSCLES OF THE HEAD AND NECK.

168. The Occipito-Frontalis elevates the eyebrows.

The Orbicularis Palpebrarum closes the eyelids. and, by pressing back the ball of the eye, it also compresses the lachrymal gland and causes a flow of tears.

The Orbicularis Oris closes the mouth, and enables the lips to embrace any substance placed between them. It receives into its periphery the fibres of the surrounding muscles, which meet here as in a common centre. It enters largely

Fig. 61

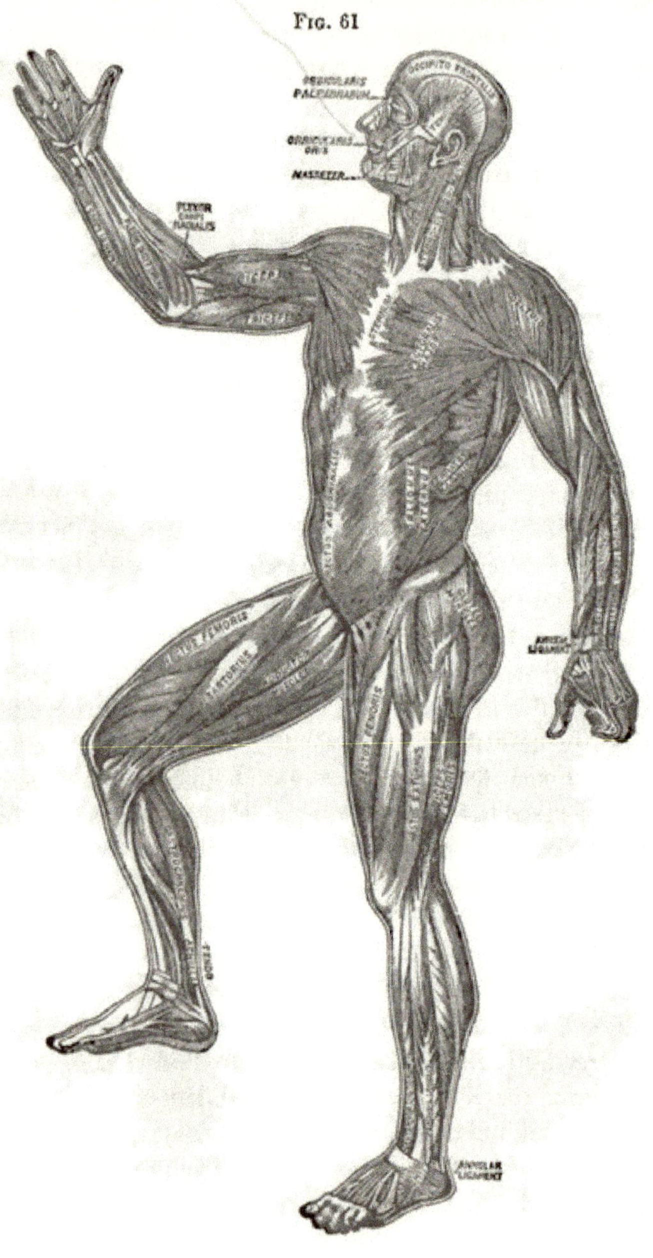

into the diversified expressions of the countenance, and in no one respect exhibits more varied adaptations than in the performance on wind instruments.

The MASSETER and TEMPORAL give motion to the lower jaw.

The STERNO-CLEIDO-MASTOID, when both sides contract, draws the head forward or elevates the sternum.

MUSCLES OF THE ANTERIOR PART OF THE TRUNK.

169. The PECTORALIS MAJOR draws the arm by the side, and across the chest, and also draws the scapula forward.

The SERRATUS MAGNUS elevates the ribs in inspiration.

The OBLIQUUS EXTERNUS and RECTUS ABDOMINALIS exert an equable pressure upon the organs contained in the abdominal cavity; when acting together they bend the body forward or elevate the hips; they also depress the ribs in respiration. When the muscles of but one side act, the body is twisted to that side.

MUSCLES OF THE POSTERIOR PART OF THE TRUNK.

170. The TRAPEZIUS, RHOMBOIDEUS MAJOR and MINOR draw the scapula back toward the spine: the two latter draw the scapula upward toward the head, and slightly backward: the former draws the head back and elevates the chin.

The LATISSIMUS DORSI draws the arm by the side and backward.

The SERRATUS POSTICUS INFERIOR depresses the ribs in expiration.

MUSCLES OF THE UPPER EXTREMITIES.

171. The DELTOID raises the arm from the side of the body to a horizontal position.

The BICEPS flexes the fore-arm on the arm, as in preparing for striking a blow.

The TRICEPS extends the fore-arm on the arm; it lies on the back of the humerus and is used in striking a blow.

Fig. 62.

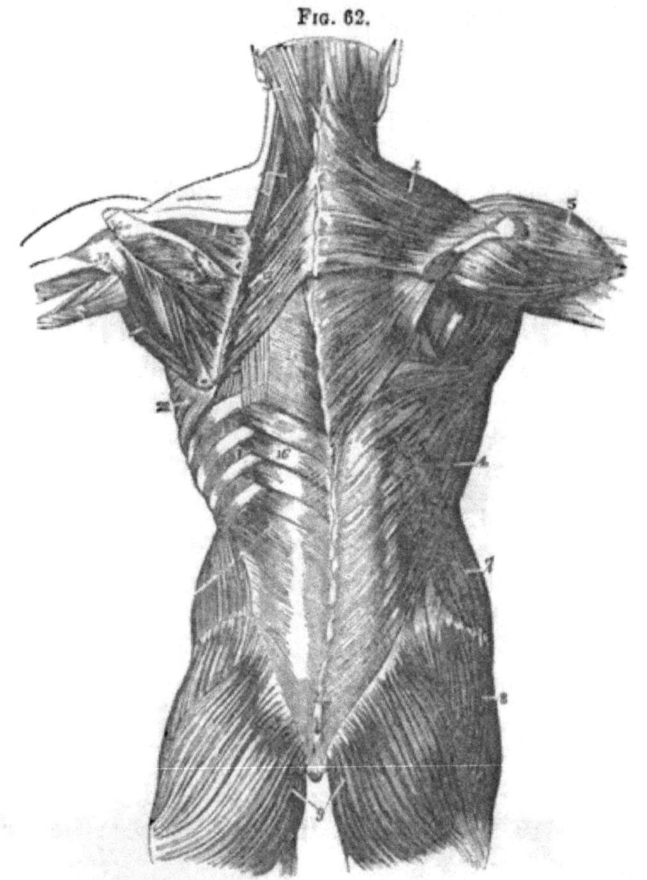

Fig. 62. The Dorsal Muscles.—The first, second, and part of the third layer of muscles of the back. The first layer is shown on the right, and the second on the left side. 1, The trapezius muscle. 4, The latissimus dorsi muscle. 5, The deltoid muscle. 7, 8, The gluteus medius muscle. 9, The gluteus maximus muscle. 11, 12, The rhomboideus major and minor muscles. 16, The serratus posticus inferior muscle. 22, The serratus magnus muscle.

The Flexor Carpi Radialis passes under the annular ligament and bends the hand on the wrist.

The Flexor Carpi Ulnaris bends the hand in the direction of the ulna.

The Flexor Digitorum bends the fingers.

The Extensor Digitorum extends the fingers.

The Extensor Carpi Radialis extends the wrist on the fore-arm.

MUSCLES OF THE LOWER EXTREMITIES.

172. The GLUTEI give power of retaining the erect position.

The SARTORIUS bends the lower extremities into the position assumed by the tailor at his work.

The RECTUS FEMORIS, VASTUS EXTERNUS, and VASTUS INTERNUS extend the leg on the thigh.

The TRICEPS ABDUCTOR FEMORIS bends the thigh on the pelvis, rotates it outwardly and acts powerfully in bending the limbs inward.

The BICEPS FEMORIS forms the outer hamstring, assists in turning the leg outward, and also flexes it upon the thigh.

The EXTENSOR DIGITORIUM splits into four tendons which pass under the annular ligament, and extend the four lesser toes and flex the foot.

The PERONEUS LONGUS extends the foot and inclines the sole obliquely outward.

The GASTROCHNEMIUS EXTERNUS raises the body in walking, and extends the foot on the leg.

The TENDO-ACHILLES (heel-cord) is formed by the conjoined tendons of the gastrochnemius externus and internus (and plantaris). It lies directly beneath the fascia and integuments.

§ **14.** HISTOLOGY OF THE MUSCLES.—*Analysis of a Muscle. Sheaths of Muscles.—Voluntary or Striated, and Involuntary or Non-Striated Muscle. Exciting Agents of Muscular Contractility. Tendons, Blood-Vessels and Nerves.*

173. A MUSCLE is separable into bundles of fibres called *Fasciculi;* each bundle or fasciculus, into smaller fibres (smaller fasciculi); each of the smaller fibres, into a multitude of filaments, or *fibrillæ* (fibrils); and each filament, or fibrilla, into cells arranged in a linear series. Hence, a single muscle is composed of some millions of these fibrillæ combined together, having the same point of attachment or origin, and concentrating in a tendon which is fixed to a movable part, or the point of insertion.

174. Each muscle is invested by a membranous covering of areolar tissue, named the *Perimys'ium;* from this, thin partitions pass inward between the large and small fasciculi, so that, were it possible to remove the muscular substance, there would remain a delicate areolar network of the exact shape of the muscle and its parts. Each elementary fibre or fasciculus is enclosed in a very thin, transparent, structureless *sheath*, called *Myolem'ma;* this sheath is entirely distinct from that of the areolar tissue; it isolates each ultimate fasciculus and probably gives off a sheath to each fibril.

FIG. 63.

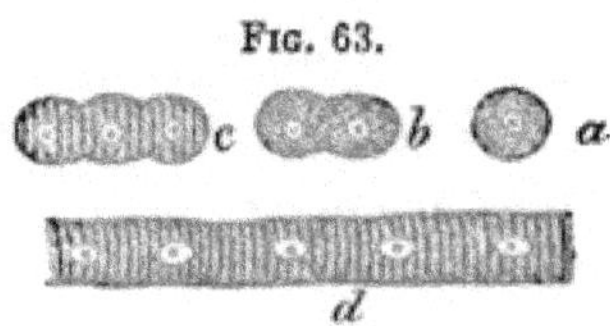

FIG. 63 (*Leidy*). DEVELOPMENT OF STRIATED MUSCULAR FIBRE, FROM CELLS.—*a*, Simple cell. *b*, A pair of cells fused together. *c*, Three cells fused, and their contents assuming the striated character. *d*, A muscular fibre, exhibiting its original composition of cells.

175. Muscles are of two classes—*Striated* and *Non-Striated;* the former are also called *Voluntary*, being, in their normal action, under the control of the will; the latter, *Involuntary*, acting independently of the will, as the heart, the stomach and the intestines. The Myolemma, or sheath, so distinct in the striated or voluntary muscle, cannot be shown to exist in the involuntary or non-striated muscle. The heart seems to be a connecting link between the two classes, as it is involuntary in action, yet composed of striated muscle; the sheath, however, is wanting.

176. The "Vis Musculosa," or contractility of the muscle, is excited on the application of certain stimuli; these may be *Mechanical*, as the touch of a sharp instrument; *Chemical*, as acids and alkalies; *Electrical*, as in shocks; and *Vital*, originating in, or acting through, the nervous system: it is by means of the latter that muscular fibre is most frequently called into action.

177. TENDONS are composed of the inelastic, white-fibrous tissue, and possess great strength. The muscular fibres do not cease immediately, but intertwine with those of the tendons, and these with those of the bone. The tendinous and muscular fibres are generally parallel; thus being straight in the sartorius, and oblique in the penniform muscles. In passing over bones or other hard parts, they are protected by synovial bursæ. In common with the muscle to which it belongs, each tendon has an envelope of very condensed areolar tissue continuous with that of the muscle: this sheath generally forms a semi-cylindrical canal, completed on the opposite side by the bone, so that the tendon itself slides into a canal partly bony and partly fibrous. This canal is lined with the synovial membrane.

178. The BLOOD-VESSELS do not enter the proper muscular substance, but everywhere abound in the areolar tissue by which the fibres are enveloped; hence the nutriment necessary for the growth and repair of muscular tissue must be absorbed through the Myolemma.

179. The NERVES seem to occupy the same position as the blood-vessels, in relation to the primitive fibres, and therefore must also exert their influence through the Myolemma. The nerves of the voluntary muscles are abundant, and chiefly of the *motor* class, or those which preside over motion, having nothing to do with sensation, and hence acting *from* the brain and spinal cord to the muscles; while the nerves of the involuntary muscles are few, and of the *sensory* class, or those which preside over sensation, having nothing to do with motion, hence conveying impressions *to* the brain and spinal cord.

§ **15.** THE CHEMISTRY OF THE MUSCLES.—*Chemical Composition of Muscle. Chemical Changes attending Muscular Action. The Muscular Current.*

180. The chemical composition of muscular tissue cannot be precisely known, because of the difficulty of isolating the fibres from the areolar tissue, blood-vessels and nerves blended

with them. We give the analysis of Berzelius, by which it appears that less than twenty-three per cent. of ordinary meat is solid matter:

Proper muscular substance	15.80
Gelatin (firm areolar tissue)	1.90
Albumen and hæmatin	2.20
Phosphate of lime with albumen	.08
Alcoholic extracts with salts (lactates)	1.80
Watery extracts with salts	1.05
Water and loss	77.17
	100.00

Inosit, or *Muscle Sugar*, exists in the juice of flesh.

181. The proper muscular substance differs from simple fibrous tissue in not being resolved into gelatin by boiling. It contains a peculiar principle called *os'mazome;* this is colored, soluble in alcohol, and gives to broth its characteristic taste and smell.

182. Muscular action is accompanied by chemical changes due to the oxidation of muscular tissue. Quiescent muscle is neutral (neither acid nor alkaline) in chemical character; but muscle after repeated contractions is acid. Heat is evolved, both by chemical action and increased capillary activity, in proportion to the amount of exercise performed. The electrical current known as the "muscular current" is probably a result of chemical action. In the entire muscle, its path lies along the outside toward the tendons. The direction of the total current of the body is from the head downward.

Observation.—In friction, or rubbing the body with the hand, the direction of the current should be followed, otherwise irritation is produced rather than the soothing influence desired. This direction is of special importance to nurses and watchers in caring for the sick, particularly nervous patients; the effect of friction is sometimes improved by moistening the inside of the hand.

§ 16. PHYSIOLOGY OF THE MUSCLES.—*Relative Uses of the Bones and Muscles. Important Functions of the Muscles. Relation of the Will and the Muscular Sense to Muscular Action. The Muscular Sense as a Source of Enjoyment. Importance of Involuntary Movements.—Of such Movements being sometimes Voluntary. Uses of Tendons. The Mechanical Powers as exhibited in Muscular Action.—Levers.—Pulley. Oblique Action of the Muscles. Deep-Seated Muscles. Minute Muscles.*

183. To give a clear idea of the relative uses of the Muscles and Bones, we quote the comparison of another: "The Bones are to the body, what the masts and spars are to the ship; they give support and the power of resistance: the Muscles are to the bones, what ropes are to the masts and spars."

184. The USES OF THE MUSCLES are manifold—they give the beautiful form and symmetry of the exterior of the body; enclose the cavities, and form a firm, defensive, but yielding wall in the trunk; invest and move the bones of the limbs; and give to some of the joints their principal protection. By means of the contractile property, and various mechanical contrivances of muscular fibres, the heart pulsates; the blood circulates; respiration is carried on; the conduits of the glands urge on their fluids; and mechanical aid is afforded in the various processes of preparing nutriment for the system. We are indebted to the same for our power of locomotion; for our ability to engage in the manifold employments of life; to enjoy its pastimes; and to hold communication with our fellow-men by speech, gesture, and the varied expressions of the human countenance.

185. The VOLUNTARY MUSCLES in their normal condition, both in their contraction and relaxation, are subject to the control of the *Will* and the guidance of the *Muscular Sense;* the will determines an act, and the muscular sense enables us to judge of the effort necessary to its performance. And here we would notice that the *motion* of a *limb* implies an *active* state, or a change in *both classes* of *muscles* (flexors and extensors), the one to *contract*, the other to *relax;* and the will

influences both classes: the relaxing muscle does not give up all effort, but is subject to as fine a sense of adjustment in its *yielding* as in its *contraction*.

186. By the aid of the *Muscular Sense*, sometimes with conscious volition, and sometimes without it, we regulate the force employed in all the movements of the body, as lifting weights; balancing the body in standing or locomotion; moving the arms in prehensile or manipulating acts; and exercising the vocal organs. The feats of the rope-dancer and trained gymnast are largely due to the cultivation of this sense. The exercise of the muscular sense is a source of positive enjoyment. The person who walks with an elastic step, holding the body easily in equilibrium, experiences a *sensible* pleasure unknown to him who moves with shuffling gait and apparent distrust of the integrity of his muscles: so in dancing, gymnastic and skating exercises, if attention is given to elegance of attitude and *harmony of motion*, there is experienced a pleasure quite distinct from that gained by the quickened activities, and which is attributable to the muscular sense.

187. The INVOLUNTARY MUSCLES perform their functions wholly independent of the will, and are essential to the action of the heart, the digestive organs, the respiratory apparatus, and various ducts, blood-vessels and lymphatics. The Divine Builder has wisely ordered that these vital operations should not be subject to the control of the individual.

188. Again, there are certain operations *generally* entrusted to the involuntary muscles that may be temporarily controlled when occasion requires, as in respiration: were these movements never under the control of the will, we should be unable to use to any advantage the vocal apparatus, either in speech or singing, and were we compelled to breathe at perfectly regular intervals, it would be exceedingly difficult to attend to the daily duties of life.

Observation.—In rare instances the action of the heart may be suspended for a short time at will. The possibility of such control would be fairly inferred from the presence of striated muscular fibre, as before described.

189. TENDONS serve to convey the contractile power of muscles to the bones; they are, in themselves, passive organs possessing no contractility. In them the evidence of care and skillful arrangement is beautifully exhibited. Wherever muscular action is wanted, and the presence of muscle would be inconvenient or mar the harmony of proportion, or where great strength is needed, there we find the small, dense, conducting tendons: An example is seen in the human hand; Suppose the large muscles of the fore-arm to extend to the hand, it would not only be destitute of beauty, but unfitted for many of the purposes of life. But the gradual blending toward the wrist of the muscles with the tendons; the band clasping them firmly at the wrist; the slits in the short tendons of the second joint to allow the long tendons to pass through to the bones of the fingers, afford the best conceivable arrangement for compactness, delicacy, beauty and utility.

FIG. 64.

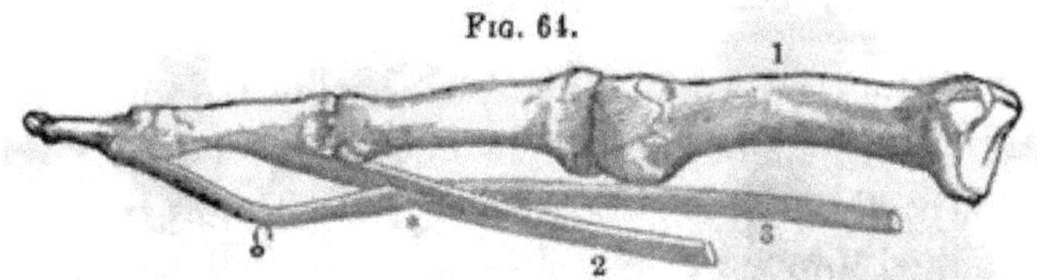

FIG. 64 (*Leidy*). METACARPAL AND PHALANGEAL BONES OF THE FINGERS, WITH THEIR TENDONS AND LIGAMENTS.—1, Metacarpal bone. 2, Tendon of the superficial flexor. 3, Tendon of the deep flexor, passing through a perforation (*) of the superficial flexor.

190. *In the action of the muscles upon the bones, we have examples of the three kinds of Levers* treated of in mechanics. A lever is a rod of wood, metal, or other substance, movable in one plane about a supported point in the rod, called a *fulcrum.* The resistance to be overcome, is called the weight; and the force used in overcoming the resistance, is called the power. The three kinds of lever are distinguished from each other, by the relative position of the power, weight and fulcrum.

191. *In the first kind,* the fulcrum is between the power

and the weight; as in scales, scissors, etc.; *in the second*, the weight is between the power and the fulcrum; as is seen in moving the common wheelbarrow, or a door; *in the third*, the power is between the weight and the fulcrum; as in using the fire-tongs. In the body, the bones are the levers; the parts attached, the weights; and the muscles, the powers. The fulcra are the joints, or extremity of the limbs in contact with the ground, or other resisting substance.

192. *The first kind of lever is illustrated* in the adjustment and movement of the skull upon the first vertebra (91); the hinge-joint is the fulcrum; the excess in gravity of the parts of the head in front of the joint over the parts behind it, is the weight; and the muscles extending from the spine to the cranium are the power. Of the same kind of leverage, are the movements of the vertebræ on each other from above downward; of the lowest lumbar vertebra on the sacrum; of the pelvis on the thigh-bone; of the thigh on the leg; of the leg on the ankle; and also the extension of most of the joints of the upper extremities—as the elbow and the joints of the fingers.

193. *The second kind of lever is illustrated* in the foot. When resting on the ground, with the heel raised, the fulcrum is at the ball of the great toe; the weight is the body transmitted through the large bone of the leg; and the power is in the muscles of the calf of the leg (Gastrocnemii) acting through the tendon of Achilles (Fig. 61). The depression of the lower jaw, when the mouth is opened very wide, is also an example of the second class. Where great weight is to be raised slowly through a short space, we find the second class of levers.

194. *The third kind of lever* is most used in animal mechanics; as in raising the lower jaw; in raising the shoulder and collar-bone, and in the flexion of all the joints of the limbs. A familiar example is the elbow. The fulcrum is at the joint; the weight is the fore-arm and hand; and the power is in the biceps and brachial muscles. This kind of lever works at what is called a mechanical disadvantage, as

the distance of the power from the fulcrum is always less than that of the weight; and the shorter the power-arm (or distance of the power from the fulcrum) in proportion to the weight-arm (or distance of the weight from the fulcrum), the greater must the power be to overcome a given resistance; but what is lost in power, is gained in velocity. In mechanics, one of two things is aimed at—either to raise a great weight slowly and through a short space, or to move a light weight quickly through a long space; the latter is most frequently needed for the purposes of life, hence, we find it accomplished by the use of levers of the third class.

FIG. 65.

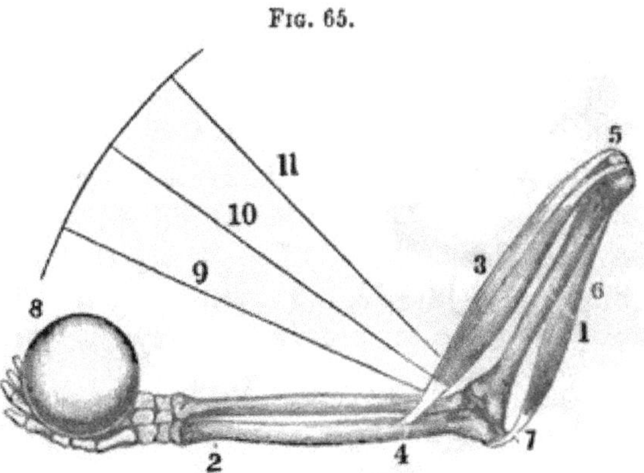

FIG. 65. DIAGRAM OF THE THIRD KIND OF LEVER.—1, The bone of the arm above the elbow. 2, One of the bones below the elbow. 3, The muscle that bends the elbow. This muscle is united, by a tendon, to the bone below the elbow (4); at the other extremity, to the bone above the elbow (5). 6, The muscle that extends the elbow. 7, Its attachment to the point of the elbow. 8, A weight in the hand to be raised. The central part of the muscle 3 contracts, and its two ends are brought nearer together. The bones below the elbow are brought to the lines shown by 9, 10, 11. The weight is raised in the direction of the curved line. When the muscle 6 contracts, the muscle 3 relaxes and the fore-arm is extended.

195. There is a loss of power, and hence mechanical disadvantage, in the *oblique action of the muscles*, since a force acting perpendicularly on the arm of a lever, operates most advantageously. It is important to notice, however, that the tendons of insertion, and sometimes also those of origin, are attached to the special eminences of bones called *processes*,

in which case the muscles gain an advantage, for their force ultimately operates in a line more or less perpendicular to the osseous surface, instead of in a line nearly parallel with it, as would happen if there were no elevations. The extensors act more obliquely than the flexors, and therefore require an additional amount of fibres, hence, they are larger than the flexors, in the ratio of eleven to five.

196. *The principle of the pulley* is also used in the arrangement of the muscles, though less frequently than the lever. The annular ligaments, which confine the tendons at the wrist, and at the ankle (61), act as pulleys. A marked example is seen in one of the muscles that pull down the lower jaw, called the *digastric* muscle; this has one extremity attached to the inside of the lower part of the chin-bone, and the other to the back of the ear; it is tendinous in its middle portion, which passes through an opening in a short muscle connecting the hyoid bone, just above the larynx, to a small process under the ear; hence, when the digastric muscle contracts, the jaw is depressed. Another beautiful example is furnished by the trochlear muscle of the eye (Fig. 181).

Fig. 66.

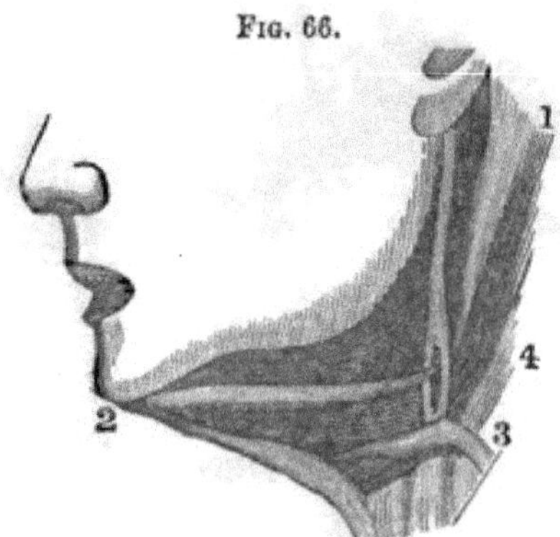

Fig. 66. Pulley Arrangement of a Muscle.—1, Digastric muscle attached to the mastoid process of the temporal bone behind the ear. 2, Its attachment to the lower jaw. 3, Hyoid bone. 4, The pulley arrangement of the digastric and stylo-hyoid muscles.

197. We have noticed only the larger of the exterior layer of muscles. The limits of this work will not allow a full view of the exquisite beauty beneath; the layers are of various sizes and forms, and crossing each other in every direction, yet the millions of fibres and multiplied millions of cells per-

form their assigned work in perfect harmony, not one interfering with the action of another. In the abdominal muscles, one layer of muscles is oblique from left to right toward the median line; another crosses this diagonally from right to left; another crosses these transversely; and still another perpendicular layer overlies all the others.

198. Infinite mechanical skill is still more wonderfully shown in the nice adjustment and accurate movements of the *minute muscles*, as those of the tongue, and the yet finer muscles of the eye and the drum of the ear, too small to be seen by the naked eye.

199. Everywhere the muscular force is one and the same, but its applications are innumerable; the instruments are constructed upon the same plan, but infinitely varied in form, size and arrangement, yet made with the greatest simplicity for effecting each its particular purpose.

> "In human works, though labored on with pain,
> A thousand movements scarce one purpose gain;
> In God's, *one single* can its ends produce,
> Yet serves to *second, too, some other use.*"

§ 17. HYGIENE OF THE MUSCLES.—*Requirements necessary to maintain a Healthy Condition of Muscle. Importance of Freedom from Compression.—Of Exercise. Conditions to be observed in Muscular Exercise. Exercise sometimes Injurious. Effect of Mental Stimulus. Regard necessary to the Age and Health.—Position of the Body.—Proper Muscular Tension. Education of the Muscles.*

200. Since so much of our happiness and usefulness in life depends upon *healthy muscle*, it is of great importance that we seek to understand the laws upon which their normal action depends. The first and great essential is, that *the muscles should be abundantly supplied with pure blood.* A pure state of the blood requires that the digestive apparatus should be in a healthy condition; that the vital organs should have ample volume; that the lungs should be plentifully supplied with pure air; that the skin should be kept warm by proper

8*

clothing, and clean by bathing, and that it should be acted upon by *air* and *sunlight*. It is also of primary importance that there be *free circulation* of the blood, which may be secured by *freedom from compression*, and by *regular and judicious exercise*.

201. *Freedom from compression* is requisite to free circulation, for, even a slight pressure upon the delicate, yielding blood-vessels checks the flow, thus preventing the necessary deposit of materials required by the waste of the system, and also the removal of the injurious products of the decomposing tissues. Again, pressure stimulates the lymphatics to undue action, thus attenuating the muscles.

A fractured limb is enfeebled, not only by inaction, but by the necessary *pressure of the dressing;* it recovers its size, tone and strength only by judicious and persistent exercise after the bandages are removed.

The pressure of *tight dresses*, under the name of a "neat fit," not only produces deformity in the general figure, but prevents the expansion of the lungs, and *stagnates the blood*, thus poisoning the whole system.

The young lady who is "horrified" at the act which takes life in a moment, adopts a plan no less suicidal. She is equally certain to take away her God-given life. And if the guilt of self-murder admits of degrees, shall he who applies the knife to his throat, or the rope to his neck, be guilty of a greater condemnation, than she who *extends the act* through months, or, if nature endures, through years, meanwhile poisoning the *soul* as well as the body?

202. *Free circulation, and, consequently, muscular power, is increased by proper exercise, and decreased by inactivity.* It is a general law of the system, that the action and power of an organ are, within a certain limit, commensurate with the demand made upon them—a law which holds good in the muscular apparatus. When the muscles are exercised, the flow of blood in the arteries and veins is increased, hence, the muscular fibre increases in size, and acts with greater force; while, on the contrary, the muscle that is little used receives

little nutriment from the sluggish blood, and decreases in size and power.

Illustration.—The muscles of the blacksmith increase in size and become firm and hard; those of the student, if not used in gymnastics or otherwise, decrease in size and become soft and less firm.

203. *Relaxation must follow contraction, or rest must follow exercise.* Exercise too long continued produces exhaustion, and in the exercise of exhausted muscle, the loss of material exceeds the deposit; also long-continued tension enfeebles, and at length destroys the contractile property.

Illustration.—The effect of over-work may be seen in the attenuated frames of over-tasked domestic animals, as the horse, or in the diminished weight of the farmer after the hurry of harvest-time. The effect of continued tension is seen in the restlessness of children at school, after sitting for a time in one position. The necessity of frequent recesses is founded upon the organic law, that relaxation of muscle must follow contraction. The younger and feebler the pupils, the greater is this necessity.

204. *Change of employment often affords the required rest,* as it brings into action a new set of muscles; hence, the person of sedentary occupation is rested by general muscular exercise, while the person of active occupation is rested by that of a sedentary character.

Illustration.—The needlewoman exhausts the muscles of the back and arm; a brisk walk or some active household employment affords rest.

205. *The muscles should be gradually called into action;* for while in action, they require more blood and nervous fluid than when at rest, and these fluids are gradually increased. In an alarm of fire, never start "on the run," but "make haste slowly" in the first instance, and then gradually increase your speed.

Observation.—If a man has a certain amount of work to perform in nine hours, his muscles having been in a state of rest, he will do it with less fatigue by performing half the work in the first five hours, and the remainder in four hours. The same principle applies in the use of beasts of burden, or in driving the carriage-horse.

206. *The muscles should be rested gradually after vigorous exercise.* If a person has made great muscular exertion, instead of sitting down immediately to rest, he should continue to exercise *moderately* for a short time, and avoid sudden *cooling* in a current of air; additional clothing is often needed. The soreness of muscles which have been severely exercised, is often prevented by bathing and thorough rubbing, followed by moderate exercise.

207. *Exercise should be regular and frequent.* The system needs this means of invigoration as regularly as it needs new supplies of food. To devote a few days to the proper action of the muscles, and then spend a day inactively, is as incorrect as to take a proper amount of food for a time, and then to withdraw the supply for a season. The industrious artisan and the studious minister suffer as surely from undue physical inactivity, as the indolent man. The evil consequences of neglect of regular exercise steal slowly but *surely* upon an individual; sooner or later they are manifested in muscular weakness, dyspepsia and nervous irritability.

208. *Every part of the muscular system should have its appropriate share of exercise.* Farming and domestic employments are superior as vocations, in respect to giving all the muscles their due proportion of action. Where the daily occupation exercises but a part of the muscles, it should be followed by some employment or recreation which will bring the others into use. Among the healthful pastimes, are those of quoits and ball-playing. Students, both boys and girls, often become chronic invalids from the want of this *general* exercise. Every institution, having no arrangement for systematic physical exercise, should be considered as wanting in one great means not only of physical, but of intellectual and moral development. If possible, there should be a gymnasium with its varied appurtenances; if not, there should, at least, be a room where light gymnastics may be practiced

daily, and this should not prevent further exercise in the open air. Experience has proved that far more intellectual vigor is gained when this practice is observed.

209. *The amount of exercise should be adapted to the age and strength of the individual.* In youth, a portion of the vital, or nervous energy of the system, is expended upon the *growth* of the organs of the body, consequently, severe labor or exercise is injurious.

Observation.—In the campaigns of Napoleon Bonaparte, his army was frequently recruited by mere boys. He complained to the French government, because he was not supplied with mature men, as the youths could not endure the exertions of forced marches.

210. *The proper time for exercise should be observed.* As a general rule, the morning is a better time for exercise than the evening; the powers of the system are greatest at that time. Severe exercise should be avoided immediately before or after a meal; the vigor of the system is then required for the digestive functions. The same rule should be observed regarding mental toil, as the powers of the system are then concentrated upon the brain.

Observation.—When an organ is in functional action, it attracts fluids (sanguineous and nervous) from other organs of the system. The vital energies are sufficient for this one work; but if two important organs are called into intense activity, injury arises to both and also to the general system. Nature can sustain in vigorous activity but one function at a time.

211. *The mind exerts a great influence upon the tone and contractile energy of the muscles.* Muscular exercise will be attended with much less fatigue, when the muscles act under a healthy mental stimulus. This we see illustrated in the ordinary vocations of life; if the mind has some incentive, the tiresomeness of labor or exercise is greatly diminished:

> "He chooses best whose labor entertains
> His vacant fancy most; the toil you *hate*
> Fatigues you soon, and scarce improves your limbs."

The effect of the mind upon the muscles is seen in the spiritless aspect of many of our boarding-school processions, when a walk is taken merely for exercise, with no other object in view. But, present to the mind a botanical or geological excursion, and the saunter will be exchanged for the elastic step, the inanimate appearance for the bright eye and glowing cheek. The difference is, simply, that in the former case, the muscles are compelled to work without the nervous impulse, which, in the latter case, is in full and harmonious operation. It must not be supposed that a walk simply for *exercise* is not beneficial; if possible it should be taken in combination with harmonious mental exhilaration; if not, let the position be erect and the pace so brisk, as to produce rapid respiration and circulation of the blood, and in a dress that shall not interfere with free motions of the arms and free expansion of the chest.

212. *The amount of exercise should be adapted to the health of the individual.* This direction is of essential importance, for what gives *vigor* to one, may bring *weakness* to another. A walk which would invigorate one in health, will quite exhaust a feebler person; hence, the *measure of strength* must be the *measure of exercise;* but a careful distinction should be made between simply *healthful fatigue*, soon removed by subsequent rest, and the *positive exhaustion* which really enfeebles.

213. *In diseases producing great muscular exhaustion, particular care and discretion are necessary regarding exercise.* In scarlet fever, typhoid diseases, etc., the muscular debility is very great; and any muscular exertion that exhausts, such as moving the patient home when he has sickened abroad, or undue exercise during convalescence, is almost sure to result injuriously, if not fatally. Exercise should be moderate, made pleasant, and followed with proper intervals of rest, and *never* at the discretion of one who is ignorant of the peculiar state of the system.

214. *In chronic diseases of the digestive organs, lungs and nervous system, well-directed and persistent exercise of the muscles*

is essential to recovery. In these ailments, the exertion of all the muscles repeated frequently, is attended with the most compensatory results. Moderation is necessary at first, but the exercise should be increased in intensity and duration. The aversion of the patient to exercise is often very great; but it should, nevertheless, be persistently taken in the same spirit in which he would perform any other part of the life-work entrusted to him. Making it a *business* to perform the labor necessary to recovery, and entering into it with the *heart* and the *will*, gives the healthy tone and stimulus so important in securing the most beneficial results.

Observation.—A patient who had suffered long from a combination of chronic ills which baffled the skill of several physicians, in extreme weakness, adopted a systematic plan of exercise, commencing with but two or three steps at a time, and adding a step or two each day, till in six months' time she walked regularly three miles a day.

215. *The muscles require an erect position of the body both in standing and in sitting.* A person can stand longer, walk farther, and perform more labor in an erect position than when stooping, since fewer muscles are then in a state of tension, and, consequently, less draught is made on the nervous system. While stooping, the muscles of the posterior part of the spinal column are kept in a state of tension, to prevent the body from falling forward; while in the erect position, the body and head are balanced upon the bones and cartilages of the spinal column. In stooping, the lower limbs are also curved at the knee, causing a constant tension of their muscles; again, the slight oscillation is wanting, which in the erect position gives alternate contraction and relaxation of the muscles of the back.

216. It is important that the muscles of the child should receive due attention, that the shoulders may be thrown back, and the chest become broad and full. Even when an adult has contracted the habit of stooping, and has become round-shouldered, it can be measurably, and generally wholly corrected, by moderate and repeated efforts to bring the shoul-

ders into proper position. This deformity should receive attention in our schools. It may be remedied as well by persistent effort on the part of a kind instructor, as under the stern military drill-sergeant, who never fails to secure the erect attitude in his raw recruits. In furnishing school-rooms with desks, care should be taken that they be of sufficient height to allow the proper attitude when pupils are using their books or the pen. This is not only essential to health, but to beauty and symmetry of form.

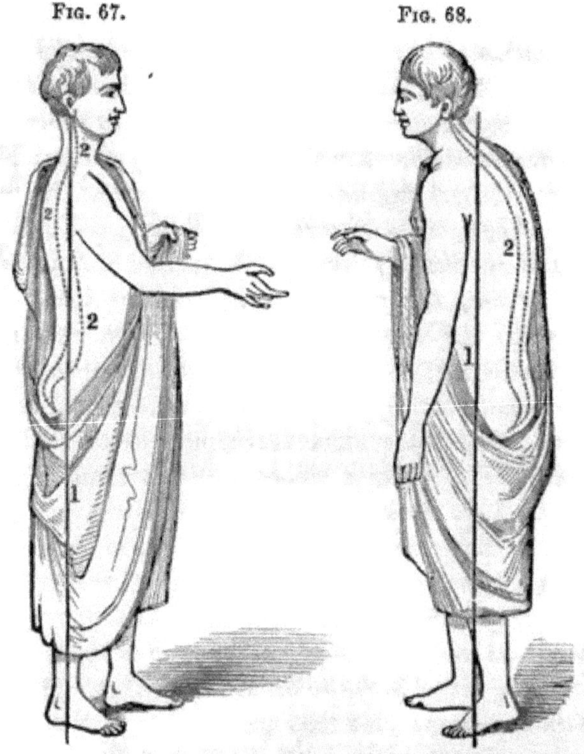

Fig. 67 represents the Erect Position of the Body. 1, A perpendicular line from the centre of the feet to the upper point of the spinal column, where the head rests. 2, 2, 2, Spinal column with its three natural curves. The head and body are so balanced that the muscles are not kept in a state of tension.

Fig. 68 represents the Stooping Position of the Body. 1, A perpendicular line. 2, Unnatural curved spinal column, and its relative position to the perpendicular line (1). The curved position of the body and lower limbs keeps the muscles in tension, which exhausts their contractile energy.

Observation.—A simple test of the erect position is to stand with the back against the wall of a room, with the heels, elbows and back of the head touching the wall. The effort required to do this will show the amount of the departure from the true attitude.

Fig. 69. Fig. 70.

Fig. 69 represents an Improper, but not an unusual, position when writing.
Fig. 70 represents a Proper position when writing.

217. *A slight relaxation of the muscles tends to prevent their exhaustion.* In walking, dancing, and most of the mechanical employments, the fatigue will be less, and the movements more graceful, if the muscles are slightly relaxed; the same condition diminishes the jar of cars or coaches. In jumping or falling from a carriage or any height, the shock will be in a measure obviated if the presence of mind is sufficient to relax the muscles, bend the limbs at the ankle, knee and hips; throw the head and body slightly forward, and fall upon the toes, instead of the heel.

Observation.—With the lower limbs firm and the muscles

rigid, jump a few inches perpendicularly to the floor, and fall upon the heels; again, slightly bend the limbs, jump a few inches and fall upon the toes, and the difference in the force of the shock will be readily noticed.

218. *The muscles require to be educated or trained.* Frequent and systematic use of the muscles at proper intervals is necessary to effective action. This education must be continued till not only each muscle, but every fibre of the muscle, is fully under the control of the will. In this way, persons become skillful in every employment. The power of giving different intonations in reading, speaking, and singing; the rapid movements in penmanship, and in mechanical and agricultural employments, depend, in a great measure, upon the education of the muscles. An individual with trained muscles will perform a given amount of labor with less fatigue and waste to the system than one whose muscles are untrained.

Observation.—It is exceedingly important that correct movements be insisted upon at the *commencement* of any muscular training, as it is very difficult to change a movement which has been long practiced. If a child holds his pen improperly during his early lessons, he will probably never become an easy and elegant writer.

§ 18. COMPARATIVE MYOLOGY.—*Compare Muscles of other Mammals with those of Man. Muscles of Birds—Of Reptiles—Of Fishes.*

219. The muscles of all *Mammals* in their general plan, resemble those of Man; the modifications in number, form, position and relative size, being only such as adapt them to the habits and necessities of the particular species. The color of the muscle is deepest in the Carnivora (flesh-eaters), and palest in the Rodentia (gnawers).

220. The muscular system of *Birds* is remarkable for the distinctness and density of their fasciculi, for the deep-red color of those employed in vigorous action, and their marked separation from the tendons, which are of a pearly-white

color and have a peculiar tendency to ossification. This high development results from the rapid circulation of warm, rich, highly oxygenated blood through the extent of the respiratory system. The energy of the muscular contraction in this class is in the ratio of the activity of the vital functions. As in Mammalia, so in birds, the muscles are varied to meet the habits, wants and condition of the several species and orders.

Fig. 71.

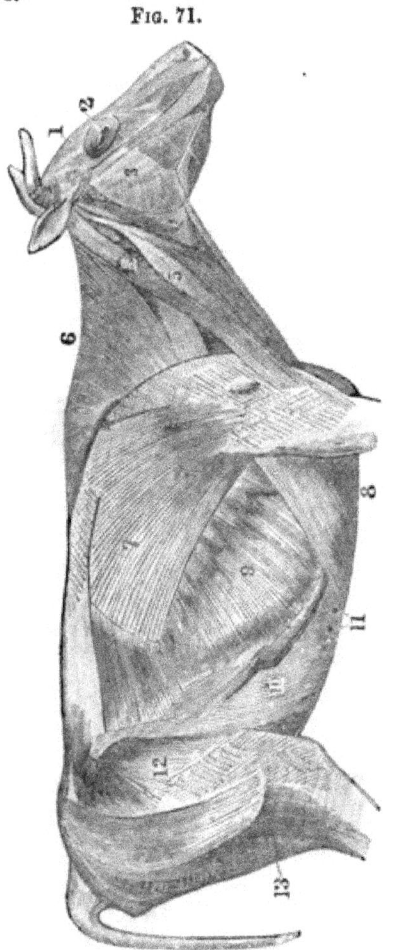

Fig. 71. Superficial Muscles of a Cow.—1, Occipito-Frontalis. 2, Orbicularis Palpebrarum. 3, Masseter. 5, Sterno-cleido-Mastoid. 6, Trapezius. 7, Latissimus Dorsi. 8, Pectoralis. 9, 10, External and internal oblique muscle. 11, Opening of the mammary artery and vein (milk-veins). 12, Glutei. 13, Rectus Femoris muscle.

221. The muscles of the air-breathing *Reptiles* are always pale in color, and the fibres are tenacious of their contractility;

the energy of their contraction in some instances and on some occasions is great, but it cannot be continuously exercised, such power being soon exhausted. The form, size and relative number of the muscles are as various as in mammals and birds.

Fig. 72.

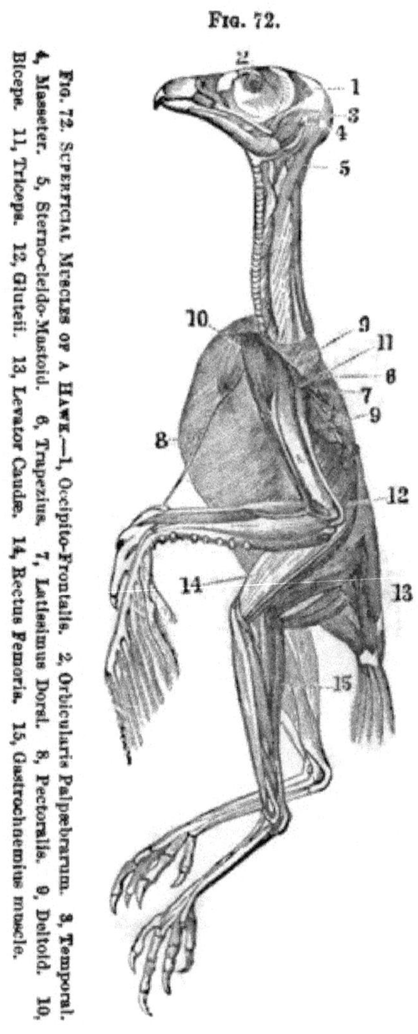

Fig. 72. Superficial Muscles of a Hawk.—1, Occipito-Frontalis. 2, Orbicularis Palpebrarum. 3, Temporal. 4, Masseter. 5, Sterno-cleido-Mastoid. 6, Trapezius. 7, Latissimus Dorsi. 8, Pectoralis. 9, Deltoid. 10, Biceps. 11, Triceps. 12, Glutei. 13, Levator Caudæ. 14, Rectus Femoris. 15, Gastrochnemius muscle.

In reptiles the muscular system of the trunk reaches its maximum development in serpents, and its minimum development in the tortoise. The mandibular development is generally large, while that of the limbs is comparatively small or entirely wanting.

Fig. 73.

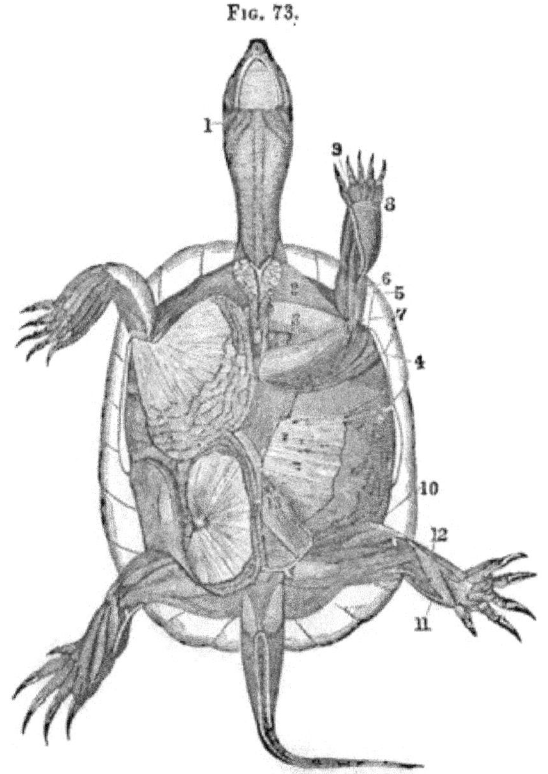

Fig. 73. **Muscles of the Tortoise.**—1, Digastricus. 2, 3, Deltoides. 4, Serratus Magnus. 5, 6, Triceps Brachii. 7, Biceps Brachii. 8, Ulnaris Internus. 9, Flexores Digitorum. 10, Sartorius. 11, 12, Gastrochnemius. 13, Triceps Adductor.

222. In *Fishes* there is a modification of the active motor organs, and a marked deviation from the fundamental vertebral type. The chief masses of the muscular system are disposed on each side of the trunk in a series of vertical plates, or *flakes*, corresponding in number to the vertebræ. Each lateral flake (myocomma) is attached by its inner border to the osseous and fibrous parts of the corresponding segment of the skeleton within; by its outer border, to the skin; and by its fore and hind surfaces, to the septum between it

and the contiguous myocommas, or flakes. The gelatinous tissue of these septa is dissolved by boiling, and the muscular segments, or plates, are then easily separated, as we find in carving fish for the table. Each flake is arranged in a zigzag manner. The muscular tissue of fishes is usually colorless, sometimes it is opaline or yellowish, but it is white when boiled.

FIG. 74.

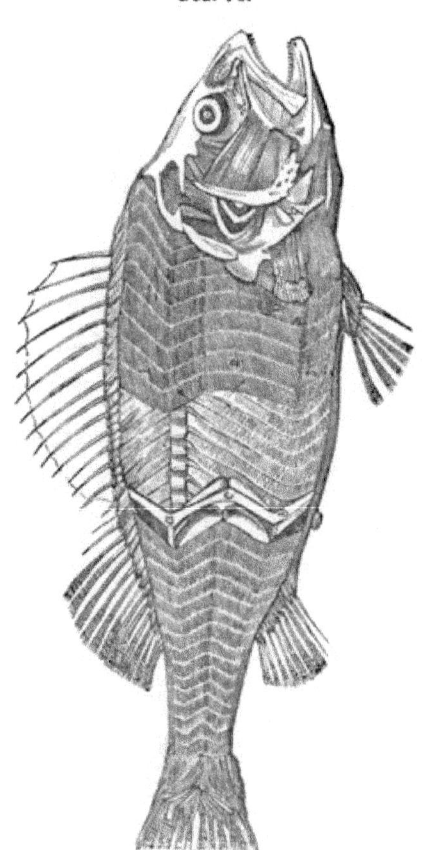

FIG. 74. MUSCLES OF THE FISH.—*a*, *b*, *c*, The zigzag arrangement of the myocomma.

DIVISION III.

THE NUTRITIVE APPARATUS.

223. In the mastication and deglutition of food; in its conversion into fluids; in its circulation in all parts of the system; in its assimilation into the various tissues and organs of the body; in its dis-assimilation, and in the excretion of useless matter;—in a word, in the building up and repairing of the system, from the earliest period of embryo life to the last moment of earthly existence, certain organs are used, which together may be termed the NUTRITIVE APPARATUS: including the *Digestive*, the *Absorptive*, the *Circulatory*, the *Assimilatory* and the *Respiratory* organs.

CHAPTER VI.

THE DIGESTIVE ORGANS.

§ **19.** ANATOMY OF THE DIGESTIVE ORGANS.—*Anatomy of the Mouth.—The Teeth.—The Salivary Glands.—The Pharynx.—The Œsophagus.—The Stomach.—The Intestines.—The Liver.—The Pancreas.—The Spleen.*

224. The DIGESTIVE ORGANS include the *Mouth*, *Teeth*, *Salivary Glands*, *Palate*, *Pharynx*, *Œsophagus*, *Stomach*, *Intestines*, *Liver*, *Pancreas* and *Spleen*.

225. The MOUTH is the space bounded by the lips in front; the soft palate behind, which separates it from the pharynx; the hard palate above, enclosed by the teeth and alveolar arch; and the floor below, upon which rests the tongue (the floor being included within the lower teeth and alveolar arch).

Fig. 75.

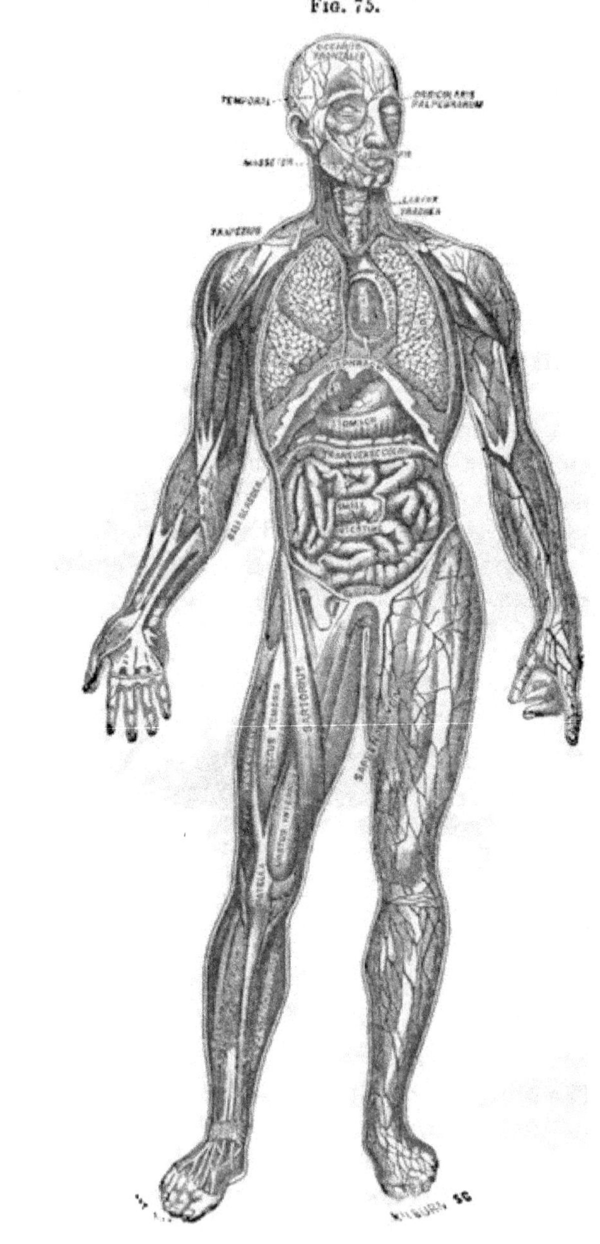

226. The TEETH are attached to the upper and the lower jaw-bone by means of bony sockets called *alve'olar* processes. The attachment is strengthened by the fibrous, fleshy structure of the gums. Each tooth has two parts—the *crown* and the *root:* the crown is that part which protrudes from the jaw-bone and gum, and is covered by the enamel; the root, or fang, is that part contained in the socket of the jaw; and the slightly constricted portion clasped by the gums is the *neck.*

FIG. 76.

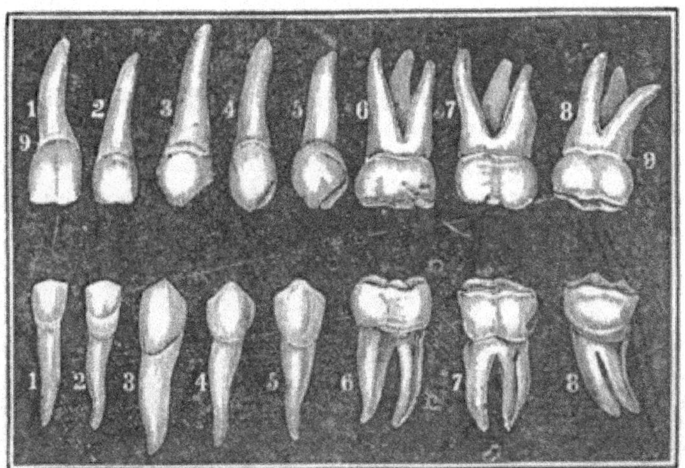

FIG. 76 REPRESENTS THE ADULT TEETH.—1, 2, The cutting teeth (incisors). 3, Eye-tooth (cuspid). 4, 5, Small grinders (bi-cuspids). 6, 7, 8, Grinders (molars). 9, 9, Neck of the tooth.

Observation.—These bony processes are absorbed after the extraction of a permanent tooth, leaving the jaw-bone covered only by the lining membrane of the gum. This gives the narrow jaw and retreating lips of old age. A piece of the alveolar process sometimes clings to a tooth when extracted, and the dentist has the credit of "breaking the jaw."

227. The first set of teeth, appearing in infancy, is called *temporary*, or the milk-teeth. They are twenty in number; ten in each jaw. Between six and fourteen years of age, they are replaced by the second set, called *permanent* teeth, numbering thirty-two, sixteen in each jaw.

The four front teeth in each jaw are called *Incisors* (cutting teeth). They are convex in front, concave behind, gently rounded at the sides, and have a broad, chisel-shaped body, based on a rounded neck, terminating above in a sharp and slightly serrated cutting edge. The next tooth on each side, the *Cuspid* (eye-tooth in the upper jaw, and stomach-tooth in the lower) is round, strong, with a very long, tapering root, and the body terminating in a point having on each side a partial serrature: the next two, *Bi-cuspids* (small grinders) have a rounded body terminating on its grinding edge in two points, one before, the other behind, with a rough groove between them: the next two, *Molars* (grinders), situated behind all the other teeth, have a crown, square or cuboid in form, with four points on the triturating surface separated by channeled depressions; the last molar is the *dens sapientiæ*, or "wisdom tooth," smaller than its fellows, late in its development, and early in its decay.

FIG. 77. FIG. 78.

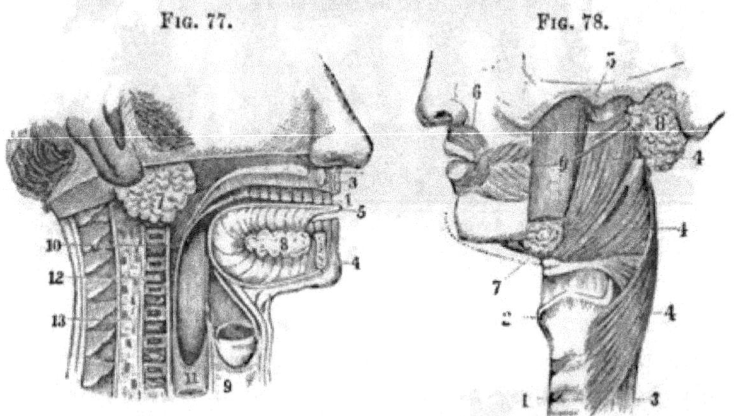

FIG. 77. THE MOUTH AND NECK LAID OPEN.—1, The teeth. 3, 4, Upper and lower jaws. 6, The tongue. 7, Parotid gland. 8, Sublingual gland. 9, Trachea (wind-pipe). 10, 11, Œsophagus (gullet). 12, Spinal column. 13, Spinal cord.

FIG. 78. A SIDE VIEW OF FACE.—1, 2, Trachea. 3, Œsophagus. 7, Submaxillary. 8, Parotid gland. 9, Duct from the parotid gland. 4, 4, 4, 5, 6, Muscles.

The incisors, cuspids and bi-cuspids have each but one root; the molars of the upper jaw have three roots; those of the lower jaw, two roots.

228. The SALIVARY GLANDS consist of three pairs—the *Parot'id*,* the *Submax'illary* and the *Subling'ual.*

The PAROTID GLAND, the largest, is situated in front of the external ear, and behind the angle of the jaw. A duct (Steno's) from this gland opens into the mouth opposite the second molar tooth of the upper jaw. The SUBMAXILLARY GLAND is situated within the lower jaw, anterior to its angle. Its excretory duct (Wharton's) opens into the mouth by the side of the *fræ'num ling'uæ* (bridle of the tongue). The SUBLINGUAL† GLAND is elongated and flattened, and situated beneath the mucous membrane of the floor of the mouth, on each side of the frænum linguæ, by the side of which are seven or eight small ducts opening into the mouth.

Observation.—The "mumps" is a disease of the parotid gland, and the swelling under the tongue, called the "frog," a disease of the sublingual gland.

229. The PHARYNX, or throat, is the funnel-like cavity about four inches in length, extending from the base of the skull to the top of the fifth cervical vertebra, where it becomes continuous with the œsophagus. The pharynx has four passages; one leading upward and forward to the nose; the second, forward to the mouth; the third, downward to the trachea and the lungs; and the fourth, downward and backward to the stomach.

230. The ŒSOPHAGUS is a large membranous tube, extending from the pharynx to the stomach. It lies behind the trachea, the heart and the lungs, and passes through the diaphragm.

231. The STOMACH is a somewhat pear-shaped dilatation of the alimentary canal. When moderately filled, it measures twelve inches in length, by four inches in diameter. It has two openings, one connected with the œsophagus, called the *car'diac* orifice; the other connected with the upper portion of the small intestine, called the *pylor'ic* orifice.

---

* Gr., *para*, near, and *ous*, ear.

† Lat., *sub*, under, and *lingua*, the tongue.

232. The INTESTINES are divided into the *Small* and the *Large* intestines. The small intestine is about twenty-five feet in length, and divided into three parts—the *Duodé'num*, the *Jeju'num* and the *Il'eum.*

FIG. 79.

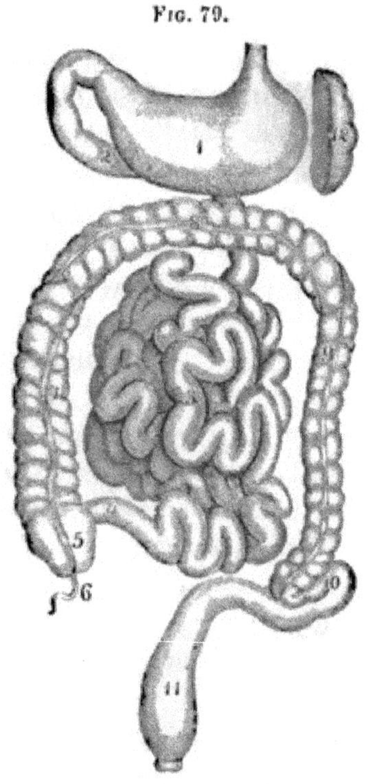

FIG. 79 (*Leidy*). THE STOMACH AND INTESTINES.—1, Stomach. 2, Duodenum. 3, Small Intestine. 4, Termination of the ileum. 5, Cœcum. 6, Vermiform appendix. 7, Ascending colon. 8, Transverse colon. 9, Descending colon. 10, Sigmoid flexure of the colon. 11, Rectum. 12, Spleen.

DUODENUM signifies *twelve,* and this part is so called because its length is about twelve fingers' breadth, or ten inches; JEJUNUM signifies *fasting,* the food passing quickly through this portion, leaving it empty; ILEUM, *twisted,* is so named from its numerous coils or convolutions.

233. The large intestine, about five feet in length, is also divided into three parts—the *Cœcum,* the *Colon* and the *Rectum:* The CŒCUM is so called from its forming a *blind*

pouch perforated at one end only; the COLON, because the excrements are arrested, for a considerable time, in its folds; and the RECTUM, from its *straight* course.

Attached to the extremity of the cœcum is the *appendix vermiformis*, a worm-shaped tube, about four inches long, and the size of a goose-quill. Its function is unknown. The COLON is divided into three parts—the *ascending*, the *transverse* and the *descending;* the lower portion of the descending colon makes a double curvature, called the *sigmoid flexure.* The RECTUM extends from the sigmoid flexure to the terminus of the intestinal canal, a distance of six or eight inches.

234. The LIVER is the largest glandular organ in the body, weighing about four pounds. It is situated in the right side below the diaphragm. It is convex above and slightly concave below; its convex surface being fitted accurately into the concavity of the diaphragm, and its concave surface in contact with the stomach, duodenum, colon and right kidney. The liver is surrounded by a peritoneal covering, which forms for it a suspensory, or broad ligament, and two lateral and triangular ligaments. It has two principal lobes, the right lobe being four or five times larger than the left. On the under side of the liver is the gall-bladder, or reservoir for the bile, which opens by the common biliary duct into the duodenum.

235. The PANCREAS* is a long, flattened organ, weighing three or four ounces, about six inches in length, and placed transversely across the posterior wall of the abdomen, behind the stomach. A duct from this organ opens into the duodenum.

236. The SPLEEN (so called because the ancients supposed it to be the seat of melancholy) is an oblong, flattened organ situated on the left side in contact with the diaphragm, stomach and pancreas. It is of a dark-bluish color, has no outlet, and its use is not well determined.

---

* Gr., *pan*, all, and *kreas*, flesh.

§ **20.** Histology of the Digestive Organs.—*Lining Membrane of the Alimentary Canal—Of the Mouth. Histological Composition of the Teeth—Of the Tongue—Of the Palates—Pharynx. The Three Coats of the Œsophagus, the Stomach and the Intestines—Their Relation to each other—Adaptation to their several Offices.*

237. The alimentary canal is lined through its entire length by the *mucous membrane*, which, with its little recesses forming tubes or sacs called *glands*, is composed of three layers—the epithelium, or surface layer, the basement membrane, and the areolar-vascular layer, or corium. The epithelium varies in different parts, both in the number of layers and in the form of its cells.

238. The cavity of the Mouth, excepting the teeth, is everywhere covered with a highly vascular mucous membrane, having a squamous epithelium, beneath which are concealed conical papillæ, excepting upon the gums and upper surface of the tongue, where they become conspicuous as organs of taste.

The Tongue is a muscular organ, composed of two symmetrical halves, separated by a median fibrous membrane. Its muscles are named *extrinsic* and *intrinsic;* the former, four in number on each side, pass into the tongue at its base and under surface, attaching it to the neighboring parts. The intrinsic are named the superior longitudinal, the inferior longitudinal, and the transverse; the first extending just beneath the mucous membrane, from the apex, through the entire length, to the hyoid bone, having some of its fibres oblique, and some branching; while the inferior longitudinal extends from the apex to the base of the tongue. The transverse muscles which give the principal body to the tongue, are connected with the median septum and pass outward, intersecting at the margins and base with the other muscles. (From this variety of arrangement, the tongue is capable of moving in all directions.)

The Palate, or roof of the mouth, comprises two parts, the *hard* and the *soft* palate. The *Hard palate* is deeply vaulted and lined with a smooth mucous membrane, excepting at the fore part, which is roughened by transverse ridges. The *Soft palate* is composed of a doubling of the mucous membrane, enclosing a muscular layer, together with several small glands. It projects as a freely movable partition, obliquely downward and backward from the hard palate between the mouth and posterior nasal orifices.

239. The TEETH are appendages developed in the mucous membrane of the mouth, and not parts of the skeleton, as is sometimes supposed. The tooth consists of a hard portion hollowed out and filled with a soft substance, called pulp. This pulp is formed of areolar tissue supplied with vessels and nerves, which enter the tooth at a minute opening at the point of the fang. It is the remains of vascular and nervous papillæ upon which the tooth was originally formed. The hard substance is composed of *ivory*, or *dentine*, *enamel* and *cement*. The *Dentine* forming the greater part of the tooth consists of microscopic tubes called *dental tubuli*. These tubuli are filled with minute processes of the pulp, affording nutrition and perhaps giving sensibility to the dentine.

FIG. 80.

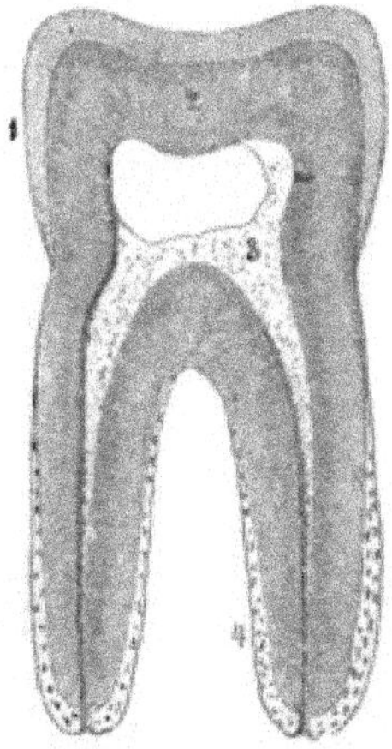

FIG. 80 (*Leidy*). VERTICAL SECTION OF A MOLAR TOOTH, moderately magnified. 1, Enamel, the lines of which indicate the arrangement of its columns. 2, Dentine, the lines indicating the course of its tubules. 3, Thin lamina of the dentine forming the wall of the pulp cavity, the dots indicating the orifices of the dental tubuli. 4, Cement.

The crown of the tooth is covered with *Enamel*, the hardest of all known animal textures, containing more earthy matter than the dentine, chiefly phosphate of lime. The enamel is composed entirely of hexagonal, prismatic fibres or rods arranged closely together upon the dentine. On the crown these fibres are vertical; on the sides, they become first oblique, then horizontal.

The *Cement* is a thin layer of true bone covering the fang, thinnest next to the enamel, and thickest along the grooves and near the point. Its outer surface is firmly attached to a fibro-vascular and sensitive membrane analogous to the periosteum, which seems to fasten the teeth in the socket of the jaw, being itself united to the periosteal membrane which lines its sockets.

240. The PHARYNX is a musculo-membranous bag, attached above to the base of the skull. Its walls consist chiefly of three pairs of constrictor muscles, supported by areolar tissue, and lined by mucous membrane which is continuous with that of the nasal cavities, Eustachian tubes, mouth, larynx and œsophagus, with all of which the pharynx communicates. The portion devoted to the passage of air has its epithelium *columnar* and *ciliated*, while that devoted exclusively to the passage of food and drinks has a *squamous non-ciliated* epithelium. Many mucous glands are found in the mucous membrane of the pharynx.

241. The walls of the ŒSOPHAGUS are composed of three coats—*muscular*, *areolar* and *mucous*. The *Muscular* coat has an external layer of longitudinal fibres and an internal layer of circular fibres. The fibres are mostly non-striated, excepting in the upper part. The *Areolar* coat is soft and distensible, supporting the mucous membrane, which lies in folds, so that no opening exists when the œsophagus is not in action. Many mucous glands are found, especially at the ends.

242. The STOMACH is the dilated portion of the alimentary canal, into which the œsophagus opens from above by the cardiac orifice, and the small intestines from below by the pyloric orifice. It is a membranous bag, consisting of mucous

membrane within, serous membrane without, with a muscular and areolar layer between. The muscular coat has three layers of fibres—*longitudinal*, *circular* and *oblique*. The *longitudinal* are continuous with the longitudinal fibres of the œsophagus, and extend divergingly over the body of the stomach, collecting into parallel bundles again at the intestinal opening, thus continuing into the intestines. The *circular* fibres correspond in arrangement with those of the œsophagus, passing in this way around the stomach, till condensed into a sphincter muscle at the pyloric orifice, which it partly closes. The *oblique* muscles are continuous with the circular fibres of the œsophagus, from which they spread obliquely for a short distance upon the anterior and posterior sides of the stomach.

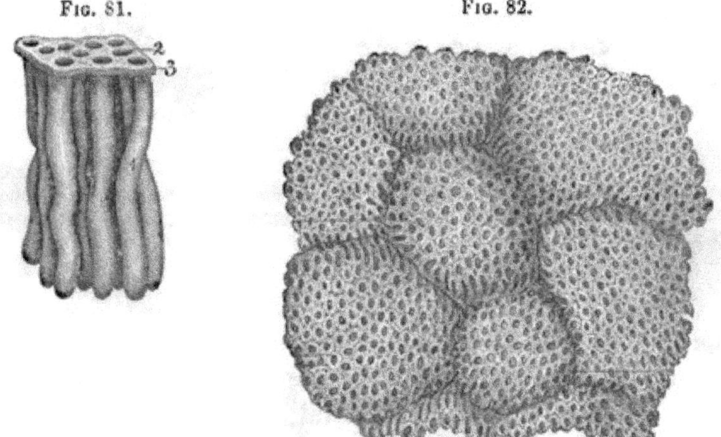

Fig. 81 (*Leidy*). Small Portion of the Mucous Membrane of the Stomach, with the Imbedded Gastric Glands. 1, The glands. 2, Orifices of the glands. 3, Epithelium of the mucous membrane; moderately magnified.

Fig. 82 (*Leidy*). Mammillæ of the Mucous Membrane of the Stomach, moderately magnified, exhibiting the orifices of the gastric glands.

The areolar coat is united to the muscular by loose areolar tissue, but the union is very firm between it and the mucous membrane which it supports. The mucous coat has numerous blood-vessels and lymphatics, also upright tubular glands, secreting the gastric juice.

10*

243. The INTESTINES have their coats and muscular fibres arranged like those of the stomach. The areolar coat, with its closely-adherent mucous membrane, projects into the interior of the small intestines, forming valves called *val'vulæ conniven'tes.* These vary in size, some being two inches long, and one-third of an inch wide in the middle, tapering at both ends; others are smaller, alternating with the layer. The intestinal mucous membrane is covered internally with thread-like processes of the membrane, which become erect when immersed in water, presenting a velvety appearance, hence called *villi.*

FIG. 83.

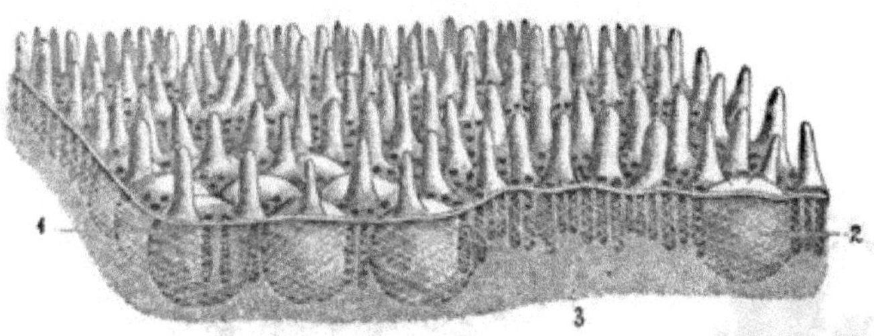

FIG. 83 (*Leidy*). PORTION OF THE MUCOUS MEMBRANE FROM THE ILEUM, moderately magnified, exhibiting the villi on its free surface, and between them the orifices of the tubular glands. 1, Portion of an agminated, or clustered gland. 2, A solitary gland. 3, Fibrous tissue.

244. The LIVER has two coats—the external serous coat, formed from the doubling of the peritoneum upon it, and the internal areolar coat. Its proper substance is composed of a multitude of compressed polyhedral masses, not larger than a small pin's head, and named *hepatic lobules.* Each lobule has one part called the base, which rests upon a hepatic vein (333), thence called the sub-lobular vein. The impress of the polygonal base may be seen within when the vein is opened. Running longitudinally down the middle of the lobule is a vein formed by six or eight veinlets coming from various parts of the circumference. This *mid-vein,* called intra-lobular, unites with the sub-lobular vein at the base of the lobule. The lobules are arranged closely side by side, their bases aiding in the formation of the membranous canals, called the

portal canal, in which the hepatic vein lies closely adherent to the canal-walls. This arrangement resembles that of sessile leaves upon the stem. The veinlets of the intra-lobular vein originate in a network of veins, upon the lobule-walls, which are continuous with the final branches of a small trunk passing between the lobules, from the *portal vein* (333) within the portal canal. The hepatic system, therefore, is continuous with the portal. Each portal vein is always accompanied by a hepatic duct and artery, and the three are enveloped in one sheath of areolar tissue, which gives off partitions to separate the vessels from each other. Outside and beyond these canals, this tissue invests the free portion of each lobule, filling the inter-spaces. If the other portions could be removed, leaving the areolar tissue intact, we should have a perfect areolar skeleton of the liver.

Fig. 84.

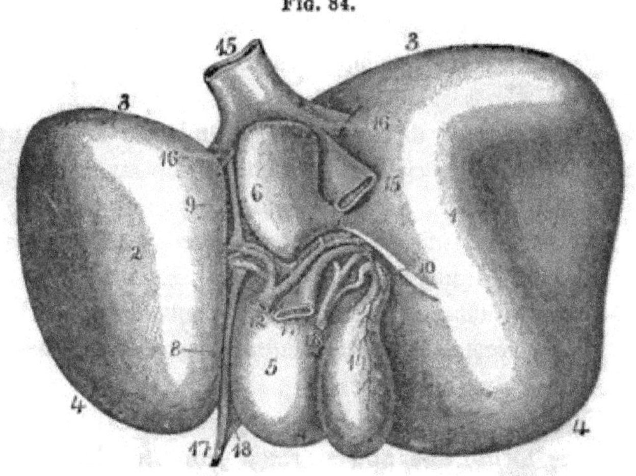

Fig. 84 (*Leidy*). Inferior Surface of the Liver.—1, Right lobe. 2, Left lobe. 3, Posterior margin. 4, Anterior margin. 5, Quadrate lobe. 6, Caudate lobe. 7, Isthmus, or caudate process, connecting the latter with the right lobe. 8, 9, Longitudinal fissure. 10, Transverse fissure. 11, Portal vein. 12, Hepatic artery. 13, Common biliary duct, formed by the union of the hepatic and cystic ducts. 14, Gall-bladder. 15, Inferior cava 16, Hepatic veins. 17, Round ligament. 18, Anterior part of the suspensory ligament.

245. The Spleen has two coats—the outer, serous coat, being a reflection of the peritoneum; the inner, fibro-elastic

coat, is composed of white fibrous tissue mingled with elastic tissue; when torn, the lacerated surfaces present a deep reddish-brown, pulpy appearance, resembling coagulated blood. The particular arrangement and relationship of its constituents have not been satisfactorily determined.

246. The PERITONEUM is a serous membrane which invests all the abdominal viscera, and is then reflected upon the walls of the abdomen. The large doubling of the peritoneum reflected from the front of the vertebral column over the small intestine is called the *mesentery*.

§ **21.** CHEMISTRY OF THE DIGESTIVE ORGANS.—*Secretions effecting Chemical Changes during Digestion. Chemical Character of these Secretions—Of Saliva—Of the Gastric Juice—Of Bile. Of the Pancreatic Juice—Of the Intestinal Juice. Relation of Acids and Alkalies in the Digestive Fluids.*

247. The chemical processes concerned in digestion consist of peculiar reactions between the food and the various secretions of the alimentary canal. These fluids are—*mucus* and *saliva*, secretions of the mucous membrane and glands of the mouth; *gastric juice*, a secretion of the stomach; *bile*, a secretion of the liver; *pancreatic juice*, a secretion of the pancreas; *mucus* and *intestinal juices*, secretions of the mucous membrane and glands of the intestines. Each of these fluids effects a special change in the constituents of food, till they are finally converted from an insoluble to a soluble condition, in which they may be absorbed.

248. MUCUS is a colorless and very viscid fluid found upon the mucous membrane, and secreted from the plasma of the blood by the epithelial cells of that membrane. It is sometimes alkaline, sometimes acid, but perhaps in its normal state, neutral. It is composed chiefly of water, holding from four to six per cent. of solids. Its characteristic constituent is *Mucin* (60), also a small amount of extractions, and salts like those of the blood. It has limited solvent powers, its chief office seeming to be to moisten the food, and thus prepare it for the action of other digestive fluids.

249. SALIVA is a transparent, watery fluid, with a specific gravity varying from 1002 to 1008. Its chemical composition, as given by Dr. Wright (who has made the salivary secretion a special study), is as follows:

Water	988.10
Salivin	1.80
Mucus (and epithelium)	2.60
Fatty matter	.50
Albumen (with soda)	1.70
Sulpho-cyanide of potassium	.90
Alkaline and earthy salts	3.20
Loss	1.20
	1000.00

When first secreted or during secretion, saliva is alkaline; in fasting, the moisture of the mouth is nearly neutral, or even acid; but it consists at that time almost entirely of mucus. Of salts, the tri-basic phosphate of soda probably gives the alkalinity to the secretion. Saliva contains a peculiar and remarkable salt, sulpho-cyanide of potassium. Besides, there are found chlorides of sodium and potassium, sulphate of soda, phosphates of lime and magnesia, and oxide of iron. The "*tartar*" of the teeth is formed by these earthy salts mixed with mucus, and minute portions of other animal matter. The chemical action of saliva is—*first*, that of a solvent; it dissolves saline substances, organic acids, alcohols and ethers, gum, sugar and the soluble albuminoid and gelatinoid bodies. *Second*, the saliva converts starch granules into dextrine, then into soluble dextrose, glucose, or grape sugar. A mixture of all the fluids of the mouth appears to form the most active combination for this purpose.

250. The GASTRIC JUICE is a colorless, or pale-yellow, transparent, slightly viscid and strongly acid fluid. Its specific gravity is 1025. Its composition as given by Schmidt is—

Water	994.4
Pepsin, with other organic matter	3.2
Salts	2.2
Free hydrochloric acid	.2
	1000.0

The small quantity of solid matter (about five per cent.) is remarkable, considering its extremely active powers. Pepsin is its characteristic constituent (58). The saline matter of the gastric juice consists chiefly of alkaline and earthy chlorides and phosphates. A small amount of lactic acid is found, but whether as a product of secretion or decomposition, is not certain. The free hydrochloric acid affords singular example of the liberation of a mineral acid from its strongly-combined base by an organic process in the animal economy; the source of this acid is probably common salt. Though the most powerful solvent known, the gastric juice seems to have no effect upon *living* animal substances; hence the membranes of the stomach remain intact as long as their vital power continues. A recent view, founded upon many experiments, attributes the non-solution to the protecting influence of the blood in the capillaries, which is supposed to maintain the alkalinity of the tissues—a chemical condition incompatible with peptic digestion. Gastric juice changes cane-sugar into glucose—albuminous substances, as albumen, fibrin, casein, etc., into substances called *peptones*. Gelatinous substances are changed chemically by the gastric juice, and lose their property of gelatinizing when cold; but this change is not necessary to their solution, which takes place so readily that these substances may be taken as food, when albuminous substances would remain in the stomach undissolved.

251. BILE is a somewhat viscid, glutinous and bitter fluid, of a dark golden-brown color. Its specific gravity is about 1026. Its acid (cholic and tauro-cholic) forms four to seven per cent. of the secretion, and is always united with soda; the coloring matter forms about five per cent.; ordinary fats, about one per cent.; salts, one per cent.; there are also traces of *choles'terine*. Bile is but slightly alkaline, and is sometimes neutral. It is an important agent in digestion, but its action does not depend upon an albuminoid like saliva, pepsin or pancreatin. It acts feebly in changing starch into sugar, and changes cane-sugar slowly, but grape-sugar rapidly, into lactic acid. It dissolves neither albuminoid substances nor

fat, but probably emulsifies the latter. Bile is said to arrest the actions of saliva and the gastric juice; it probably completes some particular part of the digestive process, but its specific action is not well understood.

252. The PANCREATIC JUICE is somewhat viscid, transparent, colorless and inodorous. Its solid constituents vary from 1.5 to 10 per cent. Its salts are chiefly chloride of sodium, and phosphates of lime and magnesia. It is more strongly alkaline than saliva; as digestion goes on, it becomes more alkaline and less viscid. Its most peculiar constituent is *pancreatin*, an albuminoid substance whose special composition is not yet determined. This juice is sometimes called abdominal saliva; as it has, like saliva, the power of converting starch into dextrine and grape-sugar. It has of itself little power over albuminoid and gelatinous substances, but is co-operative with the gastric juice. Its chief office seems to be to emulsify fatty matters, in which it probably acts with the bile.

253. The composition of INTESTINAL JUICES is not well known. They probably differ from common mucus, and have special properties. They are colorless, viscid, and contain from 2 to 3.5 per cent. of solid matter. They appear to be alkaline in the ileum, or lower part of the small intestines, acid in the cœcum, or beginning of the large intestines, and alkaline through the remainder. They convert starch into sugar, which in the small intestines passes into lactic and butyric acids; and act still more powerfully upon albuminoid substances, and also emulsify fat.

254. The changes which take place in the three *staminal principles* of food—*saccharine*, *albuminoid* and *oleaginous* substances—from their entrance into the mouth till ready for absorption, sum up as follows: The conversion of starch commences with the saliva; that of albuminoids and cane-sugar with the gastric juice; the emulsifying of fats with the bile and pancreatic juice. These processes go on independently of each other; the salivary action being unaffected by the gastric function, but both aided somewhat by the pancreatic

juice; the intestinal juice coming in as a general auxiliary agent, to complete and harmonize the several operations commenced at different points in the alimentary canal.

255. It will be noticed in the digestive fluids that there are successive alternations of alkali and acid: the saliva being alkaline; the gastric juice, acid; the pancreatic juice, bile and juice of the ileum, or third part of the small intestine, more or less alkaline; that of the cœcum of the large intestine, acid; that of the remaining portion, alkaline;—alternations giving neutralizations of great importance in the chemistry of digestion.

§ **22.** PHYSIOLOGY OF THE DIGESTIVE ORGANS.—*The Assimilation of Food. Process by which Food is transformed into Chyle. Destination of the Chyle.*

256. Food is necessary to the preservation and growth of the body, but it must first be *animalized*, or *assimilated;* that is, converted into matter having the *same characteristics* as those animal substances into which it is at length to be incorporated. We may include under the term *Primary Assimilation*, those animalizing changes necessary to the conversion of food into *chyle* and *blood:* under *Secondary Assimilation*, those necessary to the conversion of blood into integral parts of *solid tissue.* The *first* series of changes is included in the process named *Digestion*, by which food is transformed from its crude state into *chyle.*

257. The alimentary canal in which these digestive changes take place, is like a long manufacturing establishment, with many apartments;—the first room being the mouth, or masticating room, where some of the workmen cut the food; some grind it; some moisten it and supply the needed chemicals for making one of the animalizing changes. Mastication being completed, at the word of command, the obedient muscles, with greatest promptness and efficiency, convey the food onward to that wonderful laboratory—the stomach. The muscles of the soft palate raise the curtain from the base of the tongue, and incline it backward, closing the opening

into the nostrils; those of the small open lid of the trachea, the epiglottis, close the lid tightly that the food may pass safely over, while the muscles of the tongue, cheeks and floor of the mouth, force the food back into the pharynx and the œsophagus, the circular muscles of which, by alternate relaxation and contraction, urge it into the stomach. Here the food is subjected to a remarkable chemical agent, the *Gastric Juice*, which changes it from a crude state into a soft, homogeneous pulp, called *Chyme.*

258. Recent investigations show that this juice is less of a "universal solvent" than was formerly supposed—that its chemical power is limited to azotized substances; changing albuminoids into albuminose, and gelatinoids into gelatinose, the conditions best adapted to assimilation. The change in starch which continues in the stomach, is effected by the presence of the saliva, which commenced its work in the mouth. Oleaginous matters are only reduced to a fine state of division and held in suspension by the pulpy chyme.

During these processes, the mass is undergoing a churning or rotary motion, by the joint manipulations of the longitudinal, circular and oblique muscles, thus bringing part after part into the immediate presence of the Gastric Juice. While digestion is thus going on, the openings of the stomach are well guarded. A return of any part of the mass into the œsophagus is prevented by the sphincter muscles near the cardiac orifice; and the passage to the intestines is closed by the sphincter muscles of the pyloric orifice, and a valve called the *pylorus*, or "gate-keeper," which, true to its name, stands a faithful sentinel till proper chyme presents itself, showing evidence of having completed the prescribed curriculum. This sentinel-commission seems to last only during the process of digestion, as afterward many substances previously detained are allowed free egress.

259. After passing the pyloric orifice, the chyme is treated by other chemical agents—the bile, the pancreatic and intestinal juices, continuing the chemical processes commenced in other parts of the alimentary canal. The fats are reduced to

an exceedingly fine state of emulsion, but there is no proof of chemical change. The whole pulp is subjected to the constant wave-like, or peristaltic muscular action of the intestines, which forces their contents to their respective destinations. The nutritive portion is called *chyle,* and is taken up by the absorbent vessels and conveyed to the blood; while the innutritious portion is excreted from the system.

260. The absorbing surface of the intestines is enormously increased by the projecting forms and great abundance of the *villi:* they hang out into the nutritious, semi-fluid mass contained in the cavity of the intestines. as the roots of a tree penetrate the soil, imbibing the liquid portions of food with wonderful rapidity.

§ **23.** HYGIENE OF THE DIGESTIVE ORGANS.—*Suggestions relative to the Preservation of the Teeth—To their Removal. Conditions affecting the Quantity of Food demanded by the System. The Quality of Food. Directions relating to the Manner of taking Food. Conditions of the System requisite for the proper Digestion of Food.*

261. *For the Preservation of the Teeth,* the first requisite is to keep them clean. After meals they should be cleansed, to prevent the collection of *tartar,* and to remove any remaining particles of food. Such as are inaccessible to the brush, may be removed by tooth-picks made of wood, ivory or the common goose-quill. Metal injures the enamel. Night and morning, the mouth should be cleansed with pure, tepid water, after which the teeth should be thoroughly brushed on both surfaces. Occasionally, refined soap may be moderately used, if followed by thorough rinsing of the mouth.

Sudden changes of temperature crack the enamel, hence extremes of heat and cold in food and drinks should be avoided. Acid and corrosive substances should also be avoided, as acidulated drinks and mineral waters, that "set the teeth on edge." All tooth-powders containing such articles should be banished from the toilet.

Tobacco contains a "grit" which injures the enamel. It also discolors the teeth, debilitates the vessels of the gums, taints the breath and renders the appearance of the mouth forbidding.

The teeth should be frequently examined, that if enamel is removed and decay commenced, they may be filled with *gold-foil.* All amalgams, pastes and cheap patent articles should be rejected, both for the sake of the teeth and the general health.

262. *The Removal of the Teeth.* The temporary teeth should be removed at once, when loose; or before, if the permanent teeth appear. This is essential to the regularity and beauty of the second set.

Irregular or crowded permanent teeth, generally, require the removal of one or more. By pressure upon each other, the enamel is injured and the appearance rendered unsightly. With a little care the spaces left after extraction will soon be filled with the remaining teeth.

Toothache does not always indicate the necessity of extraction, as the nerve, or investing membrane, may be diseased, and the tooth sound. Relief will then be afforded by proper medication.

Observation.—When the removal of a tooth is necessary, apply to some skillful operator: something more is needed than strong muscles and a pair of forceps. Skill is as requisite in the proper extraction of a tooth, as in the amputation of a limb.

263. The health of the Digestive Organs, in general, requires the observance of certain conditions relative to their natural stimulus—*Food.* These will be considered under the following heads: 1. *The Quantity of Food.* 2. *The Quality of Food.* 3. *The Manner of taking Food.* 4. *The Proper Conditions of the System for receiving Food.*

264. The QUANTITY OF FOOD necessary to the system varies, being affected by age, occupation, temperament, habits, temperature, amount of clothing, health and mental state.

265. *The supply must equal the waste of the system.* In every department of nature, waste attends action. The greater the

amount of exercise, the more rapidly will the particles be worn out and removed, and their places need supplying with new atoms.

During the period of growth, the supply must exceed the waste, for the building of new tissues. This accounts for the keen appetite and vigorous digestion in childhood. The same is true when persons have become emaciated from famine or disease.

266. *When exercise is lessened, the quantity of food should be proportionally diminished,* otherwise the tone of the digestive organs will be impaired, and the health of the system enfeebled. This is especially applicable to students, who have been accustomed to laborious employments. Self-denial should be practiced for a few days, when the real wants of the system will generally be manifested by a corresponding sensation of hunger. It is a common remark that in seminaries and colleges, students from the country suffer more from indigestion and impaired health, than those from the cities.

267. *More food is required in winter than in summer;* hence, by diminishing the amount of food as the warm season approaches, the tone of the stomach and vigor of the system will be better maintained, thus lessening the liability to "summer complaint." In this respect, the lower animals seem to learn from instinct, what man is slow to learn from reason.

268. *The amount of food should be adapted to the present condition of the digestive organs.* Imperfectly digested food irritates the mucous membrane of the intestines and debilitates the system instead of invigorating it. In sickness, the attending physician is the person to decide respecting the proper amount. In health, the natural appetite is generally a safe guide, as to plain, nutritious food; but condiments, spices, etc., excite a morbid appetite, whose cravings it is unsafe to gratify. General languor of the body after meals, shows that undue demands are made for an increased supply of fluids to enable the overloaded stomach to free itself of its

burden. This, with the extra labor of the secreting glands, will soon be followed by debility and consequent inaction.

269. The QUALITY OF FOOD *should be both nutritive and digestible.* Substances are *nutritious* in proportion to their capacity to yield the constituents of chyle. Substances are *digestible* in proportion to the facility with which they are acted upon by the digestive fluids. Articles highly nutritive in themselves, but difficult of digestion, often yield less nourishment than those poorer in nutritive quality, but easy of digestion. If we confine our diet to easily digested articles, the digestive organs will be weakened from want of proper exercise; if to highly concentrated diet, they will be injured by over-work; hence, the necessity of choosing, in this respect, the "happy medium."

270. *Proper aliment must contain the three staminal principles of food.* These are albuminous, oleaginous and saccharine substances; the first contain carbon, oxygen, hydrogen and nitrogen; the last two are destitute of nitrogen. Various experiments have shown that if we feed upon any one of these groups, to the exclusion of the other two, or upon any two to the exclusion of the third, the health will be impaired. Milk contains all the food principles;—the albuminous, being furnished by its caseine; the oily, by the butter; and the saccharine, by the sugar of milk. Beef is rich in fat and albumen, and also contains *inosit*, or muscle sugar. Most of the cereals contain gluten (an albuminoid), starch, sugar and oil. Wheat, however, has the first three constituents without the oil. It is most nutritious in the form of "Graham flour;" by rejecting the *bran*, most of the gluten is lost. Eggs are very rich in albumen, and the yolk also contains oil. Beans, peas, etc. afford starch and much legumine (an albuminoid). Potatoes abound in starch. Sago, tapioca, rice, arrow-root, etc., are constituted almost wholly of starch. These articles or their substitutes, properly combined, will yield the necessary elements to the system.

271. *Food should be properly cooked.* However nutritious an article of food may be, if not well cooked, it is not only

unsavory to the palate, but hurtful to the digestive organs. The simplest methods of preparation by cooking are the best. Meat should be broiled, roasted, or made into soup. Fried meats are apt to be indigestible and also less nutritious. The fat used in frying is infiltrated by the heat and usually penetrates the whole mass. It is mistaken economy to *fry* meats for the laboring class; better throw the fresh steak upon coals, and add simply salt and pepper, than to deluge it in boiling fat. Much of the nutriment of beef when salted, is extracted by the brine; and during the process of boiling, still another portion remains in the boiling water, thus leaving but little more than hardened muscular fibre to grace the platter. (The liquor of boiled beef may be converted into soup.) The breakfast "hash" is too frequently unfit to be eaten by the student or sedentary person, from want of being well cooked.

The cooking of *vegetables* should be thorough and complete. The proper *combination* and *cooking* of a few articles of food (as flour, milk, eggs and butter), require skill, which in reality, assumes the importance of no inferior art.

272. *The Quality of Food should be adapted to the season and climate.* Highly stimulating food may be used almost with impunity during the cold season of a cold climate; but in the warm season and in a warm climate it is very injurious. Animal food being more stimulating than vegetable, is therefore well adapted to winter, and vegetable to spring and summer. Where the digestive organs are weakened or diseased, it is very important that a nutritious vegetable diet be adopted as the warm season approaches.

273. *Vegetable diet is most suitable for children.* The organs of a child are more sensitive and excitable than those of an adult; hence, stimulants of every kind should be strictly avoided, and the food mainly of a vegetable character. In this "fast age," this is a suggestion of vast importance. Parents mourn over many evil effects of unrestrained passion and moral deterioration in the rising generation, while in truth, these are too often but the legitimate harvest of the

seed they have themselves sown in the form of stimulating food and drinks. The old spelling-book assertion, that "Bread and milk is the best food for children" is as true now, as it was in the days of our fathers.

274. Some temperaments require more stimulating food than others. As a general rule, persons of obtuse sensations, and slow movements, are benefited by animal or stimulating food; while individuals of highly sensitive constitutions, and quick, hurried movements, require a nutritious and unstimulating vegetable diet.

275. THE MANNER OF TAKING FOOD exercises a controlling influence upon the health of the digestive organs.

276. *Food should be properly masticated.* This is essential to secure the fine division necessary to the proper action of the gastric juice and other fluids, and especially to mix the food with the requisite amount of saliva. Rapid eating should be avoided, not only as a violation of good table manners, but as a violation of the laws of our physical nature, whose penalty, in the form of dyspepsia with its numerous train of evils, will sooner or later be visited upon the transgressor.

277. *Drink should not be taken with the food.* Nature supplies the appropriate moisture, and if tea, coffee, or any other fluid be used as a substitute, indigestion will follow, from the absence of the necessary amount of saliva. Again, drinks taken into the stomach must be absorbed before the digestion of other articles is commenced. Thirst between the meals does not always arise from a demand of the system for fluids, but may be induced by fever or local disease of the parts connected with the throat. This may often be relieved by chewing a cracker, or some other dry substance, thus exciting the salivary glands. This is a safe resort when thirst accompanies a heated condition of the system, arising from over-exercise; while the practice of taking cold fluids is dangerous and should never be indulged.

278. *Regard should be paid to the temperature of food and drink.* Hot food or drink, for a short time unduly stimu-

lates the vessels of the mucous membrane of the gums, mouth and stomach; then reaction follows, bringing loss of tone and debility of these parts. This practice is a fruitful cause of spongy gums, decayed teeth, sore mouth and indigestion. On the other hand, if food or drink be taken too cold, an undue amount of heat is abstracted from the stomach, this arrests the digestive process, and thus deranges the system.

279. *Food should be taken at regular and suitable periods.* The interval between the meals should be regulated by the character of the food, and the age, health, exercise and habits of the individual. In the young, the active, and the vigorous, food is more rapidly digested than in the aged, the indolent and the feeble; consequently, it should be taken more frequently by the former class than by the latter. The average time required to digest an ordinary meal is from two to four hours. The stomach should always have from one to three hours of rest, before the next meal. Eating between meals, is a habit ruinous to the digestive organs, inasmuch as the chemical processes are by this means disturbed, and the stomach given no time for rest.

280. THE CONDITIONS OF THE SYSTEM FOR RECEIVING FOOD, are of practical importance for the healthy action of the digestive process.

281. *Food should not be taken immediately before or after severe exercise of body or mind.* The functional exercise of any organ abstracts fluids, sanguineous and nervous, from other parts of the body, thus weakening those parts for the time. Severe exercise of muscle, concentrates the forces in the muscle; severe exercise of the brain, concentrates the forces in the brain; the same is true of the vocal and other organs. After severe exercise, from thirty to forty minutes should be allowed before eating, for restoring equilibrium to the system. The student, farmer or mechanic, who hurries from his toil to his dinner to "save time," will, in the end, lose more time than he saves. After eating, the digestive organs need, for a time, the chief use of the vital forces, and if they are habitually expended elsewhere, as in study or

labor, digestion will be arrested, the chyle cheated of its proper elements, and headache, dullness and general derangement will follow. A moderate exercise of the muscles, a social chat and a hearty laugh, aid digestion, and tend "to shake the cobwebs from the brain." These directions are particularly applicable to the ambitious student who feels that he must "save time" and "must have the lesson." Let him try the experiment, and he will soon find that in the after-dinner hour, his lesson is better learned when he spends half the hour in recreation, and the other half in close application. Many students are obliged to give up their course of study, from simple neglect of these rules.

Observation.—The same principle will apply to lower animals. They will perform more labor by having a suitable period of repose after being fed. Two dogs were fed upon the same kind of food, one was kept quiet, the other sent in pursuit of game. In an hour both were killed. In the stomach of the quiet dog, digestion was nearly complete; in that of the other, the food was scarcely altered.

282. *Persons should abstain from eating at least three hours before retiring for sleep.* It is no unusual occurrence for those persons who have eaten heartily immediately before retiring to have unpleasant dreams, or to be aroused from their unquiet slumber by colic pains. In such instances, the brain becomes partially dormant, not imparting to the digestive organs the requisite amount of nervous influence; this being deficient, the unchanged food remains in the stomach, causing irritation of this organ. A healthy farmer who was in the habit of eating a quarter of a mince pie just before retiring, became annoyed with unpleasant dreams, and among the images of his fancy, he saw that of his deceased father. Becoming alarmed, he consulted a physician, who, after a patient hearing, advised the patient to eat *half* a mince pie, assuring him that then he would *see his grandfather*.

283. *The mental state exerts an influence upon the digestive process.* This is clearly exhibited when an individual receives sad intelligence. Let him be sitting at a plentiful

board with a keen appetite, and the unexpected news destroys it, because the excited brain withholds the stimulus; hence all unpleasant themes, labored discussions, or matters of business, should be banished from the table. Light conversation, enlivening wit and cheerful humor wonderfully promote digestion.

Indigestion arising from nervous prostration should be treated with great care. The food should be simple, nutritious, properly cooked, moderate in quantity and taken at regular periods. Large quantities of stimulating food, frequently taken, serve to increase the nervous prostration. Exercise in the open air, and a cheerful state of mind, are very beneficial in restoring the natural, healthy action of the brain, and thus aiding the digestive powers.

284. *After long abstinence, unstimulating food should be taken, and in small quantities.* As in case of sickness, when the appetite begins to return, the nurse must use much discretion, and the patient, often, self-denial. The popular adage, "that food never does harm, when there is a desire for it," is untrue. Too frequently, when a patient satisfies his cravings, it is to induce relapse into the former disease, and at the risk of life. The digestive organs are weak, and must be gradually brought into action. It is often better to give the food in a solid, rather than liquid form, so that the salivary and mucous glands may be stimulated to action.

285. *The condition of the skin exercises an important influence upon digestion.* Let free perspiration be checked, either from uncleanliness, chills or any other cause, and the functional action of the stomach is diminished. This is one of the fruitful causes of "liver and stomach complaints" among the filthy and half-clad inhabitants of our cities and villages. Attention to bathing and clothing would prevent many "season complaints," especially among children.

286. *Pure air is necessary to give a keen appetite and vigorous digestion.* The digestive organs must have a plentiful supply of pure blood, and to have pure blood we should breathe pure air. Poor ventilation is a frequent cause of indigestion.

Persons who sleep in ill-ventilated rooms have little or no appetite in the morning. A manufacturer stated before a committee of the British Parliament, that he had removed an arrangement for ventilating his mill, as he noticed that his men ate much more after his mill was ventilated than before, and he could not *afford* to have them breathe the pure air. Compression of the vital organs prevents the introduction of a sufficient supply of pure air, and is one of the causes of dyspepsia now so prevalent among ladies.

General Observations.—All aliment is separated into nutriment and residuum. The latter should be regularly expelled from the system, otherwise headache, dizziness and general uneasiness will ensue, and if allowed to continue, the foundation will be laid for a long period of suffering and disease. For the preservation of health, there should be in most persons a daily evacuation of residual matter. Evening is the best time, especially is this true when persons are afflicted with piles. Constipation may, in many cases, be relieved by friction over the abdominal organs, and by making an effort to evacuate the residuum at some stated period each day.

Recapitulation.—Digestion is most perfect when the action of the cutaneous vessels is energetic; the brain moderately stimulated; the blood well purified; the muscular system duly exercised; the food properly cooked and masticated, taken at regular periods, and adapted in quality and quantity to the present condition of the individual.

§ **24.** Comparative Splanchnology.—*Nutritive Apparatus of Vertebrates. Compare the Mouths and Teeth of Vertebrates.—The Digestive Fluids.—The Stomach and Intestines.*

287. In the Nutritive Apparatus of all vertebrates, as in the Motory, a general plan of parts obtains, subject to the variations required to preserve the harmony of relation between the organization and the *use* to which it is to be applied.

288. In no part do we find a greater variety or a nicer accommodation to particular wants than in the MOUTHS and TEETH of different animals. In *Mammals*, the projecting jaws, the wide mouth, the strong, pointed, sharp, enameled edges of the teeth enable carnivorous, or flesh-eating animals, to seize and hold their prey, and the hinge-like movement of the jaw, to divide it like a pair of scissors; as seen in the cat and the lion. The full lips, the rough tongue, the furrowed, cartilaginous palate, the broad, rough surface of the teeth, the central plates of enamel and the lateral movement of the jaw, qualify the herbivorous, or grain-eating animals, for grazing, and grinding their food, as the grain is crushed between the upper and nether mill-stone; as the sheep and the horse. The elongated, tapering muzzle, the

FIG. 85. FIG. 86.

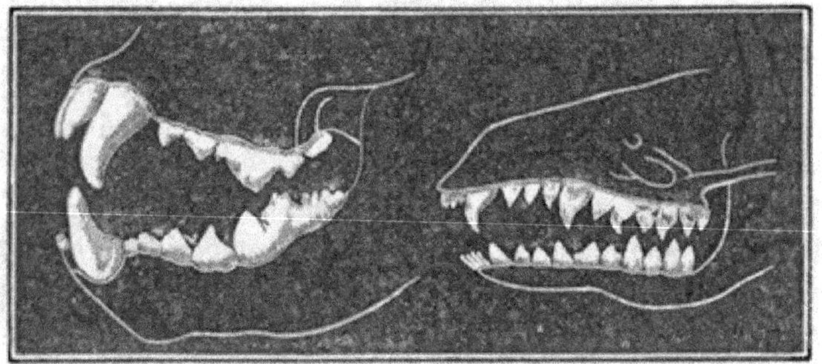

FIG. 85 REPRESENTS THE TEETH OF CARNIVORA, OR FLESH-EATING ANIMALS.
FIG. 86 REPRESENTS THE TEETH OF INSECTIVORA, OR INSECT-EATING ANIMALS.

cone-pointed, enameled molars locking into the enameled depressions of the opposite jaw, enable the insectivorous animals to burrow in the earth, for the insects and worms upon which they feed, and also to crush them; as in the mole and hedgehog. The two chisel-shaped incisors, enameled only in front, allowing more rapid wear of the posterior than the anterior part, keeping them always sharp; the bag of pulp at the base of these teeth, providing for growth equal to the wear at the top; the backward and forward movement of the jaws;

and the great size and strength of the lower jaw, adapt the rodentia, or gnawers, to their mode of life; as in the rat and the squirrel.

FIG. 87.

FIG. 87. LOWER JAW OF A SQUIRREL.—1, The enamel of the gnawing tooth. 2, The ivory. 3, The lateral furrows of the molar teeth.

289. In *Birds* the mouth receives a new character, both in substance and in form. Instead of fleshy lips and teeth of enameled bone, we have the hard and horny investment of the jaws, known as the *bill*, destitute of true teeth. This organ varies in size and form, according to the food of the species, which may be grains, insects, fishes or flesh.

290. *Reptiles* swallow their food without mastication, hence, their jaws and throats are made capable of great dilatation, and their teeth, used only for seizing and retaining their prey, all resemble each other.

291. The jaws of most *Fishes* are armed with teeth, and in many cases these are placed in all parts of the mouth, and even in the gullet.

292. In most animals, the digestive fluids are supplied by mucous follicles and glands, similar to the salivary glands in man. The simpler the function of the mouth, the smaller and simpler the arrangement for the supply of these fluids, as is seen in birds, also in reptiles, and some fishes that swallow their food without mastication, and have no organ of secretion but the liver.

293. The STOMACH and INTESTINES of vertebrates vary in size, form and relative length. They are simpler, smaller and shorter in carnivorous than in herbivorous or granivorous

animals; while the ox has intestines about twenty times the length of his body, those of the lion are but three or four times its own length.

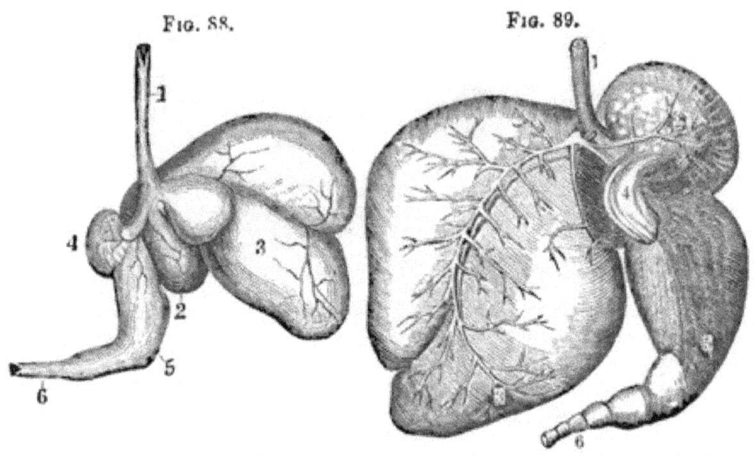

FIG. 88. STOMACH OF THE SHEEP.—1, The œsophagus. 2, The rumen. 3, The reticulum. 4, The omasum. 5, The abomasum, or rennet. 6, The intestine.

FIG. 89. STOMACH OF AN OX.—1, The œsophagus. 2, The rumen (paunch). 3, The reticulum (honeycomb). 4, The omasum (many-plies). 5, The abomasum (rennet). 6, The intestine.

294. Ruminants, as the sheep and ox, have a stomach with four cavities. The first stomach, called the *Ru'men*, or "*Paunch;*" the second, the *Retic'ulum*, or "*Honeycomb;*" the third, the *Oma'sum*, or "*Many-Plies;*" the fourth, the *Ab'omasum*, or "*Rennet:*" the latter, taken from the young calf, is used in cheese-making.

The food when first swallowed is received into the Rumen, where it accumulates while the animal is feeding. Here it is moistened by the fluids secreted by the walls of this cavity. It then passes into the Reticulum, where it receives additional secretions, and is made into little pellets, or "cuds," which, when the animal is at rest, are returned to the mouth, to be re-chewed and mixed with the saliva. This pulp passes directly into the third cavity, to be prepared for the fourth, where true digestion takes place. It is then received by the intestinal canal.

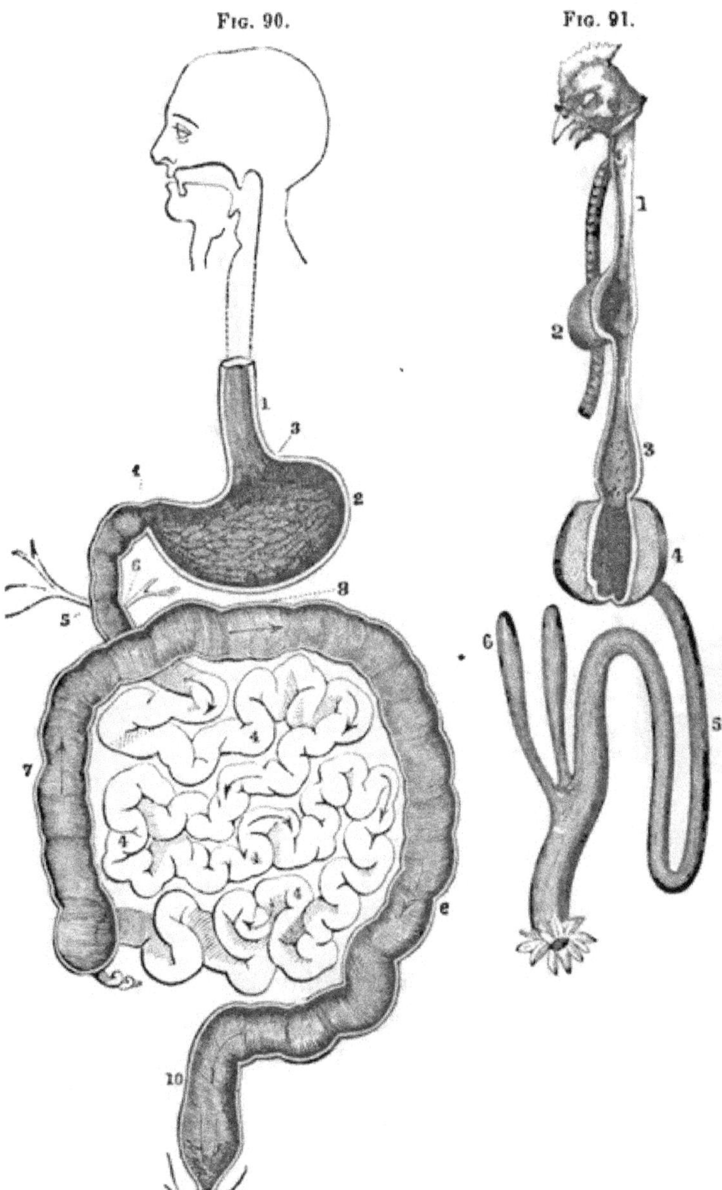

FIG. 90. THE ALIMENTARY CANAL OF MAN.—1, Œsophagus. 2, The stomach. 3, Cardiac orifice. 11, Pylorus. 5, Biliary duct. 4, 4, 4, 4, Small intestines. 6, Pancreatic duct. 7, Ascending colon. 8, Transverse colon. 9, Descending colon. 10, Rectum.

FIG. 91. THE ALIMENTARY CANAL OF A FOWL.—1, The œsophagus. 2, Ingluvies (crop). 3, Proventiculus (secreting stomach). 4, Triturating stomach (g zard). 5, Intestine 6, Two cæca.

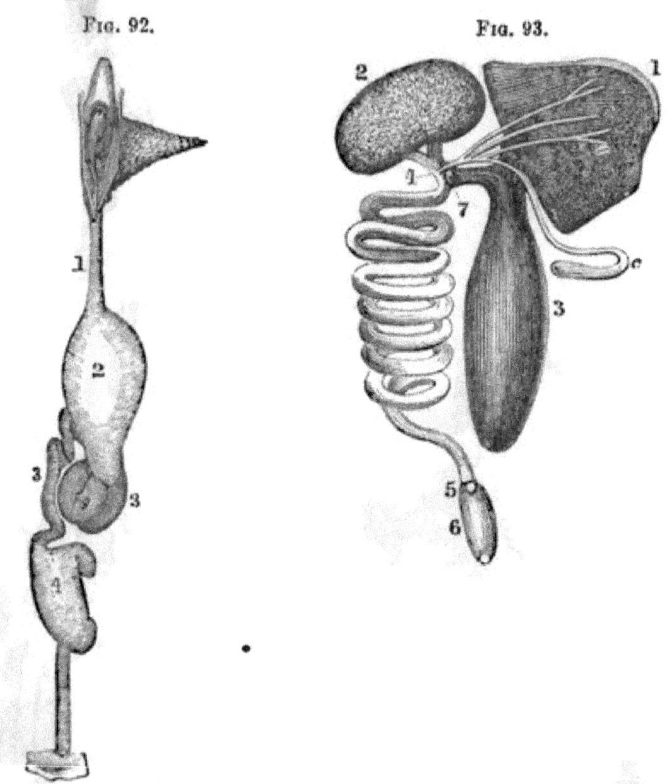

Fig. 92. The Alimentary Canal of the Flying Lizard.—1, The œsophagus. 2, The stomach. 3, 3, Small intestine. 4, Large intestine.

Fig. 93. The Alimentary Canal of the Sword-Fish.—1, Liver. 2, 3, Cæcus, or pouches, connecting with small intestine. 4, 5, Small intestine, coiled. 6, Large intestine. 7, Biliary duct.

295. In *Birds* there are usually three cavities, or stomachs; the first is a dilatation of the œsophagus, called the *Crop*, or "*Inglu'vies*," where the food is macerated and softened; the second is the true stomach, named '*Proventric'ulus*," where the mucous membrane is provided with mucous follicles, secreting an acid which acts still farther upon the food; and the third is the *Gizzard*, or *Trit'urating* cavity. The latter, in granivorous birds, has immense strength, being composed of muscular fibres running in different directions, and lined with a horny membrane. Gravel and angular stones are instinctively

swallowed to assist in the grinding process. In flesh-eating birds the gizzard is thin and membranous.

296. In *Reptiles* the alimentary canal differs much from that of mammals or birds. As a general rule, it is shorter in proportion to the trunk than in warm-blooded vertebrates. The transition from the œsophagus to the stomach is by a pouch-like enlargement; the small intestines usually have a few coils; the large intestines in most reptiles are short, simple and straight, without cæcal appendage at its beginning. The liver is relatively large.

297. In *Fishes*, the alimentary canal is more diversified in length, size and form than in reptiles; the œsophagus is a short and funnel-shaped canal; the stomach is shaped either like a syphon or a pouch (cæca). In some species of fish, the small intestines extend in a line from the stomach to their termination; in others, there are found from two to eight coils. The large intestines are short and straight. The liver is usually large, with numerous appendages. In the cod it is soft and saturated with oil, which is expressed for medicinal purposes.

Fig. 94.

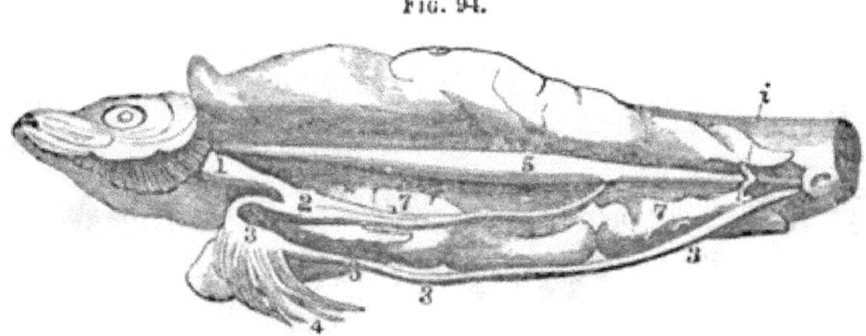

Fig. 94. The Alimentary Canal of the Herring.—1. Œsophagus. 2, Stomach. 3, 3, 3, 3, Small intestine. 4, Cæca. 5, Air-bladder. 7, Pneumatic duct.

CHAPTER VII.

ABSORPTION.

298. We have observed the changes in food till its formation into chyle—changes which have taken place in the alimentary canal, and which are included under the general term, *Digestion.* The chyle, however, is virtually *external* to the animal body. The process by which it is conveyed within is called *Absorption;* and the vessels conveying it are named *Absorbents.*

The term absorption, used in its largest sense, however, includes more than the mere taking up of nutrient material from the alimentary canal. It embraces that general process by which all external soluble substances, whether solid, liquid or gaseous, beneficial or poisonous, nutrient, stimulant or respiratory, are introduced into the tissues of the body. It also comprehends, in part at least, the process by which portions of the living tissues are themselves removed, or absorbed within the body. The former may be called *General Absorption,* and the latter, *Intrinsic,* or *Interstitial Absorption.*

§ **25.** Anatomy of the Absorbents.—*The Process of Absorption—Specific and General. The Absorbent Vessels. Lymph. Distribution of the Lymphatics. The Thoracic Duct. The Lymphatic Duct. Position of Lymphatic Glands. Absorbent Veins.*

299. The absorbents consist of certain blood-vessels, especially the venous capillaries, and the absorbents proper, viz., *Lymphatic* Vessels* and *Glands.*

The fluid conveyed by the lymphatic absorbents is a transparent, transuded portion of the blood, called *Lymph.* The lymphatic vessels of the small intestines are named *Lac'teals,*†

* Lat., *lympha,* water. † Lat., *lac,* milk.

Fig. 95.

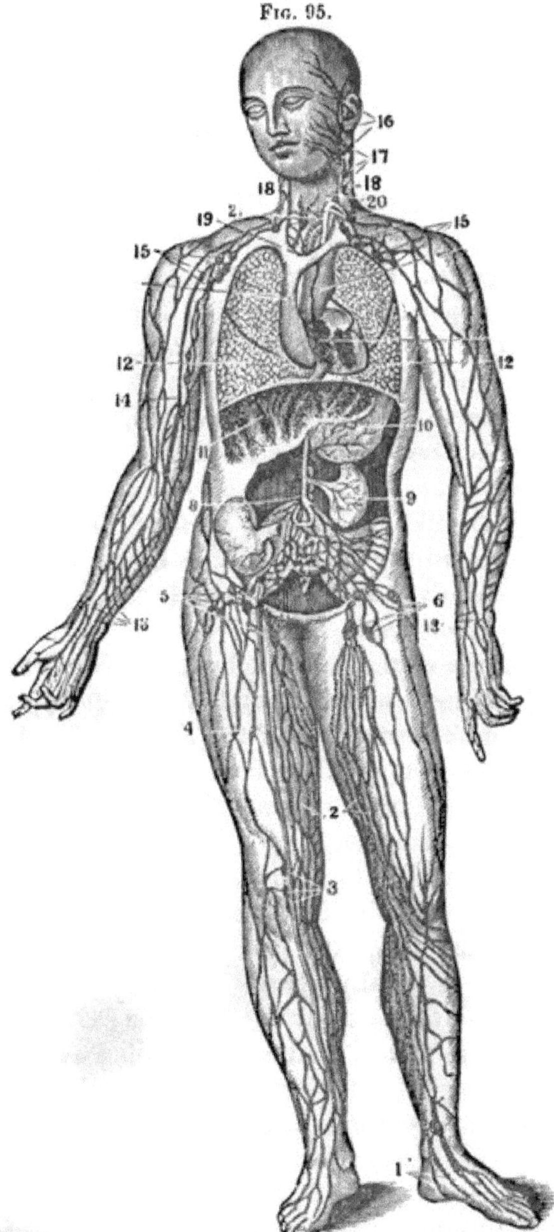

Fig. 95. A Representation of the Lymphatic Vessels and Glands.—1, 2, 3, 4, 5, 6, The lymphatic vessels and glands of the lower limbs. 7, Lymphatic glands. 8, The commencement of the thoracic duct. 9, The lymphatics of the kidney. 10, Of the stomach. 11, Of the liver. 12, 12, Of the lungs. 13, 14, 15, The lymphatics and glands of the arm. 16, 17, 18, Of the face and neck. 19, 20, Large veins. 21, The thoracic duct.

from their milky appearance during active digestion, when they are filled with chyle. In the interval of digestion, they convey lymph like the other lymphatics.

300. The LYMPHATIC GLANDS through which the vessels pass are somewhat hard, pinkish bodies, varying in size from that of a hemp-seed to that of a large pea.

301. The LYMPHATIC VESSELS are distributed through most of the system. Few are found in the muscles, and none in the brain or spinal cord, though they doubtless exist there. They abound in the secreting membranes, especially in the skin and the mucous membrane.

The finer lymphatics unite into trunks, which either accompany the blood-vessels and form the *deep* lymphatics, or run on the surface of organs or in the sub-areolar tissue, forming *superficial* lymphatics. From all parts of the body, these trunks run toward the root of the neck and unite in two main trunks which end in the venous system, viz., the *Thoracic* and *Lymphatic Ducts.*

The lymphatics of the lower limbs of the abdomen, of the left side of the head and neck, and of the left upper limb, form the *Thoracic Duct;* those of the right side of the head and neck, and right upper limb, form the *Lymphatic Duct.*

302. The THORACIC DUCT commences with a dilatation, named the "Receptaculum Chyli," or receptacle of the chyle. This vessel is formed by the convergence of lymphatics from the lower extremities, the intestines, stomach, spleen, pancreas, kidneys and the greater part of the liver. The "receptaculum chyli" is usually placed upon the second lumbar vertebra, a little to the right of the aorta (329). It soon passes behind the arch of that vessel, crossing over the œsophagus, and ascends on the left side to the root of the neck, where it curves downward and outward behind the great blood-vessels, and finally opens into the angle at the junction of two large veins (330).

303. The LYMPHATIC DUCT is about an inch long, and has a similar termination on the right side of the body.

304. The lymphatic glands are found in the axilla of the

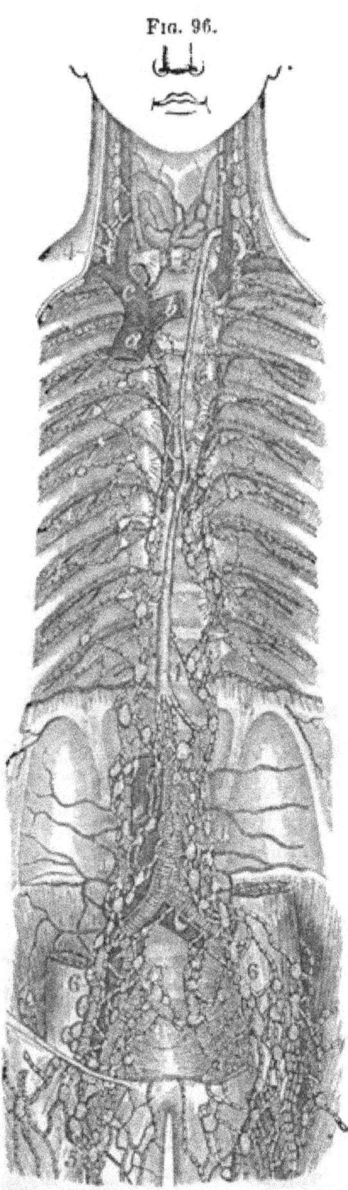

FIG. 96 (*Leidy*). VIEW OF THE GREAT LYMPHATIC TRUNKS.—1, 2, Thoracic duct. 4, The right lymphatic duct. 5, Lymphatics of the thigh. 6, Iliac lymphatics. 7, Lumber lymphatics. 8, Intercostal lymphatics. *a*, Superior cava. *b*, Left innominate vein. *c*, Right innominate vein. *d*, Aorta. *e*, Inferior cava.

arm (arm-pit) and in the groins; chains of glands are found on each side of the neck; a few in the arm; also many about the bronchi, or air-tubes; and in the pelvis or abdomen;—those of the lacteals being abundant in the *Mes'entery.**

305. The veins of the intestines acting as absorbents unite with those coming from the stomach, the spleen and the pancreas, thus forming the *Portal vein*, which enters the liver through a fissure in the concave surface.

§ 26. HISTOLOGY OF THE ABSORBENTS. — *Histology of the Lymphatic Vessels — Glands. Origin of the Lymphatics.*

306. Most of the LYMPHATIC VESSELS are long, thread-like, transparent tubes, with coats so exceedingly delicate that their structure is a matter of inference from that of the Thoracic Duct, which has three coats, like the veins. The external coat is the thickest, and consists of white fibrous tissue, with longitudinal webs of elastic tissue; the middle coat consists of unstriated muscular, elastic and connective tissues; the inter-

* Gr., *mesos*, middle, and *enteron*, the intestine.

nal coat, of a lining epithelium, an elastic basement membrane, supported by longitudinal laminæ of elastic tissue. The larger lymphatic tubes are liberally supplied with valves formed by the infolding of the inner coat. These valves are arranged in pairs, and are much more numerous in the smaller than in the larger vessels. In the thoracic duct, they are sometimes more than an inch apart. A very strong pair is placed at the opening of the thoracic duct into the large veins.

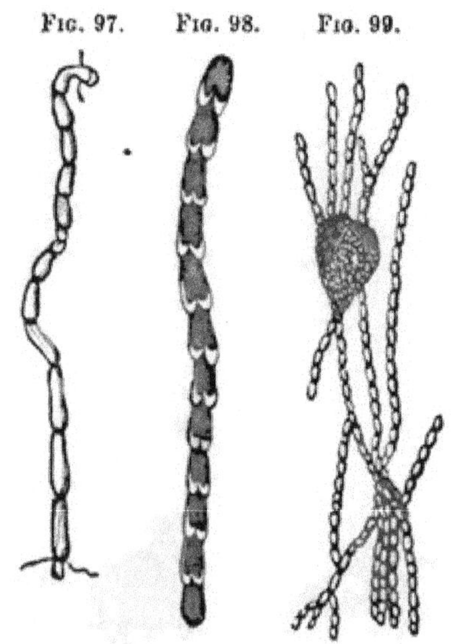

FIG. 97. A SINGLE LYMPHATIC VESSEL, much magnified.
FIG. 98. THE VALVES of a lymphatic trunk.
FIG. 99. 1, A LYMPHATIC GLAND, with several vessels passing through it.

307. The LYMPHATIC GLANDS are not well understood. They seem to be composed of a large number of vesicles, or pouches, which communicate with each other and also with the lymphatic tubes. The tubes, or vessels, entering the gland, are called *afferent* vessels, and those emerging from it, *efferent* vessels. Each vesicle of the gland seems to connect with an afferent and an efferent vessel.

308. The lymphatics are of such tenuity and transparency, it is with the greatest difficulty that they can be discovered,

hence, their origin is imperfectly known. They appear to originate in a capillary network among the sanguiferous capillaries, but not to communicate with them. The lacteals originate in the villi of the intestines, and unite more and more till their entrance into the receptaculum chyli.

309. The LYMPH consists of a fluid part containing nuclei, minute granules, and sometimes a few oily globules.

§ 27. CHEMISTRY OF THE ABSORBENTS.—*Chemical Changes in the Absorbent System—In the Portal Circulation.*

310. We know little of the chemical changes which take place in the absorbent system; but the chyle drawn from the large absorbent trunks near their entrance into the "receptaculum chyli" is very different from that just absorbed by the lacteals. During its passage through these vessels and their glands it undergoes important alterations, assimilating it to the blood.

311. The following table, by Carpenter, gives the relative proportions of the three chief ingredients of the chyle in different parts of the absorbent system.

In the afferent lacteals, from the intestines to the mesenteric glands:

Fat in *maximum* quantity (numerous fat or oil-globules).
Albumen in *medium* quantity.
Chyle-corpuscles, few or none.
Fibrin almost entirely wanting.

In the efferent lacteals, from the mesenteric gland to the Thoracic Duct:

Fat in *medium* quantity.
Albumen in *maximum* quantity.
Chyle-corpuscles very numerous, but imperfectly developed.
Fibrin in *medium* quantity.

In the Thoracic Duct:

Fat in *minimum* quantity (few or no oil-globules).
Albumen in *medium* quantity.
Chyle-corpuscles numerous, and more distinctly cellular.
Fibrin in *maximum* quantity.

312. In the portal circulation, soon after the absorbed substances are introduced into the blood, and come in contact with its organic ingredients, they become converted into other substances; the albuminose is in part changed into blood-albumen, a substance very different from albuminose or the original albumen. There is also probably some fibrin, while the sugar rapidly decomposes, losing its characteristic properties. The contents of the portal vein undergo changes in the liver before being taken up by the hepatic vein, but these are not well understood; arriving at the entrance of the general circulation, these newly-absorbed ingredients have already become measurably assimilated to those previously existing in the blood.

§ 28. PHYSIOLOGY OF THE ABSORBENTS.—*Office of the Lymphatics. Absorbent Power of Different Tissues. Absorption in cases of Disease. Imbibition of Animal Membranes.*

313. It was formerly supposed that the office of the LYMPHATICS was *excretive*—that of conveying from the system portions of waste matter no longer of use; but as these vessels are found to commence most frequently in tissues where nutritive changes are few—as there is a conformity in the nature of the fluids, chyle and lymph, the chief difference being due to the presence of fat, and a large proportion of albumen in the chyle—as the two fluids are conveyed into the general current of circulation, just before the blood is again transmitted into the system at large—the almost inevitable inference is, that lymph, like chyle, is a *nutritious* fluid. There is much evidence that the lymph is obtained from the blood, and it is not improbable that the lymphatics take up those crude materials which were absorbed directly by the veins and subject them to an assimilating agency, resembling that acting upon the nutritive substances in the lacteals.

314. The office of the lymphatics may also include another, assimilation. Disintegration of the tissues is everywhere taking place. Every respiration, every heart-beat, every

muscular movement, every thought, is produced at the expense of the life of some of the tissues; but, says Carpenter, "The *death* of the tissues by no means involves their immediate and complete destruction; and there seems no more reason why an animal should not derive support from its own dead past, than the dead body of another individual. Whilst, therefore, the matter that has undergone too complete a disintegration to be again employed as nutrient material is carried off by the excretory process, that portion which is capable of being again assimilated may be taken up by the lymphatic system." This whole lymphatic system may be looked upon as one great assimilating or blood-making gland.

315. Different membranes have different absorbent powers, and the power of the same membrane varies with change of condition. The most active is the mucous membrane; thus, in the alimentary canal, it takes up a large portion of the food; in the lungs it absorbs gases in a state of solution. In this way are introduced into the system miasmatic and contagious exhalations. Fine, solid particles are sometimes absorbed, as arsenic. Instances of poisoning are not uncommon among manufacturers of artificial flowers and green paper-hangings, arsenite of copper or "Scheele's green" being employed in the coloring.

316. Though much impeded by the cuticle, absorption takes place to a considerable extent through the skin, and the use of medicinal baths is based on this fact; shipwrecked sailors, destitute of fresh water, find that thirst is relieved by immersing the body in salt water. Life is sometimes supported for a time by immersing the patient in baths of milk or broth.

317. In serous and synovial membranes, the fluids poured out into the joint in rheumatism and other inflammations are absorbed. Absorption is shown in areolar tissue, as in taking up dropsical fluids; also by sub-cutaneous injections of a solution of morphia, to relieve suffering from neuralgic pain, from severe operations, obstinate cough and other irritations.

Observations.—1st, In cases of disease, where no food is taken into the stomach, life is maintained by the absorption of fat. In consumption, even the muscles and more solid parts of the body are absorbed. 2d, Animals living in a half-torpid state during winter, derive their nourishment from the same source.

318. There are no visible openings in the membranes for the passage of these absorbable substances, but their entrance seems to be effected by a peculiar action of animal membranes which enables certain fluids to pass directly through them by a kind of imbibition, a process called *endosmo'sis.**

§ **29.** Hygiene of the Absorbents.—*Conditions of Air affecting Absorption. Effect of Nutritious Food. Effect of the Removal of the Cuticle.*

319. *The air should be as free as possible from impure vapors and gases;* hence the importance of thorough ventilation, especially in the sleeping-room, since exhalations from the system are greater at night than by day.

Observation.—In infectious diseases, the impure air should be constantly carried from the room, and the nurse should be careful to avoid the infected air, approaching the patient *on the side* in which the currents of air are admitted.

320. *Moisture increases the activity of the absorbents;* hence, persons living in marshy districts contract miasmatic and contagious diseases more readily than those living in a drier atmosphere. In such localities the house should be plentifully supplied with fresh air, and kept dry by the use of fires. Especially is this necessary morning and evening, in spring and autumn, and often in summer.

Observation.—For the above reason, the air of the sick-room should be kept dry, otherwise the poisonous exhalations are absorbed by the lungs and skin, both of the *patient* and of the *nurse.*

321. *Nutritious food lessens the activity of the absorbents;*

---

* Gr., *endon*, within, and *osmos*, impulse.

hence, in cases of infectious diseases, due attention should be given to the food of the attendants and of the family. Some persons use alcoholic stimulants or tobacco, "to prevent taking disease," but these increase the activity of the absorbents, and the liability to contract disease. A moderate amount of nutritious food will be more efficacious.

322. *In handling poisons, care should be taken that the cuticle be unbroken*, as absorption is very rapid when the skin is removed. In contagious diseases, if the skin is broken, it should be covered with adhesive plaster while at work over the patient. In handling dead bodies, it is well to lubricate the hands with olive-oil or lard. The absorption of poisonous matter through a slight "scratch" or puncture of the cuticle, as the removal of a "hang nail," has cost several valuable lives.

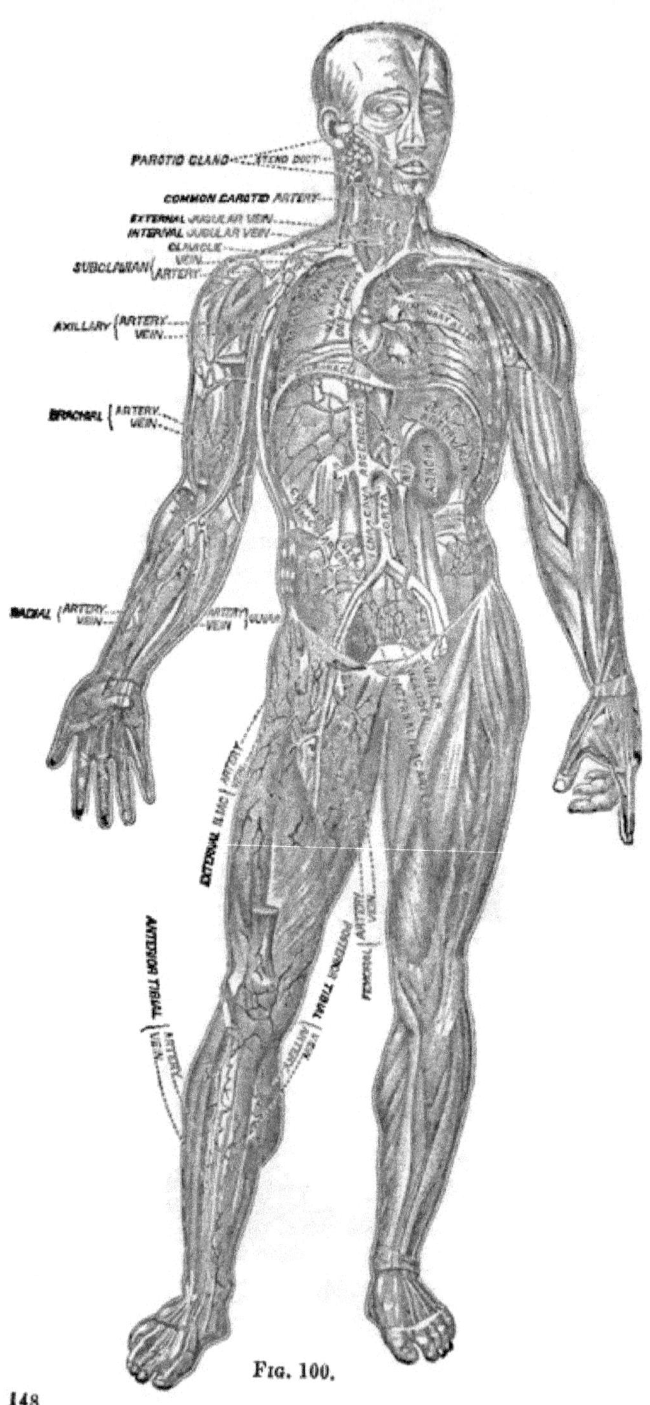
PAROTID GLAND
COMMON CAROTID ARTERY
EXTERNAL JUGULAR VEIN
INTERNAL JUGULAR VEIN
CLAVICLE
SUBCLAVIAN { VEIN ARTERY
AXILLARY { ARTERY VEIN
BRACHIAL { ARTERY VEIN
RADIAL { ARTERY VEIN
ANTERIOR TIBIAL { ARTERY VEIN
POSTERIOR TIBIAL { ARTERY VEIN
FEMORAL { ARTERY VEIN

Fig. 100.

CHAPTER VIII.

THE CIRCULATION.

§ 30. *The Blood. Composition of the Blood. Relation of the Absorbent System to the Blood.*

323. As the contents of the absorbent vessels enter the blood-vessels, they undergo their last complete change into that remarkable fluid, *the blood*, which contains all the materials for the support of every part of the animal fabric.

The blood consists of a liquid portion named *liquor sanguinis*, the *plasma*, or liquor of the blood, which holds in suspension multitudes of minute, circular bodies, called blood-corpuscles; these are of two kinds, the white, or colorless, and the red; the latter are so minute that no less than one hundred millions are said to exist in a single drop of blood; the red color is due to their accumulation, as, when in thin layers, they appear yellowish. They contain only a slightly colored fluid, while the white corpuscles have, in addition, a nucleus and indistinct granules.

324. The blood is constantly undergoing loss, from supplying material for the secretions, for nutritive changes in the solid tissues, and also in the blood itself.

Observation.—The French call blood "chair coulant," *running flesh*, and with reason, since it not only contains the same constituents as flesh, but one-fifth of its weight is solid matter.

325. In order that this blood with its cargo of supplies should fulfill its mission of nutrition, it must be kept constantly moving in a circuit, including every part of the body; this movement is called its *Circulation*, which takes place through the *Heart* and the *Blood-vessels*, which consist of the *Arteries*, *Capillaries* and *Veins*.

§ **31.** ANATOMY OF THE CIRCULATORY ORGANS.—*Construction of the Heart. The Arteries, Veins and Capillaries, and their Relation to each other. The Aorta and its Divisions. Arrangement of the Veins.*

326. The HEART is a hollow muscle enclosed in a sac, named *Pericardium.** In the male its proportion to the body is about 1 to 169; in the female, about 1 to 149. The heart is cone-like in shape, whence its triple division into base, body and apex. Its length is about five inches, and its basal diameter about four inches. It is everywhere free or unattached excepting at the base, which by means of the large blood-vessels is joined to the vertebral column, reaching from the region of the fourth dorsal vertebra to the eighth. The apex is directed downward, forward and to the left, pointing to the junction of the fifth rib with its cartilage. The interior of the heart is divided by a longitudinal muscular septum, or wall, into two chambers, named the right and the left chamber; each of these is divided by a transverse constriction into two apartments, named the *Au'ricle* † and the *Ven'tricle;* the auricle occupying the basal end of the organ, and the ventricles the body and apex. There are virtually *two hearts* placed side by side, having no communication with each other and differing in function. The right division is sometimes called the *pulmonic* heart, and the left the *systemic* heart.

327. The ARTERIES are firm, membranous, cylindrical tubes, arising from the ventricles of the heart by two trunks; that from the left ventricle, named the *Aorta,* is the systemic trunk; and that from the right ventricle, named the *Pulmonic artery,* is the pulmonic trunk.

The systemic trunk, or aorta, divides and subdivides into finer and finer arteries, like the branches from the trunk of a tree, excepting that these branches communicate with each other in a finer network, till the ultimate ramifications, too minute to be seen by the naked eye, extend to every

---

* Gr., *peri,* about, and *kardia,* heart. † Lat., *auris,* an ear.

nook and corner and atom of the body. These final branches are called *Capillaries*.

The CAPILLARIES serve to connect the terminations of the arteries with the beginning of the *veins*, so that it is impossible to tell just where the artery ends, and the vein begins.

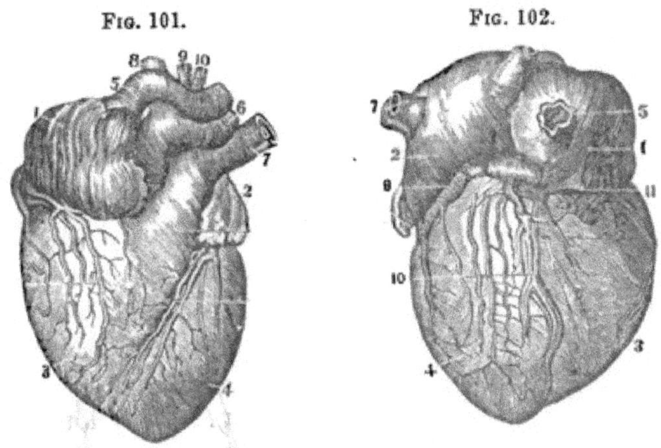

FIG. 101. A FRONT VIEW OF THE HEART.—1, The right auricle of the heart. 2, The left auricle. 3, The right ventricle. 4, The left ventricle. 5, 6, 7, 8, 9, 10, Vessels through which the blood passes to and from the heart.

FIG. 102. A BACK VIEW OF THE HEART.—1, The right auricle. 2, The left auricle. 3, The right ventricle. 4, The left ventricle. 5, 6, 7, The vessels that carry the blood to and from the heart. 9, 10, 11, The vessels of the heart.

328. The VEINS thus commencing with the capillaries unite into larger and larger veins, converging *toward* the heart, like the branches of a tree toward its trunk, till the final union in two trunks (the ascending and descending Venæ Cavæ), that connect with the right auricle of the heart. The aorta and cavæ constitute the large vessels of the Systemic, or general circulation. The Pulmonic, or lesser circulation from the right ventricle through the lungs to the left auricle, has a similar set of vessels; the trunk leaving the right ventricle is named the *Pulmonic artery*, and corresponds to the aorta; those trunks conveying the blood to the left auricle and corresponding to the venæ cavæ, are uamed the *Pulmonary veins*.

Fig. 103.

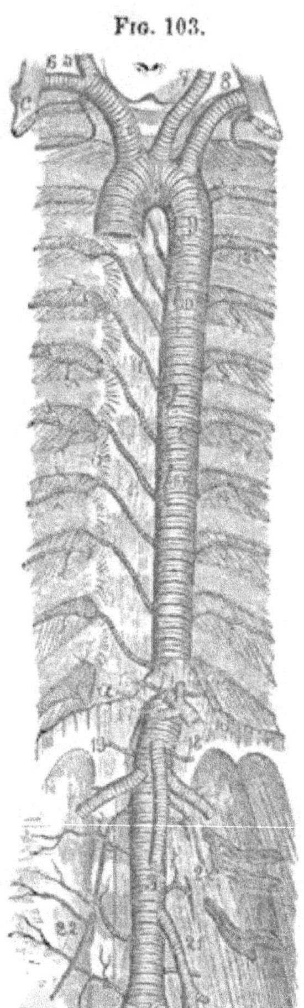

Fig. 103 (*Leidy*). The Aorta.—1, Arch of the aorta. 2, Thoracic aorta. 3, Abdominal aorta. 4, Innominate artery. 5, Right common carotid. 6, Right subclavian. 7, Left common carotid. 8, Left subclavian. 9, Bronchial artery, a small branch of the aorta. 10, Œsophageal arteries. 11, Intercostal arteries of the right side; 12, of the left side. 15, Coronary artery. 16, Splenic artery. 17, Hepatic artery. 18, Superior mesenteric artery. 19, Supra-renal arteries. 21, Inferior mesenteric artery. 22, Lumbar arteries. 23, Common iliac arteries. 24, Middle sacral artery. *a*, Aortic orifice of the diaphragm. *b*, Articulation of the head of the ribs. *c*, Anterior scalene muscle.

329. The Aorta springs from the left ventricle of the heart, is about an inch in diameter, and is the main trunk of the arterial system, supplying pure blood to every part of the body. It is divided into the arch, the thoracic and abdominal aorta. The *Arch* ascends from the heart, slightly inclines toward the right side, curves obliquely backward to the left side, and descends to the left side of the third dorsal vertebra, where it becomes the Thoracic Aorta. The arch gives off main branches as follows: the right and left *Coronary* arteries, whose branches ramify upon the walls of the heart; the three trunks going to the head and upper extremities; viz., the *right Carotid* and *right Subclavian* and the Innominata trunk, which soon divides into the *left Carotid* and *left Subclavian* arteries.

330. The Thoracic Aorta commencing with the termination of the arch, descends at the left of the vertebral column, gradually inclining toward the median line, which it nearly reaches opposite the last dorsal vertebra.

where it passes through the diaphragm and becomes the abdominal aorta. The thoracic division gives off branches to the lungs, pericardium, œsophagus, lymphatic glands, the intercostal, pectoral and serrated muscles, also those of the back.

331. The ABDOMINAL AORTA inclines a little to the left, gives off branches to the liver, stomach, spleen, pancreas, kidneys and to the abdominal muscles. Opposite the fourth lumbar vertebra, it divides into two large trunks, called the common Iliacs. These subdivide into two branches, called the external and internal Iliac arteries. The continuation of the external iliac when it reaches the groin, is named the *Femoral* artery, which passes down the groove of the thigh between the extensor and adductor muscles; after passing through the tendon of the great adductor muscle it is called the *Popliteal* artery, which divides into the anterior and posterior tibial arteries, the latter providing the fibular artery and various branches sent to several parts of the foot and toes.

332. The carotid arteries are divided into two branches, the external and the internal; the former giving off branches to the face and head, excepting the brain and orbits, which are supplied by the latter. The subclavian arteries furnish branches to the brain, spinal cord and membranes, the ears, pleura, and various muscles of the back and neck. The extension of the subclavian artery is called the brachial in the inner and fore part of the arm; the two main branches of the brachial extending down the fore-arm are named the radial, at the anterior and outer part of the fore-arm, and the ulna, situated at the anterior and inner part of the fore-arm. The radial artery toward the wrist runs near the surface, being covered only by the fascia and skin. We learn the condition of the general circulation by its throbbings, which we call the pulse. The radial and ulnar arteries divide and subdivide into the various carpal, palmar and digital ramifications, supplying the wrist, hand and fingers.

FIG. 104.

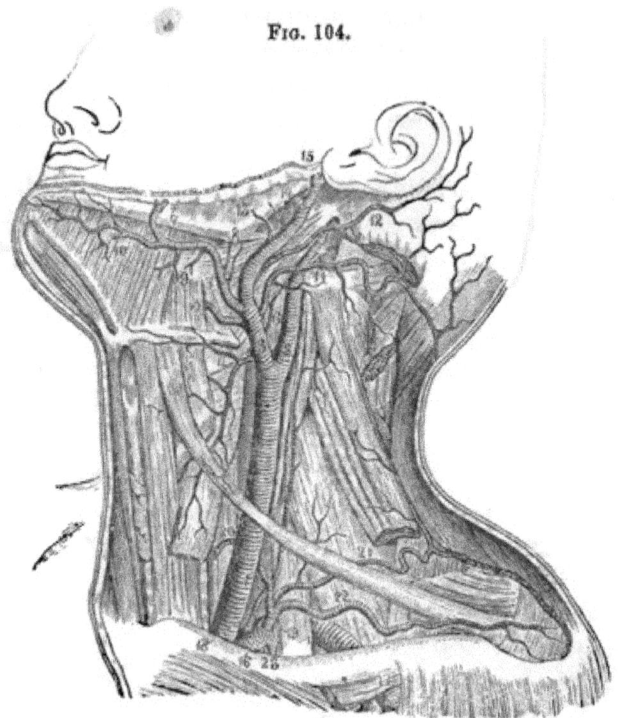

FIG. 104 (*Leidy*). LEFT COMMON CAROTID DIVIDING INTO THE EXTERNAL AND INTERNAL CAROTID ARTERIES.—1, Common carotid artery. 2, Internal carotid. 3, External carotid. 4, Superior thyroid. 5, Lingual. 6, Pharyngeal artery. 7, Facial. 8, Inferior palatine and tonsillar arteries. 9, Submaxillary. 10, Submental. 11, Occipital. 12, Posterior auricular. 13, Parotid branches. 14, Internal maxillary. 15, Temporal artery. 16, Subclavian artery. 17, Axillary. 18, Vertebral artery. 19, Thyroid axis. 20, Inferior thyroid giving off the ascending cervical. 21, Transverse cervical. 22, Supra-scapular. 23, Internal mammary artery.

333. The VEINS are arranged in two sets—the superficial and the deep-seated; the former lie immediately under the skin, possessing no corresponding arteries; the deep-seated veins directly attend the arteries, and usually take the same name. The largest arteries have one venous trunk; the medium-sized have two, called *venæ comites.* The veins unite into eight trunks with their branches; the coronary vein receives the blood from the walls of the heart, and conveys it to the right auricle; the *Superior Vena Cava* derives

its branches from the head, neck, upper extremities and walls of the thorax. It terminates at the upper back part of the right auricle of the heart.

The *Inferior Vena Cava* collects the blood from the lower extremities, pelvis and abdomen, and terminates in the right auricle.

The *Portal* vein is a short trunk about three inches in length, derived from the convergence of the veins of the stomach, spleen, pancreas and intestines; this passes into the liver, where it divides and subdivides, being distributed throughout the organ. This blood, with that of the hepatic artery, is returned to the general circulation by the hepatic veins (244).

The *Pulmonary* veins are four in number, two for each lung. They commence with the capillaries of the lungs, and converge till a single trunk is formed for each lobe, or three trunks for the right lung and two for the left; but the trunk from the middle lobe of the right lung joins that from the upper lobe of the same side, and the four mouths discharge into the four angles of

Fig. 105.

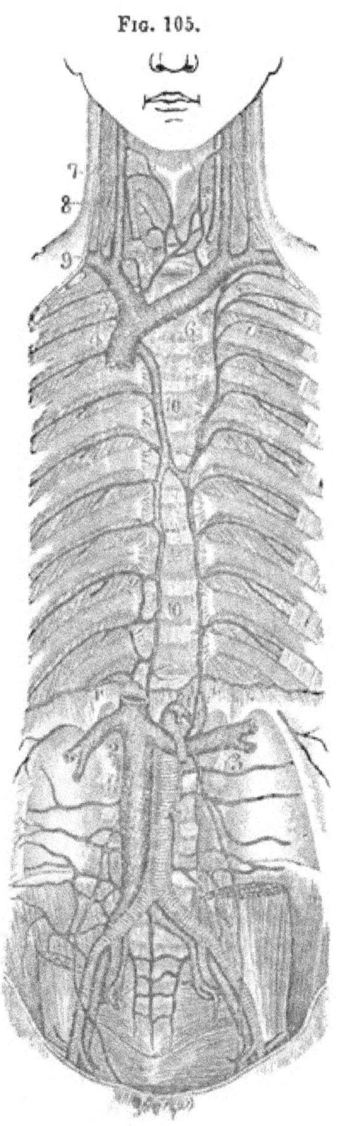

Fig. 105 (*Leidy*). Veins of the Thorax and Abdomen.—1, Inferior cava. 2, right, 3, Left renal veins. 4, Superior cava. 5, Right, 6, Left innominate veins. 7, Internal veins. 8, External jugular veins. 9, Subclavian vein. 10, Azygos vein.

the left auricle. The pulmonary veins perform the function of arteries, as they convey *pure* blood.

FIG. 106.

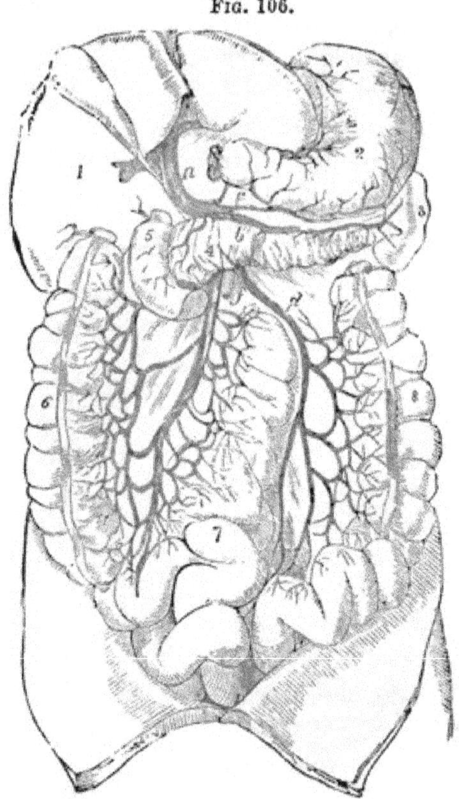

FIG. 106 (*Leidy*). THE PORTAL SYSTEM OF VEINS.—*a*, Portal vein. *b*, Splenic vein. *c*, Right gastro-epiploic vein. *d*, Inferior mesenteric vein. *e*, Superior mesenteric vein. *f*, Trunk of the superior mesenteric artery. 1, Liver. 2, Stomach. 3, Spleen. 4, Pancreas. 5, Duodenum. 6, Ascending colon: the transverse colon is removed. 7, Small intestine. 8, Descending colon.

§ **32.** HISTOLOGY OF THE CIRCULATORY ORGANS.—*The Pericardium and Endocardium. The Valves of the Heart. The Muscular Structure of the Heart. The Coats of the Arteries—Of the Veins—Of the Capillaries.*

334. The PERICARDIUM, or heart-case, is composed of two layers, one fibrous, and the other serous. The fibrous layer forms a loose sac over the heart, being connected only at the

base, from which it embraces the several blood-vessels and becomes continuous with their external coats. The *serous* layer closely invests the heart and also the great blood-vessels at its base, from which it is reflected to line the fibrous layer of the pericardium.

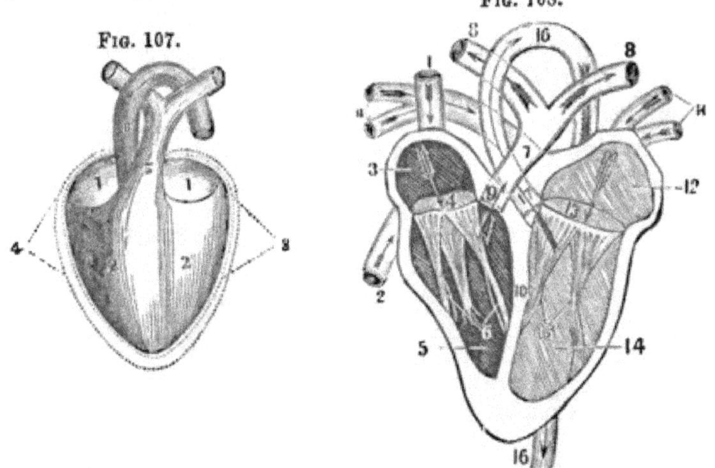

FIG. 107. DIAGRAM OF THE HEART, WITH ITS INVESTMENT.—1, 1, Right and left auricles. 2, 2, Right and left ventricles. 3, 4, Pericardium. 5, Pulmonary artery. 6, Aorta.

FIG. 108. DIAGRAM OF THE HEART AND VALVES.—1, Descending vena cava (vein). 2, Ascending vena cava (vein). 3, Right auricle. 4, Opening between the right auricle and the right ventricle. 5, Right ventricle. 6, Tricuspid valves. 7, Pulmonary artery. 8, 8, Branches of the pulmonary artery that pass to the right and left lung. 9, Semilunar valves of the pulmonary artery. 10, Septum between the two ventricles of the heart. 11, 11, Pulmonary veins. 12, Left auricle. 13, Opening between the left auricle and ventricle. 14, Left ventricle. 15, Mitral valves. 16, 16, Aorta. 17, Semilunar valves of the aorta.

335. The ENDOCARDIUM, or lining membrane of the heart, is a thin, translucent membrane continuous with the inner coats of the blood-vessels. It consists of an epithelium, an exceedingly thin basement membrane and a fibro-elastic layer closely adherent to the general muscular structure beneath. At the opening between the auricles and ventricles, at the commencement of the aorta and of the pulmonary artery, the fibro-elastic tissue forms four rings, sometimes called *fibrous zones.* It also forms valves by its little folds, enclosing muscular fibres. Those at the openings of the aorta and the pulmonary artery are named, from their shape, *Semi-*

lunar valves. They form complete pockets, three in number, and have a triangular arrangement about the orifices. Behind each of these valves is a cavity, or pouch, in the artery.

336. Between the auricles and ventricles are valves also formed by foldings of the endocardium. On the left side are two, named *Mitral* valves. They form a kind of curtain, from whose floating edge small white cords (chordæ tendinæ) pass to some of the columnæ carnæ, thus preventing the edge from being carried into the auricle. On the right side are three valves formed of three folds of membrane, called the *Bicuspid* valves.

Fig. 109. Fig. 110.

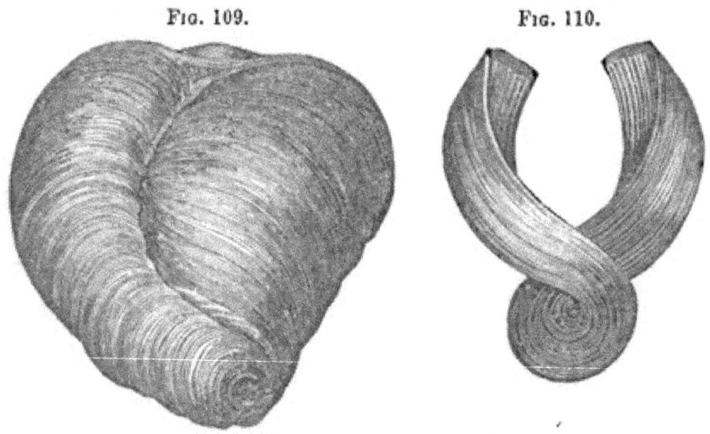

Fig. 109. Spiral and involuted arrangement of the fibres of the heart.
Fig. 110. Gyration of the fibres of the heart at the apex.

337. *The muscular structure of the heart* is based upon the four fibrous zones, which furnish a point of departure for most of the muscular fibres in the ventricles. Those of the auricles and of the ventricles are quite independent of each other. The crossing fibres form networks arranged in three circular laminæ, the superficial, middle and internal. The superficial fibres commence at the base, and pursue a spiral course to the apex; those of the right side, running from right to left; those of the left side, from left to right. These two spiral sets encircle the apex and cross each other somewhat like the lines in the figure 8, thus forming a remark-

able whorl, called the vortex. They do not stop here, but pass inward and turn upward to the auriculo-ventricular fibrous rings from whence they started, forming the deep-seated layer, or the true walls of the ventricles with their fleshy columns; hence the deep-seated and superficial layers are continuous muscles. Between these two layers is the middle stratum of fibres, more or less circular, forming a truncated cone, with its base corresponding to the auriculo-ventricular orifice.

338. The muscular fibres of the auricles consist of a superficial set investing the anterior portion of both auricles, and a deep-seated layer which in the left auricle constitutes a network of circular and oblique fibres, all traceable to the auriculo-ventricular orifice, around which they form sphincter muscles. The corresponding fibres of the right auricle intersect each other, and are traceable to the corresponding orifice.

339. The ARTERIES have comparatively thick walls, composed of three coats continuous with the endocardium and the fibrous coat of the pericardium. The external coat is chiefly of white fibrous tissue, with the spiral fibres crossing each other from opposite sides of the vessel. This coat is quite thin in the aorta and larger trunks; it forms about half the thickness of the walls in the medium-sized vessels, and disappears entirely in the smaller vessels. The middle coat is thick in the large arteries, and gradually becomes thinner till its disappearance before reaching the capillaries. This coat is, in the large trunks, chiefly composed of elastic tissue with some muscular fibres; in the smaller vessels, of muscular tissue with few elastic fibres. The inner coat is thinnest and most elastic; like the endocardium, it has an epithelium, a basement membrane and a layer of connective elastic tissue. The latter is intimately connected with the middle coat.

340. The VEINS are constructed, in general, like the arteries, but their coats are much thinner. Many of the larger veins, particularly in the limbs, have crescent-shaped valves, usually arranged in pairs and opposite each other. These are

formed by the doublings of the lining membrane, strengthened with intervening fibro-elastic tissue. Behind each valve there is a dilatation of the vein, forming a little pouch. The pulmonary veins have no valves: the same is true of the venæ cavæ, the portal vein and its branches, the hepaticæ, renal and spinal veins, and most of those of the head and neck; they are more abundant in the lower than in the upper extremities. The walls of both arteries and veins are furnished with nutritive vessels and with nerves.

Fig. 111.

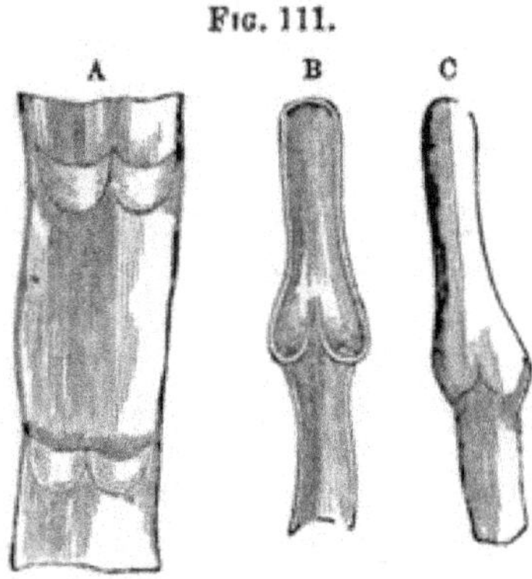

Fig. 111 (*Leidy*). Diagrams exhibiting the Arrangement of the Valves of Veins. —A, Vein laid open, showing the valves in pairs. B, Longitudinal section of a vein, indicating the mode in which the valves, by apposition of their free edges, close its calibre. The dilated condition of the walls behind the valves is also seen. C, Vein distended, showing how the sinuses behind the valves become dilated.

341. The Capillaries are exceedingly delicate tubes, which are continuous with the basement membrane of the internal coat of the arteries and veins. The network of the capillaries varies, adapting itself to the particular tissue in which it is found; thus, in the lungs it takes the form of the air-cells; in the muscles the meshes are elongated.

§ **33.** Chemistry of the Blood.—*Analysis of the Blood. Relative Proportions of Different Chemical Substances in the Corpuscle and the Plasma.*

342. The analysis of blood by different chemists gives very different results, due chiefly to the variable composition of

this fluid under different conditions connected with health, age, temperament, etc. The following table from Lehman shows the composition of 1000 parts of blood, calculated from the analysis of venous blood by Lecanu:

	Corpuscle.	Plasma.	Total.
Water	344.000	451.45	795.45
Hæmatin	8.375	......	8.375
Globulin	141.11	......	141.11
Fat	1.155	.86	2.015
Extractive matter	1.3	1.97	3.27
Salts	4.06	4.275	8.335
Fibrin	......	2.025	2.025
Albumen	......	39.42	39.42
	500.000	500.000	1000.000

343. According to this estimate, blood contains about eighty per cent. water and twenty per cent. solid matter. In round numbers, of the 205 solid parts, 156 belong to the red and white corpuscle, and 141 are globulin (modified albumen), 8½ parts hæmatin; the red coloring substance, 1 part fat, 1½ extractive matters, and 4 parts salts, chiefly salts of potash. The remaining 49 parts of solids belong to the liquor sanguinis, or fluid portion of the blood, and include rather more than 2 parts blood-fibrin; the rest of these solids are proper to the serum of the blood, and consist of 39½ parts of albumen, 1 fat, 2 extractive matters, and 4½ salts, chiefly soda.

344. Other mineral substances are found in small quantities. The distribution of mineral substances in the blood is peculiar. Thus, the red corpuscles contain ten times as much potassium as the liquor sanguinis, but only one-third as much sodium; the corpuscles contain five times as much phosphoric acid as the liquor sanguinis, but only about half as much chloride. The chloride of sodium (common salt) is, therefore, chiefly contained in the fluid plasma of the blood, and the phosphoric acid principally, and the potassium almost entirely, in the corpuscles, which also contain a large share of the fatty matters.

Blood charged with gases, especially oxygen, nitrogen and carbonic acid, has a saline taste, and is an alkaline fluid. When blood is exposed to the air, the fibrin coagulates, carrying down with it mechanically the corpuscle; this leaves an amber-like fluid, called serum, in which the clot floats.

§ 34. Physiology of the Circulatory Organs.—*Necessity for Circulation—For the Double System of Circulation. Plan of Systemic Circulation—Of Pulmonic Circulation—Their Relation to Each Other. Provisions necessary in a Circulatory Apparatus. The Circulatory Impulse. Prevention of a Re-flow. Additional Forces for maintaining the Current in the Arteries—In the Veins. Equalization of the Current. Supply of a due Proportion to each Organ. Provision for Contingencies. The Mechanism of the Body Compared with Works of Art.*

345. The tissues are so constructed that their vitality depends upon their activity, and their activity upon the amount of oxygen and nutritive material supplied; the oxygen being essential to the chemical combinations, without which there could be no new deposit of tissue particles, and also to furnish a stimulus, especially to the nervo-muscular system; and the nutritive matter being necessary to supply the waste produced by these chemical and vital activities. Hence, the necessity of a pneumatic apparatus for providing a constant and sufficient supply of oxygen; and of a hydraulic apparatus for conveying the prepared nutriment to every atom of the body, and also to remove the waste, worn-out particles. The former need is met by the exquisite mechanism of the lungs, and the latter by the no less refined mechanism of the heart and blood-vessels. The two apparatuses are brought into use and harmonious co-working, by the *double circulation* of the blood, hence the necessity of the double heart.

346. From the left ventricle the blood is forced into the aorta, to be diffused through the arteries to the capillaries in every part of the body; thence it is returned by the veins, through the venæ cavæ, to the right auricle, which delivers it to the right ventricle; this completes the *Systemic Circulation.* From the right ventricle it is thrown into the pul-

monary artery, and through its branches to the pulmonary capillaries, thence, returned by the pulmonary veins, which coalesce into four trunks, and finally enter the left auricle, which immediately pours it into the left ventricle. This completes the *lesser*, or *Pulmonic Circulation*, and the two constitute one complete circuit of the double circulation.

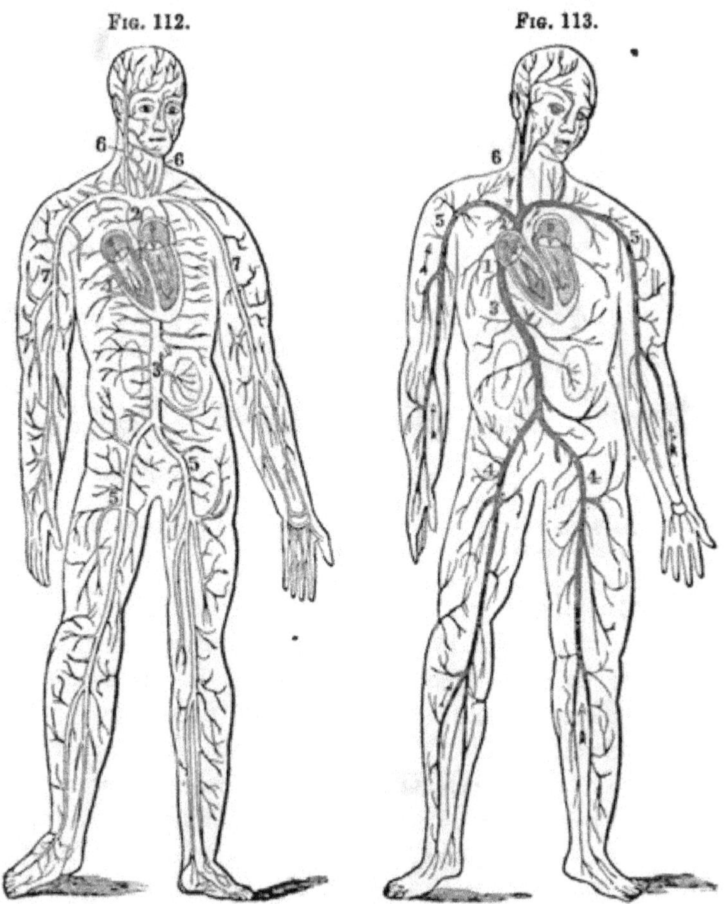

Fig. 112. A Diagram.—1, Left ventricle of the heart. 2, 3, Aorta. 5, 5, Arteries that extend to the lower extremities. 6, 6, Arteries of the neck. 7, 7, Arteries of the arms.

Fig. 113. A Diagram.—1, Right auricle of the heart. 2, 3, Large veins that open into the right auricle. 4, 4, Veins of the lower extremities. 5, 5, Veins of the arms. 6, Veins of the neck. The arrows show the direction that the blood flows.

347. Both circulations are carried on at the same time, that is, the auricles contract and dilate simultaneously; the same is true of the ventricles, whose action immediately follows that of the auricles. Hence, at the same instant, by the action of the ventricles, pure blood is thrown into the body, and impure blood into the lungs; and at the same instant, the auricles receive impure blood from the body, and pure blood from the lungs.

Fig. 114. Fig. 115.

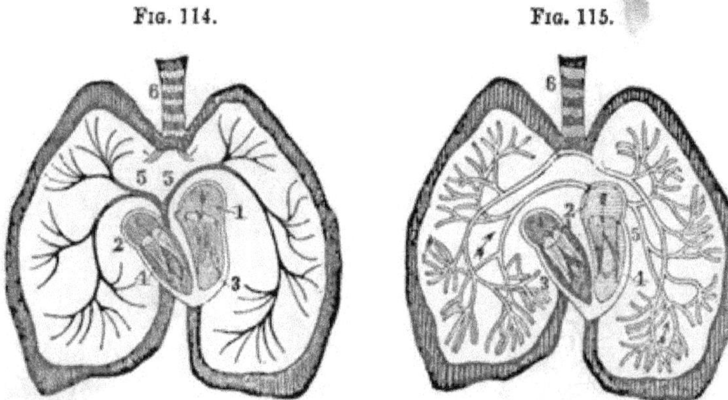

Fig. 114. A Diagram.—1, Left auricle. 2, Right auricle. 3, Left ventricle. 4, Right ventricle. 5, 5, Pulmonary artery. 6, Trachea.

Fig. 115. A Diagram.—1, Right auricle. 2, Left auricle. 3, Right ventricle. 4, Left ventricle. 5, 5, Right and left pulmonary veins. 6, Trachea.

348. How to construct and keep in successful operation an apparatus which should secure the free circulation of the blood, was no easy mechanical problem. It was necessary to provide the requisite motor-power at the starting-point; to prevent a backward flow; to protect the arteries against the force of the heart; to maintain a ceaseless current; to equalize the pressure, especially in the capillaries; to ensure the proper relative quantity of blood to each organ; and to provide for contingencies arising from accident, or other abnormal action.

349. For giving the proper circulatory impulses, we find in each heart, instead of a single cavity, the auricle and ventricle affording a far more powerful impulse. The auricle

is gradually filled by steady streams from the veins, hence the contraction and consequent force is moderate; but as the ventricle receives the whole quantity at once, there is a sudden energetic contraction or *jerk*, hence, a powerful thrusting of the blood into the aorta and pulmonary artery. Also the peculiar spiral and circular arrangement of the muscular fibres of the ventricles is most effective in producing the greatest projectile force. Here comes in a beautiful example of the adaptation of each part to its destined use. The walls are much thinner in the auricles than in the ventricles; and of the two ventricles, they are thinner in the right than in the left, inasmuch as the right sends the blood only to the surrounding lungs, the left, to the remotest part of the body. The power and the required impulse exactly correspond.

350. Though the arrangement of fibres in the heart is such as to give the blood a decidedly forward impulse, yet the danger of a backward flow is evident. This movement is prevented in the auricles, by the contraction of the muscular fibres about the mouths of the veins; by the contraction of the vein-walls; and also in the right auricle, by valves in the mouths of the inferior vena cava and the cardiac sinus, and by the valves in the veins at the base of the neck. The re-flow from the ventricles is prevented by the mitral or bicuspid valves of the systemic heart, and by the tricuspid valves of the pulmonic heart. By the contraction of the muscular columns of the ventricles, the *chordæ tendinæ*, or little cords of the valves, are stretched, bringing the delicate membranes together and into the ventricle, thus effecting a closure. The reflux from the aorta and the pulmonary artery is obviated by the semi-lunar valves. The slightest re-flow fills the little pouches behind the valves, thus closing them till the next contraction of the ventricle. The valves of the right side are more delicate than those of the left, their strength and form being, in each case, exactly adapted to their specific work.

351. The arteries are protected against the sudden action of the heart, by the elastic fibres of their middle coat, which yield easily, thus preventing the liability to rupture.

352. The maintenance of the circulatory current, though largely due to the original impulse of the heart, or the "*vis a tergo*," is aided in various ways. The smooth, glassy surface of the inner arterial coat lessens the friction; the recoil of the elastic fibres of the middle coat after distension, and the contraction of the muscular fibres of the same coat, urge the blood forward. These fibres increase in number, according to the distance from the heart. The respiratory movements also aid the arterial flow.

353. The capillaries have probably no contractility, and though the heart-impulse may be sufficient to inject the blood *into* them, it can hardly effect the passage *through*. Hence other means are employed. We think the following physical principle, as applied by Prof. Draper, will account for the capillary circulation. If two liquids communicating with one another in a tube, have for that tube different chemical affinities, movements will ensue, and that liquid having the strongest affinity will move most rapidly, often driving the other liquid before it. Now, these are the exact conditions in the capillaries of the systemic circulation; the arterial blood, as it contains oxygen, with which it is ready to part and take in exchange carbonic acid which the tissues set free, must have a greater affinity for these tissues than has the venous blood in which these changes have already taken place. Hence, the arterial blood entering at one end of the capillaries must drive before it, and expel at the other end, the blood which has become venous in passing through them. The same principle holds in the pulmonic circulation, but the affinities are opposite. The venous blood has a strong affinity for the oxygen in the air-cells of the lungs, and contains carbonic acid which it is ready to give up; hence, the exchange takes place, and the arterialized blood, having no longer an affinity for the air, is driven by the venous blood, and thus the circulation goes on as long as the blood continues to be aërated.

The portal current is accounted for in the same way. The bile-secreting cells of the liver are made up of materials conveyed by the portal veins and capillaries, and hence have an affinity for them. The supply having been deposited and the affinity thus destroyed, the fluid will be driven into the hepatic capillaries, thus maintaining the portal current.

354. The flow through the veins is continued by the combined action of several forces; viz., the capillary impulse; the suction-power of the dilating auricles, drawing the blood to the heart, or the "*vis a fronte;*" the presence of valves, single in the small veins, double in the larger trunks, and sometimes composed of three flaps; and the thoracic respiratory movements.

355. The intermittent pressure caused by the action of the heart is equalized by the frequent branching and the anastomosing of the arteries as they approach the organs to which they are distributed, since the more points of entrance, the less will be the pressure; and by the elastic coat of the arteries, whose after-distension *gradually* converts the separate impulses into a continuous motion, otherwise the capillaries of many delicate structures would doubtless be ruptured. We find the elastic tissue most abundant in the vessels near the heart, just where it is most needed.

356. The proper relative amount of blood is secured to each organ primarily by the adaptation of its main artery, and it is interesting to notice how the size of the artery everywhere corresponds to the need of the organ. Again, the calibre of the arteries is susceptible of variation within certain limits; hence, the supply of blood to any organ may be in some measure regulated by the contractility of its arteries, which is itself controlled by the nervous system.

357. Contingencies are also provided for, by the frequent anastomoses of the arteries, by their capability of distension, and also by their capability of positive enlargement by the increased nutrition of their walls. Hence, though obstructions should exist in some part, the organ may be measurably supplied with blood.

358. Though our knowledge is so imperfect, our tracing so indistinct, our souls must be dead indeed if they do not

respond to the exclamation of him of old, "I am fearfully and wonderfully made"—fearfully, for often, as in the heart-valves, there is but a gossamer web, a tendinous cord, between the life here and the life beyond: wonderfully, for in all the round of human arts, we find nothing which can at all compare, in perfect simplicity, in faultless skill, in matchless beauty, in the refinements of philosophy and in the subtleties of chemistry, with this vital workmanship, which can be none other than that of God. Till we reach our utmost range of vision, it is ever the same unfolding of the care, the wisdom, the benevolence of Him to whom nothing is great and nothing small; and beyond our finiteness, His eye alone surveys the work of busy legions of artificers, ever building up what the wear and tear of life is ever breaking down; His ear alone listens to the music of the million life-rills, as they murmur on in their ever-ceaseless flow.

§ 35. Hygiene of the Circulatory Organs.—*Conditions favoring Free Circulation. Treatment of Divided Arteries.*

359. *A natural and equal temperature should be preserved.* The blood-vessels are contracted by cold, hence, a chill in any part of the body drives the blood to other parts. The chilled part is thus weakened, while the over-burdened parts suffer from congestion. If the surface is chilled, the blood is thrown upon the internal vital organs, hence the necessity of warm clothing, and also frequent bathing, which favors the free action of the cutaneous vessels.

360. *The clothing should be loosely worn.* Compression of any kind impedes free circulation. Pressure about the vital organs is especially injurious. Ligatures used to retain in place any article of apparel should be elastic. Tight dressing of the neck deprives the brain of its due amount of blood, and retards the free return of venous blood from this organ; an item of particular importance to students, public speakers and persons predisposed to apoplexy or any brain disease.

361. *Exercise promotes the circulation of the blood.* By the action of the muscles, the blood is propelled more rapidly

through the blood-vessels, thus promoting a vigorous circulation in the extremities and skin. The best stimulants for a pale skin and cold extremities are a union of vigorous muscular exercise with agreeable mental action, and systematic bathing attended by thorough friction.

362. *The quality and quantity of the blood modify the action of the heart and blood-vessels.* If this fluid is abundant and pure, the circulatory vessels act with more energy than when it is deficient in quantity or defective in quality.

Illustrations.—1st, If blood in large quantities is drawn from the veins of an athletic man, the heart will beat feebly and the pulse become weak. 2d, A similar effect is produced when the blood becomes vitiated by the inhalation of impure air.

Fig. 116.

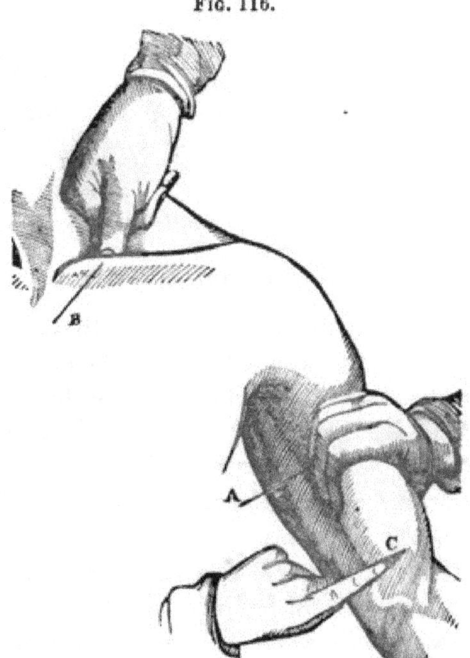

Fig. 116. The Manner of Compressing Divided Arteries.—A, Compressing the large artery of the arm with the thumb. B, The subclavian artery. C, Compressing the divided extremity of an artery *in the wound* with a finger.

363. *Hemorrhage from divided arteries should be arrested.*

otherwise the heart soon ceases its action, and the person faints. If a large artery be wounded, every beat of the pulse throws out the blood in jerks. Until surgical help can be summoned, the flow of blood may be stopped either by compressing the vessel between the wound and the heart, or by compressing the end of the artery next the heart in the wound.

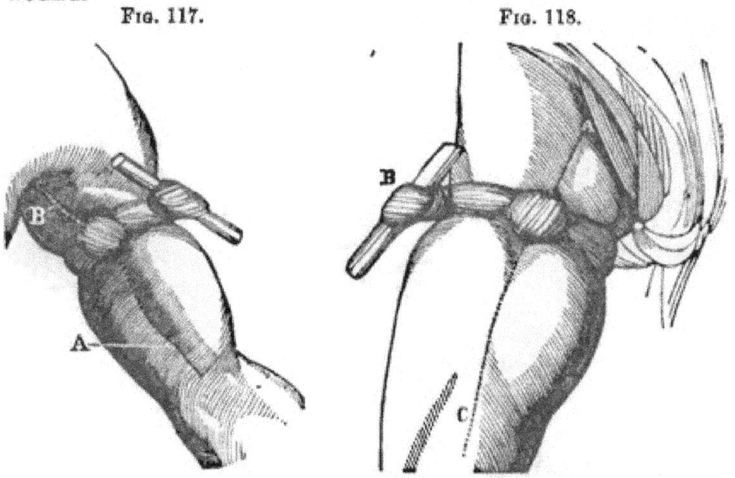

Fig. 117. The Method of Applying the Knotted Handkerchief, to compress a divided artery. A, B, Track of the brachial artery.

Fig. 118. A, C, The track of the femoral artery; the compress applied near the groin.

After compression as described and illustrated, take a square piece of cloth, or handkerchief, twist it cornerwise, and tie a hard knot in the middle. Place the knot over the artery between the wound and the heart, carry the ends around the limb and tie loosely. Place a stick under the handkerchief near the last tie, and twist till the fingers can be removed from the compression without a return of the bleeding. When an artery in a limb be cut, elevate the limb as far as possible, till the bleeding ceases.

364. In *flesh wounds*, when no large vessel is divided, wash the parts with cold water, and when bleeding has ceased, draw the incision together, and retain it with strips of adhesive plaster, not more than a quarter of an inch in width. Then apply a loose bandage, and avoid all ointments, "heal-

ing salves" and washes. In removing the dressing from the wound, both ends of the plaster should be raised and drawn *toward* the incision. To lessen the liability of a reopening, a proper position for the union should be regarded. If the wound be between the knee and ankle, and on the anterior part, extend the knee and bend up the ankle; if on the posterior part, reverse the movement, and, in general, suit the position to the case.

Fig. 119.

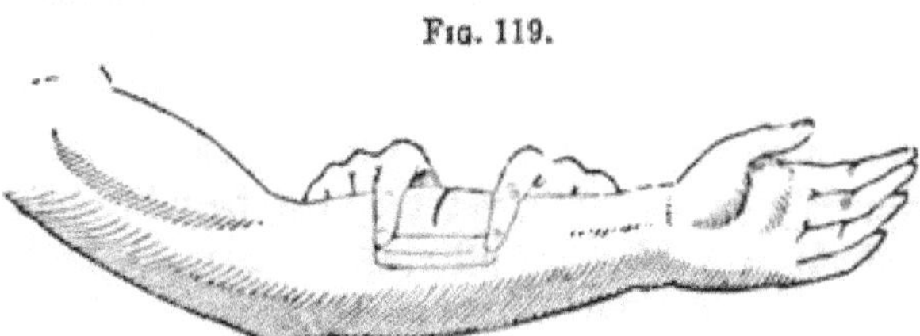

Fig. 119 represents the Manner of applying adhesive strips to wounds.

Observation.—The union of the divided parts is effected by the action of the blood-vessels, and not by salves or ointments. The only object of the dressing is to keep the parts together and protect the wound from air and impurities. Nature performs her own cure. Such wounds seldom need a second dressing, and should not be opened till the incisions are healed.

Fig. 120.

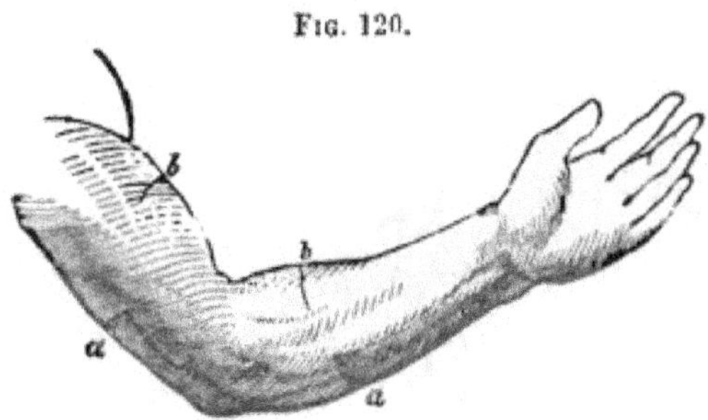

Fig. 120. *a, a*, Representation of Wounds on the back part of the arm and fore-arm. *b, b*, Wounds of the anterior part of the arm and fore-arm. By bending the elbow and wrist, the incisions at *a, a*, are opened, while those at *b, b*, are closed. Were the arm extended at the elbow and wrist, the wounds at *a, a*, would be closed, and those at *b, b*, would be opened.

Wounds made by blunt instruments do not admit of direct

and immediate union. In these cases, a soothing poultice, as of linseed-meal, may be applied, and the limb should be kept still. A physician should be consulted, as dangerous diseases may be induced by such wounds. Wounds from poisonous bites may be treated at first by suction, either by cupping-glasses or the mouth, thus preventing the absorption of the poisonous matter into the system. When this is effected, cover the wound with a poultice, as one made of ground slippery-elm bark.

Observation.—Although animal poisons, when introduced into the circulatory fluid through the broken surface of the skin, frequently cause death, yet they can be taken into the mouth and stomach with impunity, if the mucous membrane lining those parts is unbroken.

§ 36. COMPARATIVE ANGIOLOGY.—*The Composition and Circulation of the Blood of other Mammals, of Birds, of Reptiles and of Fishes, as compared with the same in Man.*

365. In most *Mammals*, the blood is similar to that of man, but the largest animals, as the elephant, have very small corpuscles. All mammals have four cavities in the heart, as in man. Its form, however, is more rounded and less elongated. The heart in quadrupeds lies on the median line of the body, and not a little to the left of it, as in man. There is a marked peculiarity in the distribution of the

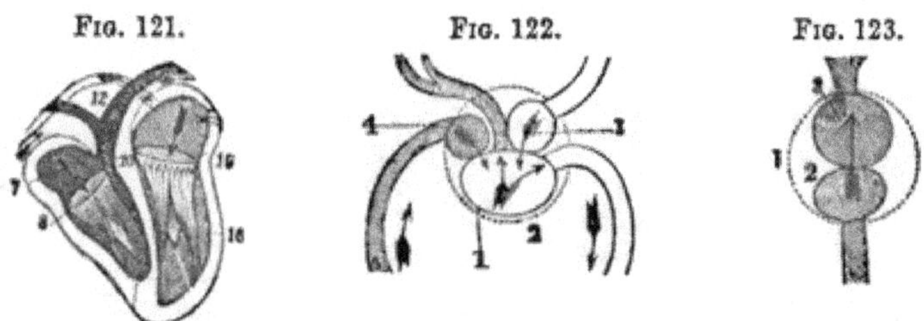

FIG. 121. DIAGRAM OF THE HEART OF THE MAMMAL.—7, Right auricle. 8, Right ventricle. 10, Pulmonic artery. 12, Pulmonic vein. 15, Left auricle. 16, Left ventricle.

FIG. 122. DIAGRAM OF THE HEART OF THE REPTILE.—1, Pericardium. 2, Single ventricle. 3, Left auricle. 4, Right auricle. The arrows show the direction of the blood.

FIG. 123. DIAGRAM OF THE HEART OF THE FISH.—1, Pericardium. 2, The ventricle that receives the blood from the body. 3, The ventricle that sends blood to the gills.

arteries of quadrupeds. In the long necks of grazing animals, there is found a large number of small arterial trunks, which are termed "Wonder Nets." Were these trunks few and large, as in man, the life of the animal would be endangered by the constant dependent position of the head.

366. The blood of *Birds* has the highest temperature of the vertebrate animals. It is richer in corpuscles than that of Mammals, and these corpuscles are elliptical in form, instead of globular. The heart of birds is highly muscular, and of large size in proportion to the bulk of the body. The aorta, at its commencement, divides into three large branches, of which the first two convey the blood to the head and neck, wings, and muscles of the chest; while the third, curving downward around the right bronchus, becomes the descending aorta. There are "Wonder Nets" in various parts of the

Fig. 124.

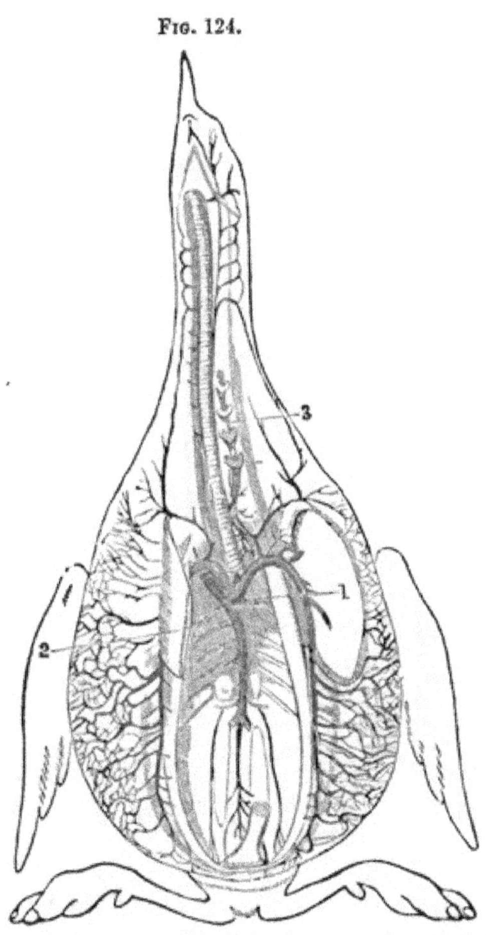

Fig. 124. Arteries of the Trunk of a Bird (the Grebe).—1, The aorta. 2, The vena cava. 3, A cerebral artery. The small lines on each side represent the arteries and veins of the lungs.

body, especially in the arteries supplying the brain, eyes and legs.

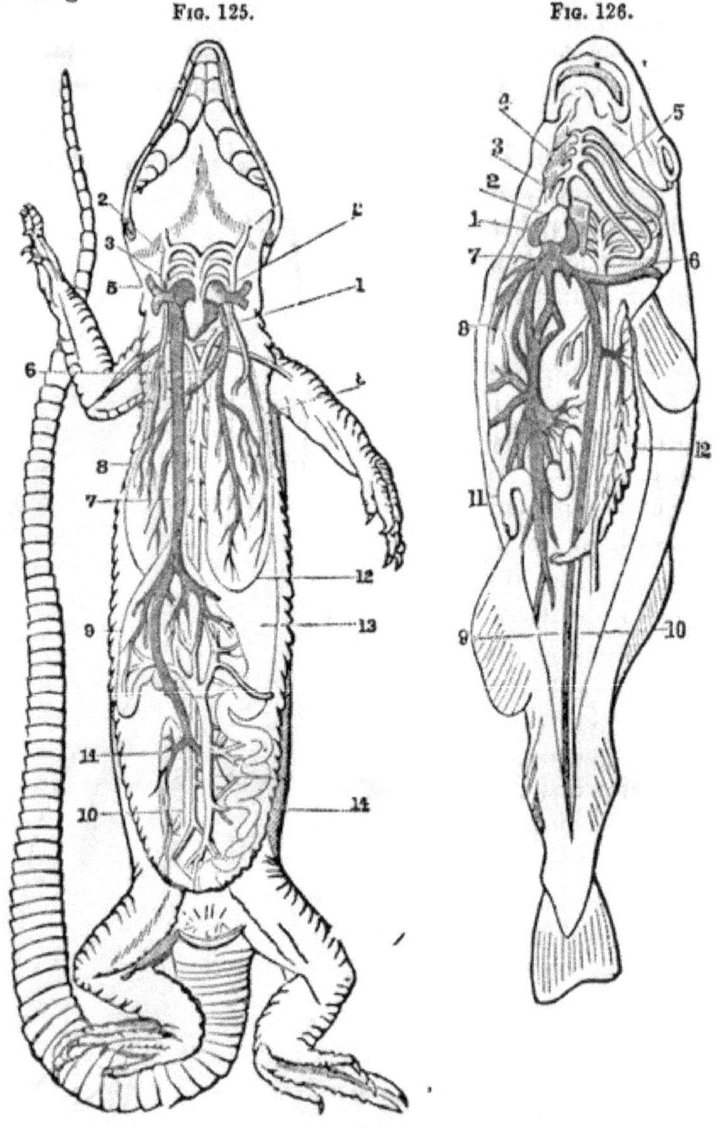

Fig. 125. Circulation of a Reptile (a Lizard).—1, Heart. 2, Left auricle. 3, Right auricle. 4, Arches of the aorta. 5, Superior vena cava. 6, 10, Abdominal aorta. 7, Inferior vena cava. 8, Pulmonary artery. 9, Portal veins. 12, Lungs. 13, Stomach. 14, Intestines.

Fig. 126. Blood-Vessels of a Fish.—1, Auricle. 2, Ventricle. 3, Arterial bulb. 4, Bronchial artery (gill). 5, Brónchial vessels. 6, 10, Dorsal artery. 7, Venous sinus. 8, Portal vein. 9, Vena cava. 11, Intestines. 12, Kidneys.

367. In *Reptiles*, the blood is much cooler than in mammals and birds, and, having fewer globules, is lighter in color. The heart has only three cavities instead of four, viz., two auricles and one ventricle. The arterial blood coming from the lungs is received into the left auricle, and the venous blood from all parts of the body into the right auricle; both are poured into the single ventricle, thus mixing the pure and impure blood, which will account for the sluggishness of these animals. A portion of this mixture returns by the aorta into the different organs it is intended to nourish, while another part proceeds to the lungs by vessels springing from the ventricle or the aorta. The arrangement of the blood-vessels of different classes of reptiles greatly varies, as some breathe by gills and others by lungs; the frog in its early condition is furnished with the former, but in its later growth with the latter.

368. In *Fishes*, the blood is cold, usually red, and the corpuscles small and bi-concave. The heart has but two cavities—one auricle and one ventricle, containing only impure blood; this blood is sent to the *gills*, which answer the purpose of lungs, and, after being there exposed to the oxygen of air contained in the water and purified, it is distributed immediately to the different parts of the body, without the interposition of a heart.

CHAPTER IX.

ASSIMILATION.

§ **37.** *Assimilation, General and Special. Formation of Different Portions of the Blood. Changes included under Secondary Assimilation. Secretion, or Special Assimilation. Excretion, characteristic of all Secretory and Excretory Glands. The Kidneys.*

369. In the human body, as elsewhere, the essential condition of physical *life* is *death.* While the vital force holds the mastery over the chemical forces, the more frequent the death-knell of the particles, the more abounding is the life. They perform their mission, yield up their vitality and pass away, while their places are supplied with new material. This new material is obtained from the food after its proper assimilation. As before stated, the processes by which food is converted into chyle, and then into blood, may be included under *Primary Assimilation;* while the changes which convert portions of the blood into solid tissue may be termed *Secondary Assimilation;* both of these we will include under the head of *General Assimilation,* and the processes of secretion under *Special Assimilation.*

370. The formation of chyle has already been fully noticed, and also its general relations to the blood. The white corpuscles of the blood are supposed to be replenished from the corpuscles of the lymph and chyle, which enter the blood and are identical with its white corpuscles in size, form, structure and general composition. Some suppose the red corpuscles are developed from the white.

The albuminous portion of the liquor sanguinis, or blood-plasma, is supplied from that of the lymph and chyle, and by the venous absorption of digested food; but it may also contain more highly elaborated albuminoid materials derived from the corpuscles, whose elaborative office is undoubted.

the

371. Secondary Assimilation, or Nutrition of the Organs and Tissues, consists of the following stages:

First, A nutritive fluid, or plasma, exudes from the blood, through the coats of the capillaries, filling the finest interstices of the tissues between the capillary networks, and bathing all the elementary parts of those tissues. The nature of this plasma is the same in all parts of the system, and it is sometimes thought to be identical with the liquor sanguinis of the blood, but this is doubtful; it is more probable that the exuded plasma destined for the nutrition of the tissues is of a purer nutrient material.

The *second* stage of the nutritive process consists in the exercise of a certain selective act by the elementary parts of tissues and organs, enabling them to appropriate to themselves such portions of the nutritive fluid as are suitable, either with or without further change, to renew, molecule by molecule, their worn-out substance. "The nucleated cells of the epithelium and epidermis; the corpuscles of the gray matter of the brain; the tubular fibres of the white nervous tissue; the complex fibres of the striated muscles; the simple fibrous forms of the contractile non-striated muscles; the fibres of the fibrous and areolar tissues; and lastly, the consolidated substance, with the remnants of cells imbedded in it, as in cartilage and bone—each derives from the exuded plasma of the blood, and assimilates its required constituents." This assimilating power of the tissue-elements is the persistent, primitive, nutritive force inherited from the germ-cell. It is probably possessed by every cell, however modified or remote in its descent from the parent cell. This power is greatest at the commencement of the life of any animal, and declines till the power to maintain the body is overcome by the forces which lead to its decay.

Third, The result of the act of assimilation is to leave a residual fluid in the interspaces of the tissue-elements outside the capillary vessels. The nature of this fluid must differ in the different tissues, inasmuch as different tissues make different appropriations. This fluid is not worthless, but only defective, and portions of it are probably taken up by the lymphatics, re-assimilated and returned to the blood through the absorbent system.

Fourth, The final residue of the exuded plasma, that which is not taken up by the tissues nor lymphatics, is probably taken up by the venous capillaries.

Fifth, With the final residuum are mingled the effete particles of waste from the tissues, which also enter the venous blood, through the walls of the venous half of the capillaries and of the minute veins. These processes, though separately described, are, of course, in the living body, all going on at the same time, and continuously, and, in a healthy condition, with a perfect balance of action.

372. Nutrition not only supplies the waste, but in *new growth*, new cell-elements, or germinal centres, are constantly reproduced and developed. This process occurs after the body has attained maturity, in the epidermis, nails, hair, the epithelial tissues, and probably the gray nervous substance, and perhaps in some of the other tissues.

373. Special Assimilation, or Secretion, is the separation from the blood of materials in a more or less fluid condition, through a *gland* or *membrane*. After assimilation, or secretion, the products are discharged from the ducts of the glands, or the surfaces of the membranes, and are used for certain purposes in the living economy.

374. The secreting glands are the liver, the pancreas, the salivary and the lachrymal glands; the true mucous glands of the nose, mouth, fauces, pharynx, œsophagus and duodenum; the simple tubular glands of the stomach and intestines; the sebaceous and the mammary glands. The secreting membranes are the mucous, serous and synovial membranes. The serous and synovial fluids are little more than transuded materials of the blood-plasma, unaltered in chemical character, but modified in their relative proportions. By other secreting processes, substances are formed which do not exist in the blood, but resemble its constituents, being albuminoid in character; as pepsin, pancreatin and salivin, etc. Others

differ from the blood in chemical constitution, and are very complex in character, as certain acids of the bile, and the fat of the sebaceous secretions. Extreme examples of special secretive power, by which compounds not existing in the blood are formed from it, are afforded by the appearance of sulpho-cyanogen in the saliva, and of hydrochloric acid in the gastric juice; so also soda is withdrawn from the normal soda salts of the blood, by the agency of the liver, to combine with the fatty acids of the bile.

375. EXCRETION is effected by glands only, and the *educts* are eliminated from the blood and thrown out of the system. The excretory glands are the kidneys, the sweat glands of the skin; to a certain extent, the liver, and, perhaps, the intestinal tubuli, especially of the large intestine, also the sebaceous glands of the skin, and lastly, the lungs, which eliminate carbonic acid from the blood.

In excretion, the substances eliminated from the blood pre-exist in that fluid as the result of decomposition, and are sometimes completely oxidized, and always to a greater extent than the secretions. The successive stages of oxidation remove substances more and more from an organizable character and necessitate their removal from the system.

376. In all cases of Secretion and Excretion there is invariably found, even in the ultimate ramifications of the gland-ducts, a basement membrane covered by a layer of epithelial cells. All glands are very vascular, and receive large quantities of blood. In many secretory processes the epithelial cells are ruptured, and their contents, and sometimes the cells themselves, escape as an essential part of the secretion itself; as in the saliva, pancreatic fluid, gastric juice, the sebaceous and mucous secretions, and perhaps the bile; but the lachrymal and excretory processes simply withdraw their substances from the blood, and convey them from the body without themselves undergoing dissolution or decay.

377. The numerous glands and membranes have been

noticed in other relations, with the exception of the kidneys and the glands of the skin.

378. The KIDNEYS lie one on each side of the spinal column, in a line with the lowest dorsal and the two or three

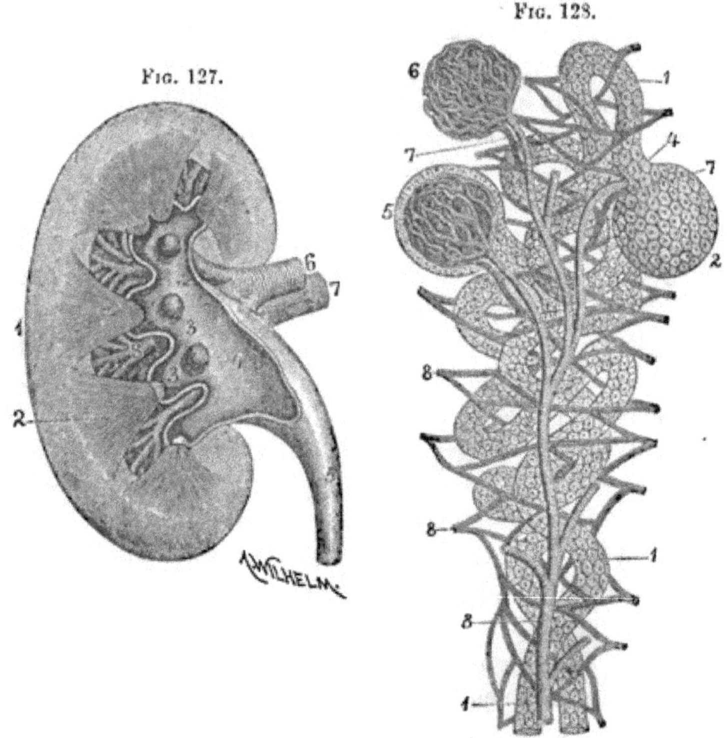

FIG. 127 (*Leidy*). LONGITUDINAL SECTION OF A KIDNEY.—1, Cortical substance. 2, Renal pyramid. 3, Renal papillæ. 4, Pelvis. 5, Ureter. 6, Renal artery. 7, Renal vein. 8, Branches of the latter vessels in the sinus of the kidney.

FIG. 128 (*Leidy*). DIAGRAM OF THE STRUCTURE OF THE KIDNEYS.—1, Two uriniferous tubules of the cortical substance lined with a pavement epithelium. 2, Dilatation of a tubule at its extremity. 3, Branch of the renal artery ending in vessels which enter the dilatations as seen at 4, 5. 6, Knot of blood-vessels freed from its investment. 7, Veins emerging from the vascular knots. 8, Plexus formed by the latter veins among the uriniferous tubules, from which plexus originate the branches of the renal vein.

upper lumbar vertebræ; the right kidney is a little lower than the left. Their shape is that of a bean, and their color a brownish red. They are made up of two very different substances, one covering the whole organ, called the *Cortical* substance; the other is called the *Medullary* substance, and consists of a series of pyramids, with their bases toward the surface of the organ, and their summits, or renal papillæ, toward the fissure. The substance of the kidney is mainly composed of secretory tubes, named *Uri'niferous tubules,* and blood-vessels, with little connective tissue. These tubules are convoluted in the cortical substance, and straight in the medullary, where the terminal orifices are seen by hundreds at the summit of each renal papilla. The tubes are lined with an epithelium which secretes the urine. This secretion is conveyed to the bladder by a cylindrical tube about eighteen inches in length, called the *Ureter.*

Observation.—The retention of the secretion of the kidneys should never be allowed by the young or the old, the healthy or the diseased, as suppression of the secretion of these glands immediately affects the whole system, especially the nervous centres. Both the quantity and color of this secretion indicate the condition or health of the body.

379. The glands of the skin will be described in Chapter XII.

Fig. 129.

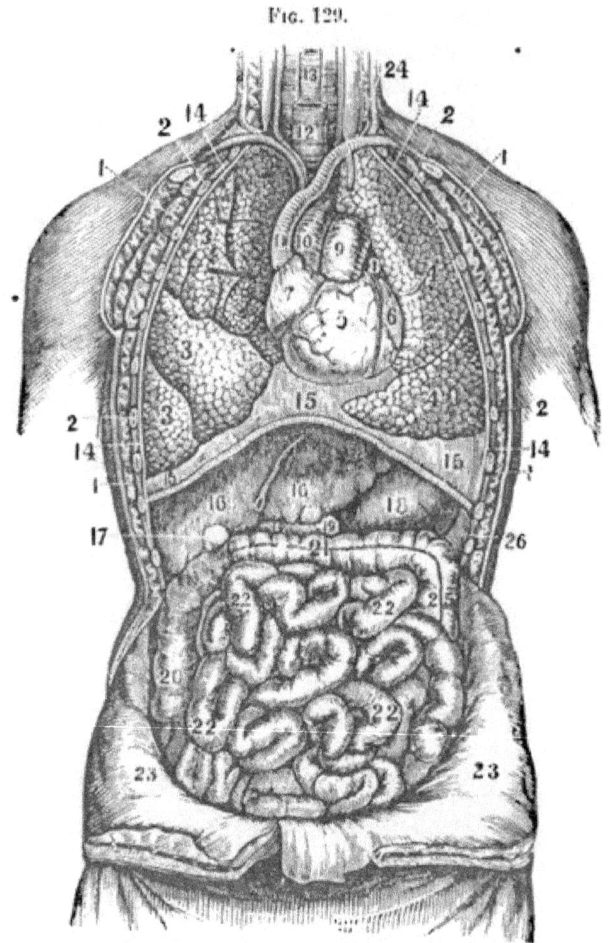

Fig. 129. A Front View of the Organs of the Chest and Abdomen.—1, 1, 1, 1, The muscles of the chest. 2, 2, 2, 2, The ribs. 3, 3, 3, The upper, middle and lower lobes of the right lung. 4, 4, The lobes of the left lung. 5, The right ventricle of the heart. 6, The left ventricle. 7, The right auricle of the heart. 8, The left auricle. 9, The pulmonary artery. 10, The aorta. 11, The vena cava descendens. 12, The trachea. 13, The œsophagus. 14, 14, 14, 14, The pleura. 15, 15, 15, The diaphragm. 16, 16, The right and left lobes of the liver. 17, The gall-cyst. 18, The stomach. 26, The spleen. 19, 19, The duodenum. 20, The ascending colon. 21, The transverse colon. 25, The descending colon. 22, 22, 22, 22, The small intestine. 23, 23, The abdominal walls turned down. 24, The thoracic duct, opening into the left subclavian vein (27).

CHAPTER X.

THE RESPIRATORY AND VOCAL ORGANS.

§ 38. Anatomy of the Respiratory and Vocal Organs.—*The Organs of the Voice and of Respiration—The Larynx—Trachea—Bronchi—Lungs.*

380. The Respiratory and Vocal Organs consist of the *Larynx*, the *Trachea*, the *Bronchi* and the *Lungs*, the whole being acted upon by a complicated series of muscles.

381. The Larynx, the organ of the voice, is a short, quadrangular, cartilaginous cavity, extending from the root of the tongue and the hyoid bone, to the trachea, with which it becomes continuous below. It is separated from the spinal column by the pharynx, into which it opens above by a triangular and oblique aperture.

The Larynx is composed of five principal parts—the *Thy'roid*, the *Cri'coid*, the two *Aryte'noid* cartilages, and the *Epiglot'tis*. The *Thyroid** is the largest cartilage. It consists of two lateral, quadrangular, wing-like plates, which meet in front and form the prominence called *pomum Adami* (Adam's apple). This cartilage is connected with the hyoid bone above, and with the cricoid cartilage below.

The *Cricoid*† cartilage is about one-fourth of an inch wide in front, and one inch behind. This cartilage connects above with the thyroid cartilage by an articulation permitting the latter to move downward and forward, and also in the reverse direction; below, it communicates with the first ring of the trachea.

The *Arytenoid*‡ cartilages are two in number, small, tri-

* Gr., *thureos*, a shield. † Gr., *krikos*, a ring.
‡ Gr., *arutaina*, a pitcher.

angular and curved. They are placed upon the summit and back part of the cricoid cartilage, forming articulations.

The *Epiglottis* is oval-shaped, having its convex surface toward the mouth. It stands in a vertical position above the aperture of the larynx, which is closed by it in the act of swallowing.

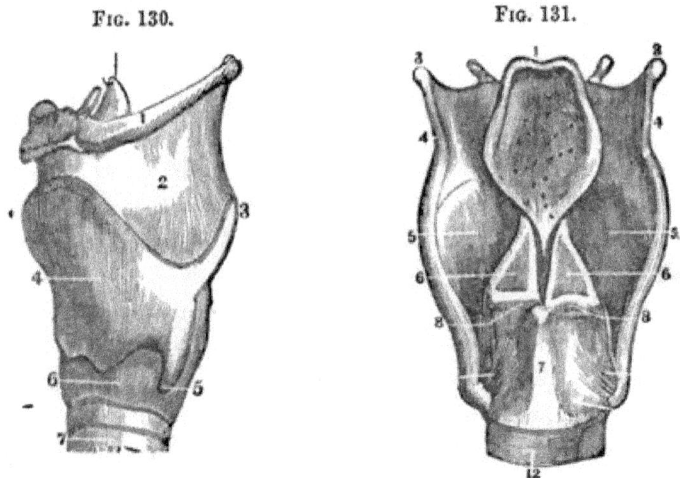

Fig. 130. A Side View of the Cartilages of the Larynx.—*, The front side of the thyroid cartilage. 1, The os hyoides (bone at the base of the tongue). 2, The ligament that connects the hyoid bone and thyroid cartilage. 3, 4, 5, The thyroid cartilage. 6, The cricoid cartilage. 7, The trachea.

Fig. 131. A Back View of the Cartilages and Ligaments of the Larynx.—1, The posterior face of the epiglottis. 3, 3, The os hyoides. 4, 4, The lateral ligaments which connect the os hyoides and thyroid cartilage. 5, 5, The posterior face of the thyroid cartilage. 6, 6, The arytenoid cartilages. 7, The cricoid cartilage. 8, 8, The junction of the cricoid and the arytenoid cartilages. 12, The first ring of the trachea.

382. The Trachea is a vertical tube about an inch in diameter and four inches in length. It is continuous with the larynx and extends to the third dorsal vertebra, where it divides into two branches, called *Bronchi.* The trachea is separated from the spinal column by the œsophagus.

383. The Bronchi* carry air to their respective lungs, and again divide, sending a branch to each lobe. These

* Gr., *brogchia*, the windpipe or throat.

divisions, called bronchiæ, are repeated, until each ultimate ramification terminates in a dilatation, called an *air-cell*.

384. The LUNGS, consisting of two divisions, are situated in the cavity of the chest, enclosing between them the heart and the great blood-vessels. They accurately fill the cavity, adapting themselves to the varying size attending respiration. They have the form of a double, but very irregular cone, with the apices above, and the basal ends below. The outer surfaces are convex, fitting the form of the chest; the inner surfaces are concave, conforming to the shape of the heart; the basal portion is also concave, owing to the upward pressure of the diaphragm. They are everywhere unattached, excepting at the root, where they are firmly secured by the pulmonary ligaments, the pulmonary artery, the pulmonary veins and nerves, and the bronchial tubes. The lungs are closely invested with a serous membrane, named *pleura*. The right lung is shorter than the left, but wider, and of somewhat greater bulk. It is divided into three lobes; the middle lobe being the smallest, and the lowest one the longest. The left lung has two lobes, of which the lower is the larger.

FIG. 133.

FIG. 132.

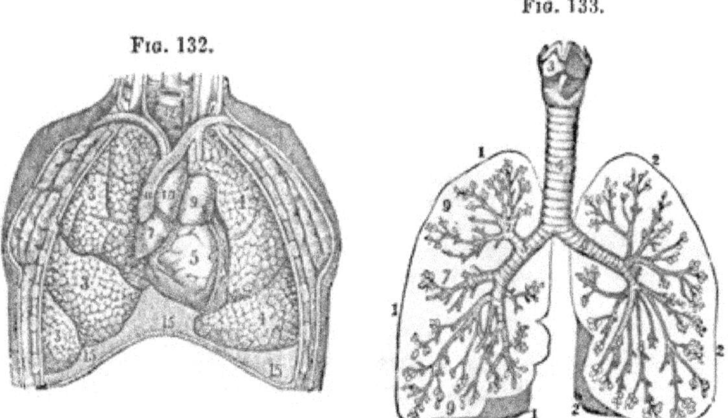

FIG. 132. THE LUNGS.—3, 3, 3, The lobes of the right lung. 4, 4, The lobes of the left lung. 5, 6, 7, The heart. 9, 10, 11, The large blood-vessels. 12, The trachea. 15, 15, 15, The diaphragm.

FIG. 133. THE BRONCHIÆ.—1, Outline of right lung. 2, Outline of left lung. 3, 4 Larynx and trachea. 5, 6, 7, 8, Bronchial tubes. 9, 9, Air-cells.

§ **39.** Histology of the Respiratory and Vocal Organs.—*Minute Structure of the Larynx.—The Trachea—The Bronchi—The Lungs and Pleura.*

385. With the exception of the epiglottis, the so-called cartilages of the Larynx are true cartilage, and in advanced life are strongly disposed to ossify. They are invested with a *perichon'drium.** The articulations of the cricoid cartilage are lined with synovial membrane and covered with capsular ligaments. The epiglottis is of a soft, elastic nature, fibro-cartilaginous in structure, and invested with mucous membrane.

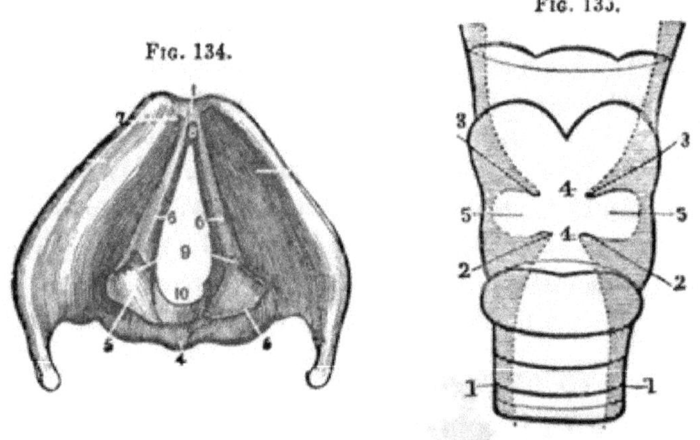

Fig. 134. A View of the Larynx, showing the Vocal Ligaments.—1, The anterior edge of the larynx. 4, The posterior face of the thyroid cartilage. 5, 5, The arytenoid cartilages. 6, 6, The vocal ligaments. 7, Their origin within the angle of the thyroid cartilage. 8, Their termination at the base of the arytenoid cartilages. 8, 10, The glottis.

Fig. 135. An Ideal Section of the Larynx.—1, The trachea. 2, 2, The lower vocal cords. 3, 3, The upper vocal cords. 4, 4, Rima glottidis, or glottis. 5, 5, Cavities between upper and lower vocal cords.

386. In the cavity of the larynx, the mucous membrane is reflected at each side, outward and upward, forming a pair of pouches, called the ventricles of the larynx. Just below these ventricles, are the true *vocal cords,* extending from a small process on the fore part of each Arytenoid cartilage to the recessed part of the Thyroid cartilage. They are com-

* Gr., *peri*, around, and *chondros*, a cartilage.

posed of yellow elastic tissue, covered by mucous membrane, and form two ridges, having very fine, smooth edges turned toward each other, and placed accurately on the same level.

387. The TRACHEA is made up of cartilage, fibrous tissue, muscle and mucous membrane. The cartilaginous part consists of flattened rings, or rather segments of circles, as they are wanting in that part of the tube next to the spine. The last ring is so modified as to accommodate it to the two first rings of the bronchi. The fibrous part is of yellow elastic tissue. It commences at the cricoid cartilage, and not only covers the rings in front, but forms for each a distinct sheath, thicker in front, and gradually losing itself with the termination of the rings. The posterior third of the trachea has a basis of strong, elastic fibrous tissue, arranged in longitudinal bands. The muscular portion has a simple layer of fibres running transversely, being attached to the ends of the cartilaginous rings and to the connecting tissue. The trachea is lined with mucous membrane.

388. The BRONCHI are constructed like the trachea, excepting in the ultimate bronchial ramifications, where the cartilages are composed of several pieces distributed around the tube, and the muscular fibres form a continuous layer. The cartilaginous element finally disappears, when the tubes consist only of fibro-elastic membrane with muscular fibres and a lining mucous membrane.

389. The LUNGS are made up of numerous small, polyhedral, primary lobules, or clusters of air-cells, which unite into larger secondary lobules. The latter give rise to the polyhedral markings seen upon the external surface of the lungs. The lobules seem to have no communication with each other, each primary lobule being in itself a miniature lung, performing independent functions. It has been calculated that no less than eighteen thousand of these air-cells group around each terminal tube, giving a sum-total of not less than six hundred millions.

The air-cells are connected together by fibro-elastic tissue, which renders them highly elastic. The cells are surrounded by fine networks of capillary vessels, the terminations of the branches of the pulmonary artery which accompany the branches of the bronchi. The trachea, bronchial tubes and air-cells are lined with a mucous membrane having a *ciliated epithelium*.

Fig. 136.

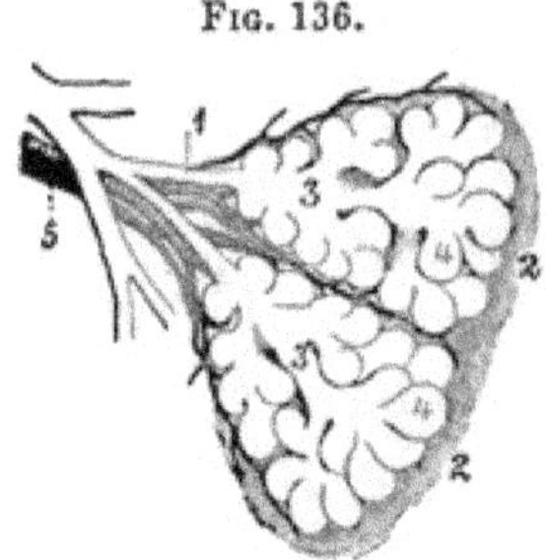

Fig. 136 (*Leidy*). Diagram of Two Primary Lobules of the Lungs, magnified.—1, Bronchial tube. 2, A pair of primary lobules connected by fibro-elastic tissue. 3, Inter-cellular air-passages. 4, Air-cells. 5, Branches of the pulmonary artery and vein.

390. The Pleura is a serous membrane which lines the thorax and then is reflected from the root of each lung over its surface. A fold of this membrane extends from the root downward to the diaphragm, and is called the pulmonary ligament. The pleural cavity is lubricated by the serous secretion, thus preventing friction during the respiratory movements (43). By the approximation of the two pleuræ in the median line, they form the *medias'tinum*, or partition of the thorax, which contains the heart enclosed within its pericardium.

§ 40. Chemistry of the Respiratory and Vocal Organs.

391. Respiration consists of two conjoint processes: that of supplying to the body the requisite amount of vitalizing oxygen, by inspiration; and that of removing from the body the deleterious carbonic acid, by expiration. The source of the oxygen is the air; the sources of carbonic acid are the blood and the tissues.

392. Some carbonic acid is generated in the blood, both from the respiratory or heat-giving elements of food, which chiefly enter the blood and are there oxidized, and from the changes of growth and decay to which the corpuscles of the blood are themselves subject. It is also probable that some intermediate or partly oxidized products of the decomposition of solid tissues undergo further oxidation in the blood.

393. We find the main source of carbonic acid, however, in the tissues. It appears both as a product of their natural decay, and of muscular and nervous activity. The sum of all the *chemical changes* of the body is *oxidation*, and the chief product of this oxidation is carbonic acid.

394. The proportions of oxygen and carbonic acid in venous and arterial blood are—

	Oxygen.	Carbonic Acid.
100 vols. venous blood	5 vols.	25 vols.
100 vols. arterial blood	10 vols.	20 vols.

It has also been found that the proportions of oxygen and carbonic acid in venous blood returning from muscles *at rest* are—oxygen, 7.5 vols.; carbonic acid, 31: from muscles *in action:* oxygen, 1.265 vols.; carbonic acid, 34.4.

395. The exchange of oxygen and carbonic acid in the capillaries is effected partly by physical and partly by chemical processes. The physical process is in accordance with the law of the "diffusion of gases." Two gases of different densities, and having no chemical affinity for each other, will intermix when brought into contact, and also when separated by a porous septum, provided they have no chemical affinity for that septum. These are the exact conditions in the capillaries; the oxygen and carbonic acid are the two gases, the capillary walls the porous septum. In addition to this physical process, there is a chemical process; the venous blood has a strong affinity for oxygen, hence readily unites with it in the pulmonic capillaries. When the arterial blood reaches the systemic capillaries, it yields its oxygen to the elements of the decomposing tissues which surround them. The carbon and hydrogen in their *nascent* state, or at the *moment of liberation,* seize the oxygen with great avidity, and give in exchange carbonic acid and water.

396. The air of expiration differs from that of inspiration, not only in its increase of carbonic acid, but in that of moisture and of temperature. As a rule, the expired air is saturated with moisture. The drier the external air, the greater the pulmonary exhalation, for in breathing air already saturated,

only so much more can be added as the higher temperature of the body will enable it to dissolve. The pulmonary exhalation has, besides water and carbonic acid, traces of ammonia, chlorides, urates, and even some albuminous substances; it readily undergoes decomposition.

397. The heat of the body is the result of the various chemical actions. The temperature of the tissues generally ranges from 98° to 100°; that of blood, from 100° to 102°. The blood varies in temperature in different parts, being warmest in the hepatic veins.

§ 41. PHYSIOLOGY OF THE RESPIRATORY AND VOCAL ORGANS.—*Objects of Respiration. Two Modes of Respiration. Renovation of the Air in the Lungs. Amount of Air concerned in each Respiration. Conditions affecting the Number of Respirations. Modifications of Respiratory Movements. Double Function of the Larynx. Resemblance between the Action of the Vocal Cords and Reed Instruments. Conditions affecting the Tone and Strength of the Voice.*

398. The FUNCTION OF RESPIRATION has for its immediate object, the *purification of the blood*, and for its ultimate uses, the *production of heat, motion and nervous energy.* The blood which becomes impure in the systemic capillaries, is carried to the pulmonary capillaries, which everywhere surround the air-cells. Through the thin walls, the poisonous carbonic acid passes from the capillaries into the air-cells, and is expelled from the body; at the same time, the oxygen of the external air passes from the air-cells into the capillaries, and the blood is changed from a dark maroon, to a bright red color.

The chemical changes in every part of the body caused by the union of this oxygen with carbon, hydrogen and other elements of the blood and tissues, maintain the temperature of the body, and are the source of its nervous power and electricity.

399. Respiration consists of two acts—taking air into the lungs, or *inspiration*, and expelling air from the lungs, or *expiration.* An act of inspiration is effected by the enlarge-

ment of the chest, which is done by elevating the ribs and sternum, and depressing the convex surface of the diaphragm. To elevate the ribs, two sets of muscles are used; those which are attached to the upper rib and sternum, contract and elevate the anterior extremities of the ribs; this enlarges the cavity between the spinal column and the sternum. The central portion of the ribs are raised by the intercostal muscles. The second rib is elevated by the contraction of the muscles between it and the first; the third rib is raised, by the combined action of the muscles between the first and second, and between the second and third.

FIG. 137.

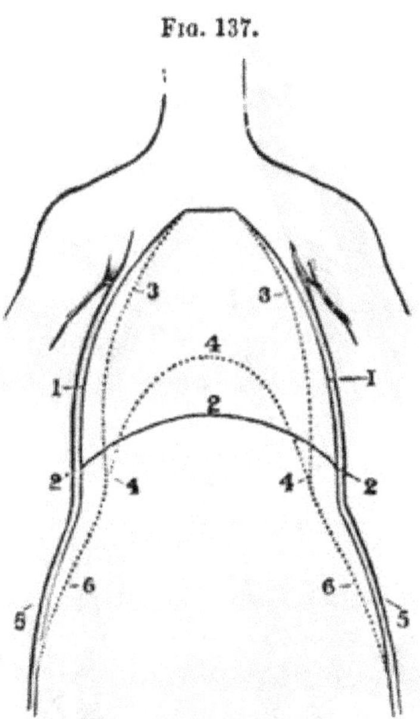

FIG. 137. A FRONT VIEW OF THE CHEST AND ABDOMEN IN RESPIRATION.—1, 1, The position of the walls of the chest in inspiration. 2, 2, 2, The position of the diaphragm in inspiration. 3, 3, The position of the walls of the chest in expiration. 4, 4, 4, The position of the diaphragm in expiration. 5, 5, The position of the walls of the abdomen in inspiration. 6, 6, The position of the abdominal walls in expiration.

The motion of each succeeding rib is increased in the same way, so that the movement of the twelfth rib is very free, as it is elevated by the contraction of eleven sets of intercostal muscles. Simultaneously with the elevation of the ribs, the central portion of the diaphragm is depressed by the contraction of its muscular margin and the relaxation of the muscular walls of the abdomen. By these combined movements the chest is enlarged in every direction. This enlargement of the thorax tends to produce a vacuum between the thoracic walls and the lungs, hence, the pressure of the external air fills the air-cells, forcing the elastic lungs to expand and fill the cavity; as when an elastic membrane is fitted over an open-mouthed vessel connected with an air-pump; exhaust any portion of the air within, and the pressure of the external air will cause the membrane to assume a convex form, the convexity being within the vessel.

Fig. 138.

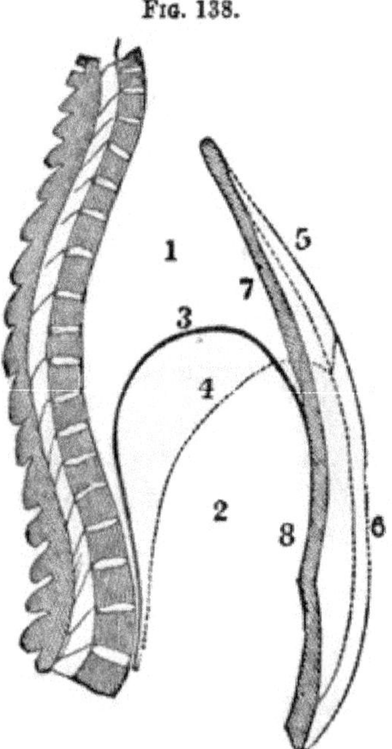

Fig. 138. A Side View of the Chest and Abdomen in Respiration.—1, The cavity of the chest. 2, The cavity of the abdomen. 3, The line of direction for the diaphragm when relaxed in expiration. 4, The line of direction for the diaphragm when contracted in inspiration. 5, 6, The position of the front walls of the chest and abdomen in inspiration. 7, 8, The position of the front walls of the abdomen and chest in expiration.

The elastic walls of the air-cells yield in every direction, so also do the surrounding areolar tissue and the pleura;

the air-tubes yield both in a circular and a longitudinal direction.

400. In *expiration*, the movements are of a more passive character, depending mainly on the relaxation of the inspiratory muscles and the elastic *resilience* of the tissues concerned. When the muscles relax, the sternum and ribs descend; the diaphragm vaults upward; the elastic walls of the air-cells diminish their size; the longitudinal and circular fibres of the bronchi and bronchiæ shorten and narrow their tubes; and the entire elastic lungs rebound like an extended spring let loose, while the interlobular and sub-pleural tissues aid powerfully in compressing them on all sides.

Ordinary expiration is undoubtedly aided by the action of proper expiratory muscles, especially the *internal intercostals* and the *infracostals*, small muscular bundles having the same direction as the former, but reaching over two or three spaces, instead of a single space between two ribs; also by a portion of a thin layer within the sternum and the cartilages of the true ribs.

The auxiliary expiratory muscles are the serratus magnus, the serrated muscles passing from certain dorsal and lumbar vertebræ upward to the last four ribs, and those which ascend from the pelvis and lumbar vertebræ to the lower ribs, certain portions of the long muscles of the back, and, lastly, the abdominal muscles. In difficult respiration almost every muscle in the body is made in some way subservient to the distension of the chest.

401. When respiration is performed chiefly by the diaphragm, it is called *abdominal* respiration; when chiefly by the action of the ribs, *pectoral* respiration. The former is the characteristic mode in men and children; the latter, in women.

402. The ordinary respiratory movements alone would not renovate the air in the smaller air-tubes and air-cells. Additional aid is rendered in two ways: 1st, By the *diffusion of gases*, causing the carbonic acid and the oxygen to mix equally in all parts of the lungs; and 2d, By the *epithelial*

air-current. In the lining mucous membrane of the trachea and the bronchial tubes, the cilia of the epithelium are always directed from below upward, and, like all ciliary motion, it has the effect of producing a current in the fluids of the mucous membrane (41). Now, the air in the tubes must move, to a certain extent, with this current, hence, a double stream of air is established in each bronchial tube; one current passing from within outward, along the walls of the tube, the other passing from without inward, along the central part. Thus a kind of aerial circulation is maintained, which, together with the mutual diffusion of the gases and the ordinary respiratory movements, ensures a complete renovation of the air in all portions of the pulmonary cavity.

FIG. 139.

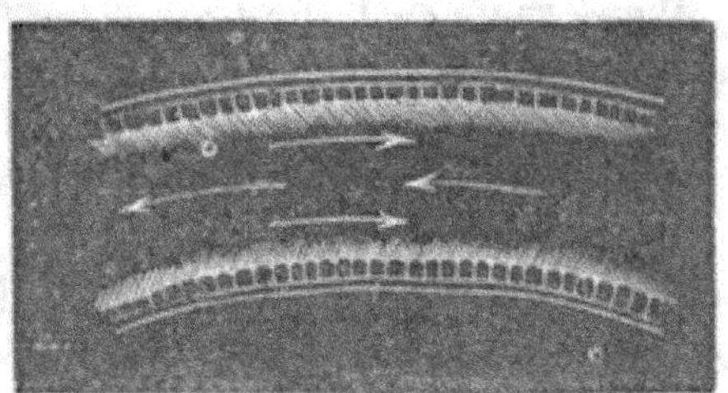

FIG. 139. DIAGRAM OF A SMALL BRONCHIAL TUBE, showing outward and inward current, produced by ciliary motion.

403. The amount of air taken in and given out in a respiratory movement must vary with different individuals and different conditions of the system. The volume of air ordinarily received by the lungs in a single inspiration is about one pint; the volume expelled, a little less than a pint. In the mutual action that takes place between the air and the blood, every twenty-four hours, the air loses about thirty-seven ounces of oxygen, and the body fourteen ounces of carbon.

404. Respiration is more frequent in women and children than in men. Persons of small stature breathe more frequently but less deeply than taller people. In health, the smallest number of respirations in a minute, by an adult, is not less than fourteen, and they rarely exceed twenty-five; eighteen may be considered the average number. The number of respirations is increased by exercise, food, stimulants and moderate cold; while it is diminished by inactivity,

moderate heat, starvation and general weakening influences, especially mental depression.

405. The actions of *sighing, yawning, sobbing, laughing, coughing* and *sneezing* are simple modifications of the ordinary movements of respiration, excited either by mental emotions, or by a stimulus arising in the respiratory organs themselves. Sighing and yawning often occur as simple results of deficient aëration; sometimes the former results from depression of the feelings; the latter from mere imitation. Laughter and weeping seem to be always either expressions of the emotions, or simple results of sensations. Coughing and sneezing are occasioned by irritation in the air-passages, and the sudden expiratory movement has a tendency to remove all intruding substances.

406. The LARYNX performs a double function, one part being concerned with respiration; the other with the voice.

In inspiration, the vocal cords separate, allowing the air to pass in freely; in expiration they relax. The former movement is active; the latter, passive. Both co-operate with the other respiratory movements. The extreme sensibility of the vocal cords and the posterior part of the epiglottis, causes them to throw off any foreign substances happening to come in contact with them, by a sudden, expulsive cough.

The larynx, however, is the special organ of the voice; sounds being produced by the vibratory action of the vocal cords. During ordinary, tranquil breathing, the cords are widely separated, the glottis being of triangular shape, but when a vocal sound is to be produced, the arytenoid cartilages are said to become erect, and almost to touch each other; the cords are made suddenly tense, closing the posterior portion of the glottis, while the anterior two-thirds opens a very fine fissure; and the air, driven by an unusually forcible expiration through the narrow opening in passing between the vibrating vocal cords, is itself thrown into vibrations which produce the sound required.

407. The vibrations of the vocal cords take place according

to the laws which regulate the action of the stretched membranous tongues, or reeds, in reed instruments. If one extremity of a short tube be covered by two portions of elastic membrane, leaving a small chink between them, a form of double membranous tongue is obtained, which resembles the vocal cords in man. The narrower the chink, the more easily are sounds produced. The size, however, in no way affects the *pitch*, which is somewhat determined by the length, tension and thickness of the tongues, but chiefly by the *tension*.

408. The *tones* of different individuals are doubtless modified by the shape and size of the vocal apparatus. Thus, a large larynx usually gives a deep-toned voice; a smaller one gives a comparatively high pitch. The difference in the tone of the *male* and female voice is due largely to the great difference in the walls of the larynx. In the female, the cavity is smaller, the angle in front less acute, and the cartilage softer.

Vocal sounds are further modified by the elevation and depression of the larynx, for when the voice is raised from a low to a high pitch, the whole larynx is elevated toward the base of the skull, drawing with it the trachea; the vocal tube is thus slightly lengthened; the diameter of the trachea lessened and variations are produced in the tension of its walls, enabling it to accommodate itself to the different vocal tones.

The general *strength* of the voice depends upon the capacity of the chest; the development of the muscles used in vocalization; the extent to which the vocal cords can vibrate; and the power of communicating resonance possessed by the air-passages and neighboring cavities.

§ 42. HYGIENE OF THE RESPIRATORY AND VOCAL ORGANS.—*Importance of Proper Respiration. Effect of Carbonic Acid Gas upon Respiration and Combustion. Sources of this Gas. Location of Dwellings. Danger of Impure Air within the House. Importance of Ventilation in Public Buildings—In Sleeping Rooms—In Sick Rooms. Means of Securing Warm and Pure Air in Winter. Importance of Moisture in the Air. Effect of Compressing the Respiratory Organs. Means of Enlarging the Chest. Influence of the Nervous System upon Respiration.*

409. In the circulating system, we have seen the minutest care manifested in supplying each organ, tissue and cell with blood; if the blood be pure, this is the best conceivable arrangement for securing health and vitality; if impure, the means is equally effective for poisoning every part of the system.

410. That pure blood can be obtained only by a healthy action of the respiratory organs, and this action only by a constant and sufficient supply of pure air, is evident from what has already been said. Limit this supply, and a double evil ensues; the stimulus furnished to the tissues, especially the nervous and muscular, is withdrawn, and the carbonic acid is retained in the blood; hence, the brain works sluggishly; the muscles become inactive; the heart acts imperfectly; the secretions are deteriorated; the food is not properly assimilated; and the whole body becomes *weak*.

411. Pure air is composed of oxygen and nitrogen in about the proportion of 21 to 79. The air is most frequently rendered unfit for vital purposes by the presence of carbonic acid gas, and volatile particles of corrupted animal matter. Carbonic acid gas will not support combustion, as may be seen by introducing into it a burning taper, which is as readily extinguished as if dipped in water; neither will it support life; if a small animal be placed in a jar of the gas, life soon becomes extinct.

412. The sources of this deleterious gas are mainly—decaying animal and vegetable matter; combustion; and the respiration of animals.

Plants in their healthy state take up carbonic acid gas, and give out oxygen, thus maintaining, under ordinary circumstances, a pure and respirable atmosphere.

In wells, mines and caves, where the circulation is obstructed, this gas often accumulates in quantities sufficient to cause death to those who enter. Hence, before entering them, the air should be tested by a lighted taper. If it will not burn, respiration cannot be maintained.

413. *The location of dwelling-houses should be chosen with reference to free circulation of air, and the avoidance of marshes, stagnant pools, slaughter-houses, and other sources of vegetable and animal decay.* Careful attention should also be given to the drainage of a house, and to the *cellar.* These underground store-rooms should always be well ventilated, and all vegetables removed from them in early spring. A little neglect in these and like respects has not unfrequently prostrated a whole family with typhoid disease.

414. The chief danger, however, is *within* the house proper, and from the breaths of its inmates. Unless ventilation receives proper attention, the carbonic acid gas from the lungs, and the effete particles of animal matter thrown off from the system, will soon render the air poisonous.

415. *School-rooms, churches, concert-halls and all rooms designed for public purposes should be amply ventilated.* The child at school becomes listless and uninterested: why? *Because he is stupefied by foul air.* When a pupil continues to breathe such air, month after month, his brain is injured, and often consumption or other fatal disease destroys his young life, and then we wonder at the "mysterious providence" that takes from us the gifted and beautiful.

The good man at church feels that he *ought* to be interested in the services, and yet, powerless to fix his attention, he sits nodding: why? *Because he is stupefied by foul air.* The air breathed over and over again last Sabbath, and shut in during the week, is all the poor man can obtain. How can acceptable worship be offered by those who are, at the very moment, violating the plainest laws of the Being worshiped?

416. The lamps of the concert-hall burn dimly long before the closing hour: why? *Because they are bedimmed by the foul air;* and just in proportion to the decrease of light is the increase of dullness in the audience. Let in the pure air, and how soon will the light perceptibly brighten, and the audience become animated. The air of a well-filled school-room, church or hall, will be rendered unfit for respiration in a few minutes.

417. The influence of habit, in accustoming us to foul air is strikingly expressed by Birnan, in the "Art of Warming and Ventilating Rooms": "Not the least remarkable example of the power of habit is its reconciling us to practices which, but for its influence, would be considered noxious and disgusting. We instinctively shun approach to the dirty, the squalid and the diseased, and use no garment that may have been worn by another. We open sewers for matters that offend the sight or the smell and contaminate the air. We carefully remove impurities from what we eat and drink, filter turbid water, and fastidiously avoid drinking from a cup that may have been pressed to the lips of a friend. On the other hand, we resort to places of assembly, and draw into our mouths air loaded with effluvia from the lungs, skin and clothing of every individual in the promiscuous crowd—exhalations offensive, to a certain extent, from the most healthy individuals; but when arising from a living mass of skin and lungs in all stages of evaporation, disease and putridity, prevented by the walls and ceiling from escaping, they are, when thus concentrated, in the highest degree deleterious and loathsome."

418. *The sleeping room should be thoroughly ventilated.* Proper ventilation would often prevent morning headaches, want of appetite and general languor, so common among the feeble. The impure air of sleeping rooms probably causes more deaths than intemperance. Those who live in open houses little superior to the sheds that shelter the farmer's flocks, are usually the most healthy and robust. Headaches, liver complaints, coughs, and a multitude of nervous affec-

tions, are almost unknown to them; not so with those who spend their days and nights in rooms with double or calked windows, breathing over and over again the confined air; disease and suffering are their constant companions.

Observation.—1st, By many, a sleeping apartment twelve feet square and seven feet high, is considered spacious for two persons, and "good accommodation" for four. This room contains one thousand and eight cubic feet of air; allowing ten cubic feet to each person per minute, two occupants would vitiate the air in fifty minutes, and four in twenty-five minutes.

2d, Among children, convulsions, or fits, often occur when they are sleeping, and not unfrequently in consequence of impure air. In such cases, by carrying the sufferer into the open air, relief is afforded. Children should not sleep in *low beds,* while adult persons occupy a higher bed in the same unventilated room, as carbonic acid is most abundant near the floor, nor is it advisable that the young sleep with the sick or aged.

419. *The ventilation of the sick room should receive special attention.* It is no unusual practice, when the patient is suffering from acute disease, for the attendants to prevent the ingress of pure air, simply from the apprehension that the sick person will take cold; and caution is indeed necessary; the patient should not *feel the current.* No room is suitable for sickness that is not so arranged that pure air may be constantly admitted without inconvenience or injury to the patient; and here we would say that *cool* air should not be mistaken for *pure* air. A very little sound judgment in this matter would doubtless save much suffering, and lengthen life in a multitude of cases.

The custom of having several persons sit in the sick room vitiates the air and delays the recovery of the patient.

420. *The great means of ventilation, in summer, are open windows and doors. Motion* is at that season the great desideratum. On a hot summer's day we go into a cool room that has been shut up, and at first it is grateful; but in a

short time, the cool, stagnant air becomes oppressive, and we select the open window with its *circulation* of air, even if it is a little warmer. Windows should be made to lower from the top.

421. *In winter, ventilation may be obtained by properly constructed flues.* As cold weather approaches, we must close the windows, excepting when in bed; but good flues secure a good circulation of air. Leeds, in his "Lectures on Ventilation," in speaking of the value of an open fireplace for ventilation, says, "Thousands of lives are thus saved, and many more would be, if all fireplaces were kept open. If you are so fortunate as to have a fireplace in your room, paint it when not in use; put a bouquet of fresh flowers in it every morning, if you please, or do anything to make it attractive, but never *close it;* better use the fire-boards for kindling-wood. It would be scarcely less absurd to take a piece of elegantly-tinted court-plaster and stop up the nose, trusting to the accidental opening and shutting of the mouth for fresh air, because you thought it spoiled the looks of your face to have two such great, ugly holes in it, than to stop your fireplace with elegantly-tinted paper because it looks better."

422. For heating a small room, where the occupants may change position at pleasure, an open fire is the healthiest known means, for the air cannot become stagnant, as the fire is continually drawing a considerable amount from the room to support combustion, the place of which is supplied by other air. Just here comes in the greatest inconvenience of the open fire; if the cold air comes in at the cracks of a door or window on the opposite side of the room, it will flow across to the fire, chilling the feet and backs of those sitting in its track.

423. A stove is a very economical mode of heating ordinary sitting rooms, offices, etc.; but there should be an air-chamber, or box, on or near the top of the stove, and communicating with this should be a pipe for introducing fresh air from the external atmosphere. If this supply of fresh air

is abundant, with a constant evaporation of moisture, and an opening into a heated flue near the ceiling, to be opened when the room is overheated or the lights are burning in the evening, and kept closed at other times, with another opening into a heated flue on a level with the floor, which should be *always open* to carry off the cold, heavy, foul air from the floor; if a stove be thus arranged, for many small, isolated rooms, it is one of the most economical, comfortable and wholesome means of heating at our command.

424. For the general warming of a house, heating the air by steam is one of the most healthy arrangements, and a very good mode is thus given by Leeds: "Where a steam furnace is used, two-thirds of the heating surface should be below the floor, and fresh air be brought into it, and thence conducted to the rooms through large pipes. The warmed air should be let into the room at the floor, and an opening into an exhaust-flue, two-thirds the size of the inlet, should be provided at the floor for the escape of the foul air. The remaining one-third of the heating surface should be exposed in the halls and some in other parts of the house, to heat by direct radiation, but *under no circumstance should a room or office be occupied which is heated exclusively by direct radiation from exposed steam-pipes.* It is one of the worst, most unhealthy, *killing systems* in existence." "Probably one of the very best arrangements is to have a good steam furnace, with a large fresh-air box, letting in an abundance of air moderately warmed, and overflowing the house with this, also to have some direct radiation in the halls, and a bright, cheerful, open fire in the family sitting room." Two things are indispensable in every furnace—a large fresh-air box communicating with the *external atmosphere,* and a large *evaporating* vessel. Few persons realize the necessity of supplying a proper amount of moisture in our stove and furnace-heated rooms. If it is not furnished by other means, the heated air will have it from the natural moisture of the skin and lungs, thus producing a dry, parched, feverish condition of the system.

425. *The conditions of proper respiration require not only that the air be pure, but sufficient in quantity.* Hence the *chest and lungs must not be reduced in size.* In children who have never worn close garments, the circumference of the chest is generally about equal to that of the body at the hips; and similar proportions would exist through life if there were no improper pressure of the clothing. Such is the case with the Indian woman, whose blanket allows the free expansion of the chest. The symmetrical statues of ancient sculpture bear little resemblance to the "beau ideal" of American notions of elegant form. The chest is often contracted in infancy, because of the mother's ignorance of the pliant character of the ribs and cartilages; thus she sows the seeds of disease and shortens the life of her offspring. In later years the same result is produced by a steady and *moderate* pressure. It is in this way that the "genteel," *tapering* waist is produced. The style of dress adopted at the present day is a prolific cause of deformity, for deformity it certainly is, since the design of the human chest is not simply to form a connection between the upper and lower parts of the body, like some insects.

426. The Chinese, by compressing the feet of female children, prevent their growth, so that the foot of a *Chinese belle* is not larger than the foot of an American girl of five years; the American women *compress their chests,* so that the chest of an *American belle* is not larger than the chest of a Chinese girl of five years. In these respects, which country exhibits the greater intelligence?

427. *Individuals may have small chests from birth,* this being, to the particular individual, natural. That like produces like is a general law. If the mother has a small, tapering waist, either hereditary or acquired, the form may be impressed on her offspring, thus illustrating the truthfulness of Scripture, which declares that the sins of the parent shall be visited upon the children unto the third and fourth generations.

428. The question is often asked, Can the size of the chest

and the volume of the lungs be increased when they have been once compressed? Yes. The means to be used are, a full inflation of the lungs at each act of respiration, and a judicious exercise of them by walking in the open air, reading aloud, singing, sitting erect, and practicing appropriate gymnastic exercises. Unless these exercises are systematic and persistent, they will not afford the beneficial results desired.

Fig. 140. Fig. 141.

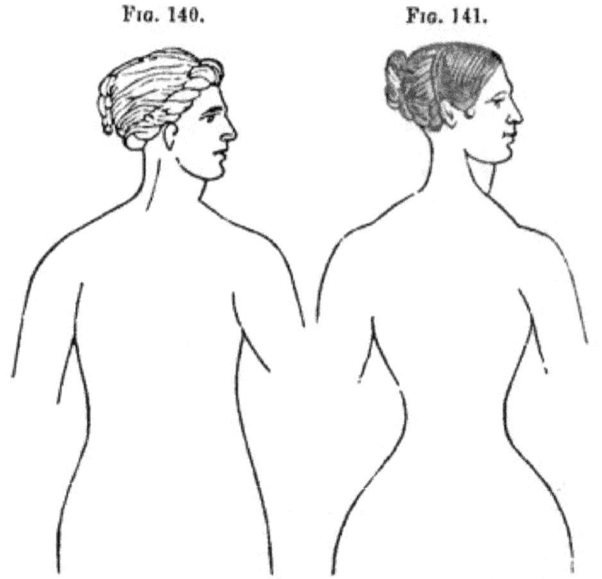

Fig. 140. A Correct Outline of the Venus de Medici, the *beau ideal* of female symmetry.

Fig. 141. An Outline of a Well-Corseted Modern Beauty.

One has an artificial, insect waist; the other, a natural waist. One has sloping shoulders, while the shoulders of the other are comparatively elevated, square and angular. The proportion of the corseted female below the waist is also a departure from the symmetry of nature.

Observation.—Persons of sedentary habits should often, during the day, take full, deep breaths, filling the smallest air-cells with air; the shoulders should be thrown back, and the head held erect.

429. *Respiration is much influenced by the condition of the nervous system.* Abstract thought, anxiety and the depressing passions diminish the contractile energy of the diaphragm

and the muscles that elevate the ribs, thus preventing the full inflation of the lungs. Cheerfulness, joy and all the exhilarating emotions favor free respiration, and consequently promote health.

430. *To resuscitate persons asphyxiated from drowning, strangulation, electricity or breathing poisonous gases,* the chest should be suddenly and forcibly pressed downward and backward, then the pressure suddenly discontinued. This should be continually repeated till a pair of bellows or some other means of artificial respiration can be obtained. When bellows are used, introduce the nozzle well upon the base of the tongue, and closely surround the mouth and nose with a towel, press upon the part of the neck called Adam's apple while introducing the air, then press upon the chest to expel it, thus imitating breathing. If other means of artificial respiration cannot be immediately obtained, let the lungs of the sufferer be inflated by air from the lungs of other persons present. That this air may be as pure as possible, the lungs should be quickly filled, and the air instantly expelled into the lungs of the asphyxiated person. The patient should be placed in pure air, and a physician procured immediately. In case of drowning, wrap the body in warm flannel and place near the fire; *use no friction till breathing is restored.*

Observation.—Inhaling the gas from burning charcoal placed in an open pan to warm a room, or gas from a furnace or coal stove, when the draught is imperfect, is deleterious, often producing death. Care should be taken, when gas is used for lighting, that it is completely turned off before retiring to sleep.

§ 43. Comparative Pneumonology.—*Respiratory Apparatus of Mammalia—Of Birds—Of Reptiles—Of Fishes.*

431. The Respiratory Apparatus in all the *Mammalia* is similar to that of man, both in structure and function. There are similar arrangements and movements of the ribs, sternum, intercostal muscles and diaphragm. The lungs fill

the cavity of the thorax, and have the same general composition of lobes, lobules and air-cells.

Fig. 142.

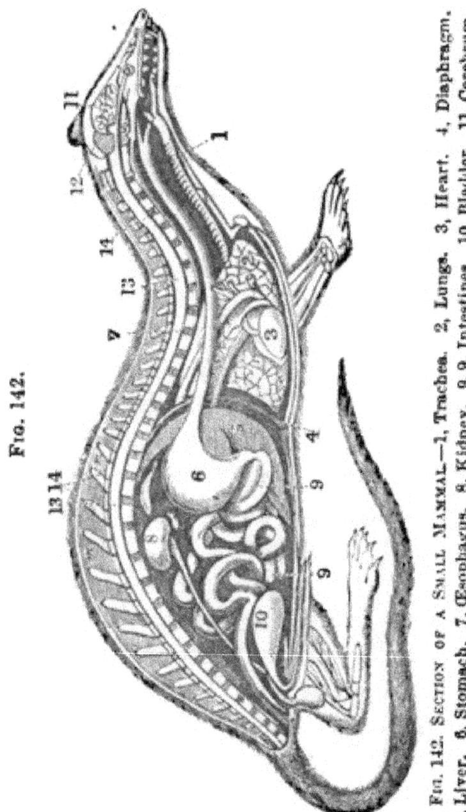

Fig. 142. Section of a Small Mammal.—1, Trachea. 2, Lungs. 3, Heart. 4, Diaphragm. 5, Liver. 6, Stomach. 7, Œsophagus. 8, Kidney. 9, 9, Intestines. 10, Bladder. 11, Cerebrum. 12, Cerebellum. 13, 13, Medulla spinalis. 14, 14, Vertebræ.

432. In *Birds* the lungs are confined to the back part of the thoracic abdominal cavity, being firmly attached to the ribs in their interspaces. They are not separated into lobes, as in the mammalia, but are lengthened, oblong and flattened in shape, and connected with large membranous cells scattered through every part of the body. They have the larynx, trachea, bronchia, pulmonary arteries, veins and capillaries, although much modified.

433. The ultimate pulmonary capillaries do not form a network lining definitely-bounded air-cells, as in mammals,

but each capillary crosses an open air-space of its own. They interlace in every direction, forming a mass of capillaries, permeated everywhere by air (B, fig. 144).

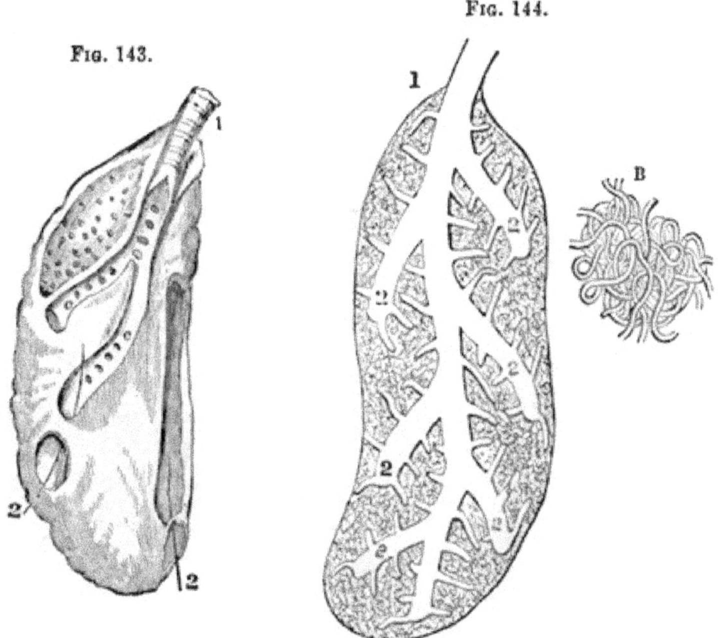

Fig. 143 (*Owen*). The Right Lung of a Goose.—1, A bronchus which divides into two tubes that open into the abdominal air-receptacles at 2, 2.

Fig. 144 (*Owen*). Ideal Section of a Bird, magnified two hundred and sixty times.—1, A primary bronchus dividing into secondary bronchi that end in cæca, 2, 2, 2, 2, 2, 2. These secondary bronchi give off smaller penniform branches that ramify among the lobules. B, A plexus of capillary vessels.

434. A marked modification of the respiration of birds is the connection of the pores of the bones and feathers with the bronchial tubes and air-spaces of the lungs, so that there is an interchange of air between the lungs, the bones and the investing plumage. The walls of the bones of birds are more cancellated than those of the mammalia. Birds consume more air in a given time, proportionally, than any other class of animals, and they soonest become asphyxiated when deprived of it.

435. In *Reptiles*, respiration is more simple than in mammals or birds. The lungs are less lobular and more bag-like, extending into the abdominal cavity. Upon the walls of these sac-like lungs the pulmonary vessels ramify. In the turtle, the tortoise and the frog, the thorax is not so formed as to act like a suction-pump, and accordingly these animals swallow the air by a sort of deglutition.

Fig. 145. Fig. 146.

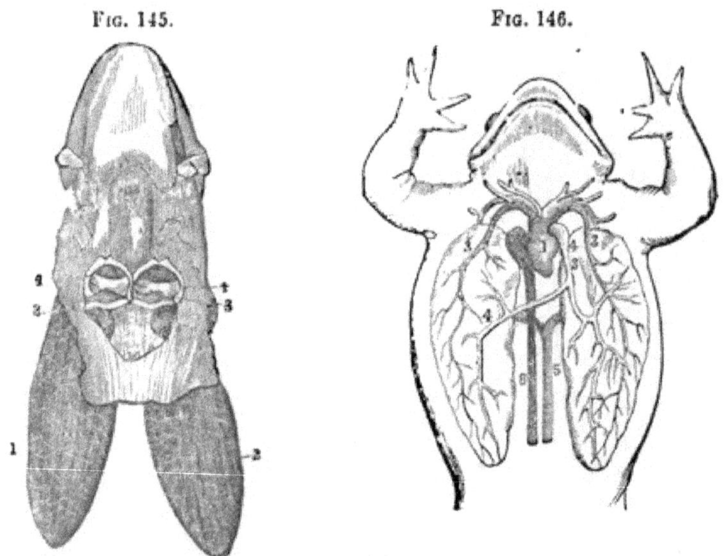

Fig. 145 (*Owen*). Tongue, Larynx and Lungs of a Frog.—1, 2, Lungs. 3, 4, Larynx.
Fig. 146 (*Owen*). Heart and Lungs of a Frog.—1, Heart. 2, Arch of the aorta. 3, 3, Pulmonary artery. 4, 4, Pulmonary veins. 5, 5, Aorta. 6, Vena cava.

436. In *Fishes*, the respiration is still more modified and more complicated than in reptiles. Instead of lobular or bag-like lungs, there are found only a series of slit-like openings, or arches on each side near the head, called the branchiæ, or gills.

437. The bony and cartilaginous frames of these arches, on the convex side, support processes. On these are many plates, or leaflets, covered by a delicate tessellated membrane, or epithelium, on which the microscopic capillary blood-vessels ramify. By this arrangement of extensive epithelial surface, the blood-particles are more minutely separated and acted upon by the air in the water. In breathing, the mouth

and gills of a fish open alternately; the water entering the mouth escapes by the openings of the gills.

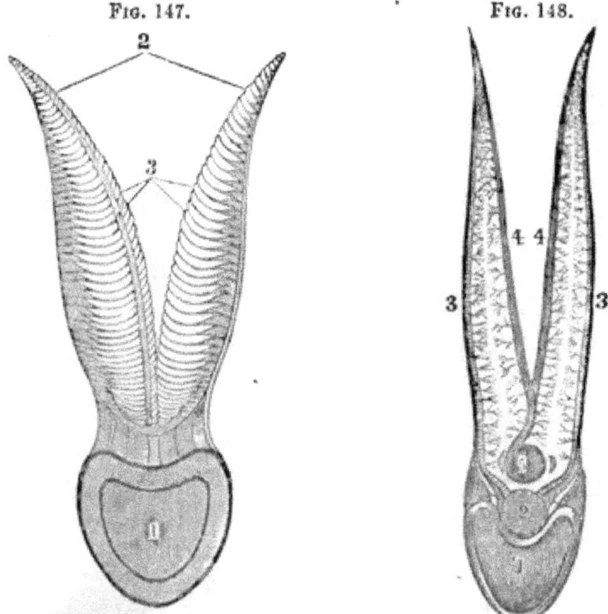

Fig. 147 (*Owen*). Section of a Branchial Arch, with a pair of processes supporting leaflets or plates, from a cod, magnified two hundred and sixty diameters. 1, A section of a branchial arch. 2, A pair of processes. 3, Branchial leaflets, or plates. The number of leaflets in one process of the cod is about one thousand; in the salmon, fourteen hundred; in the sturgeon, sixteen hundred.

Fig. 148 (*Owen*). Circulation of the Blood through the Branchial Leaflets (a diagram). 1, A section of a branchial arch. 2, A section of a branchial artery. 3, An artery sent along the outer margin of the processes, giving off capillary vessels to the leaflets. 4, A vein that receives the blood from the capillaries on the inner margin of the process after the respiratory change has been effected and returns it to the branchial vein (5).

438. A remarkable feature in the organization of some fish is the swimming, or air-bladder, placed in the abdomen under the dorsal spine, communicating often with the œsophagus or stomach by a canal, permitting the escape of air from its interior. By a movement of the ribs, the air-receptacle is acted on, so that by diminishing the quantity of air, the specific gravity of the fish alters according to circumstances. Fish that swim near the bottom have no air-bladder; as the eel, sole and turbot.

DIVISION IV.

SENSORIAL APPARATUS.

In the two preceding Divisions, the tissues and organs directly involved in the movements of the body, and those most intimately connected with the preparation and assimilation of nutrient material, have been briefly described. In the present Division, we consider the organs through which is manifested the subtle power that controls these motions and processes, establishes telegraphic communication between the several parts of the body, and brings it into important relations with the external world. These, taken collectively, we name the *Sensorial Apparatus.*

CHAPTER XI.

NERVOUS SYSTEM.

§ **44.** Anatomy of the Nervous System.—*Two Forms of Nervous Tissue. Classification of the Ganglia, Nerves and Commissures. Spinal Cord. Medulla Oblongata. Peduncles of the Cerebellum—Of the Cerebrum. Corpora Striata. Optici Thalami. Corpora Quadrigemina. Corpus Callosum. Ventricles. Hemispheres of the Cerebrum. Convolutions of the Cerebrum and Cerebellum. Classification of Cerebro-Spinal Nerves—Of Cranial Nerves—Spinal Nerves. Sympathetic System. Distribution of Sympathetic Nerves.*

439. Nervous Tissue presents two formal characters, one, cell-like and gray in color; the other, fibrous and white. The former is arranged in masses called *Centres* or *Ganglia*, being the originating, active centres of nerve-force; the latter, in threads, which are simple conductors of nerve-force, and

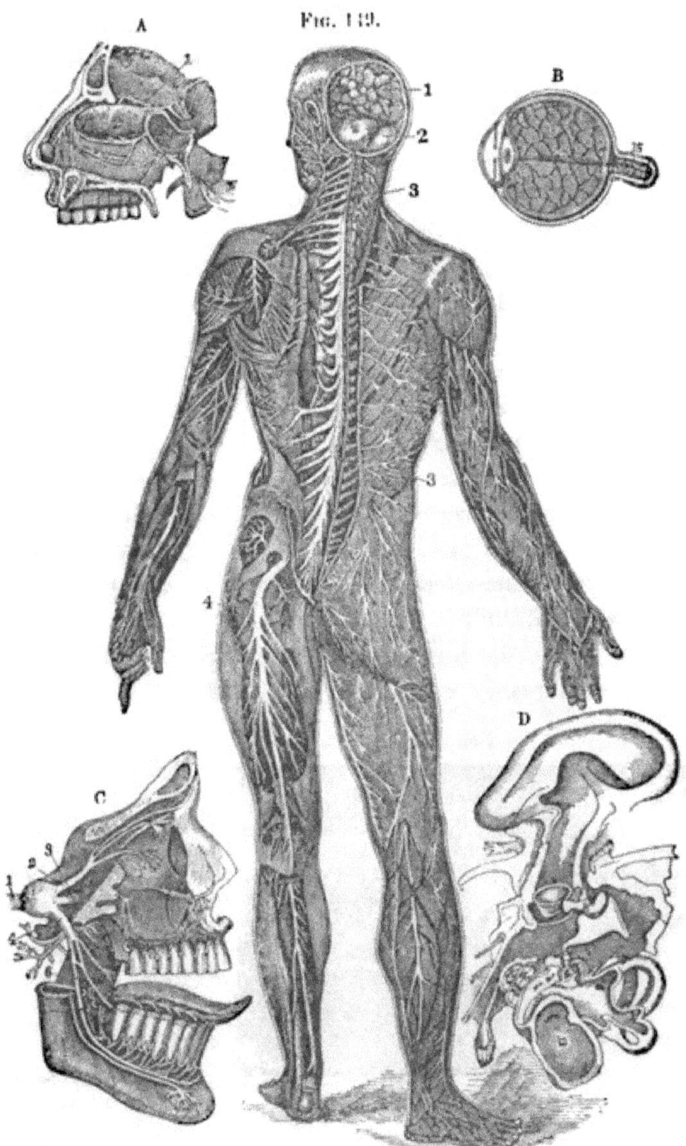

Fig. 149. A Representation of the Brain, Spinal Cord and Spinal Nerves.—1, The cerebrum. 2, The cerebellum. 3, 3, Spinal cord. 4, The sciatic nerve.

A. Distribution of the Olfactory Nerve.—1, 2, Nerve of smell.

B. Optic Nerve.—15, The nerve of vision.

C. The Gustatory Nerve.—1, 2, 3, 4, Branches of the nerve of taste.

D. Auditory Nerve.—13, Nerve of hearing.

are named *Nerves* when they connect the ganglia with the various parts of the body; and *Commissures* when they connect the ganglia with each other.

440. For convenience in study, the numerous Ganglia, Nerves and Commissures may be arranged in two great and closely-connected systems—the *Cerebro-Spinal* and the *Sympathetic:* the Cerebro-Spinal system including the series of ganglia within the skull and spinal column, their nerves, commissures and the lesser ganglia in the nerve-tracts: the Sympathetic system including the long chain of ganglia lying in front of the spinal column, their nerves, commissures and additional ganglia found chiefly in the abdominal cavity.

441. The CEREBRO-SPINAL AXIS commences with that portion of nervous matter which lies within the spinal column, extending from the second lumbar vertebra to the base of the skull, and known as the *Spinal Cord.* It contains within itself the filaments of all the nerves of the external parts of the trunk and limbs. It is soft, and white externally, but grayish within, forming the longest ganglion in the system. The cord is nearly cylindrical and double, the two halves connected by a narrow commissure or bridge of the same substance as the cord, having within, through the entire length, a minute *central canal.* On each half are two slight, longitudinal lines, serving to distinguish it into *Anterior, Lateral* and *Posterior columns.* As it enters the cavity of the skull, the cord becomes enlarged and receives the name of *Medul'la Oblonga'ta.* This enlargement is due to the presence of an important ganglion imbedded

FIG. 150.

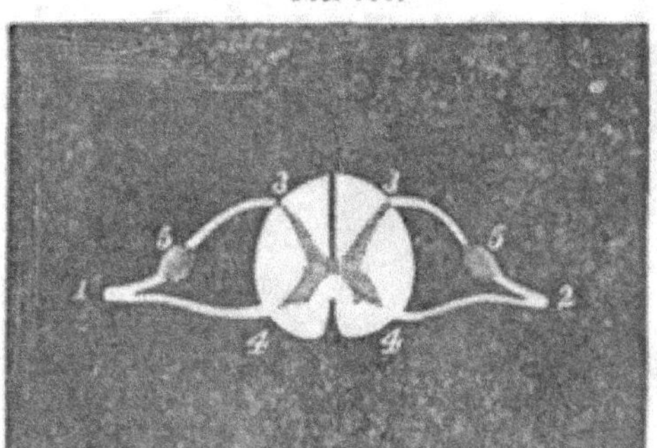

FIG. 150. TRANSVERSE SECTION OF SPINAL CORD.—1, 2, Spinal nerves of right and left sides, showing their two roots. 4, Origin of anterior root. 3, Origin of posterior root. 5, Ganglion of posterior root.

within, named the *Ganglion of the Medulla Oblongata,* and also to the accession of the fibres of most of the cranial nerves. In each of the lateral halves of the medulla oblongata may be seen four principal bundles of nerve-fibres, ranging backward from the middle line in front as follows: 1st, *Anterior Pyramids;* 2d, the *Olivary Bodies;* 3d, the *Restiform Bodies;* and 4th, the *Posterior Pyramids.* These bodies are continuous with their corresponding portions of the columns of the spinal cord. Many of the fibres of the anterior pyramids cross each other, bringing each side of the column into communication with the opposite side of the brain; this crossing forms the *Decussation of the Anterior Pyramids.* Some of the fibres of the posterior pyramids also cross a little above. By the divergence of the restiform and posterior pyramidal bodies, a somewhat broad cavity is left, which may be considered a widening of the central canal, and which receives the name of the *Fourth Ventricle.*

FIG. 151.

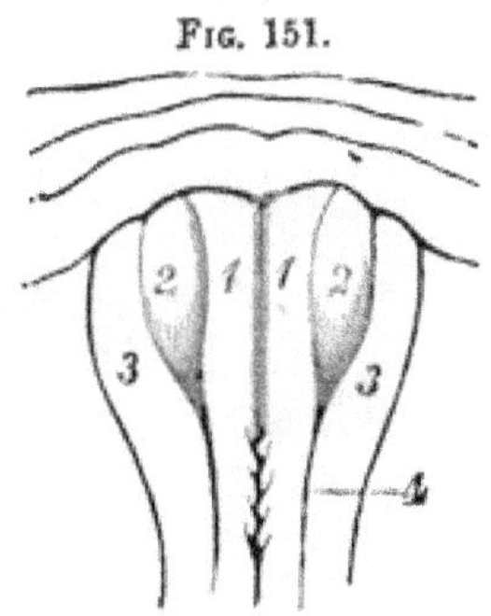

FIG. 151 (*Dalton*). MEDULLA OBLONGATA OF HUMAN BRAIN, anterior view. 1, 1, Anterior pyramids. 2, 2, Olivary bodies. 3, 3, Restiform bodies. 4, Decussation of the anterior columns. The medulla oblongata is seen terminated above by the transverse fibres of the Pons Varolii.

442. Overshadowing this ventricle is a mass of nerve-substance, called the *Cerebel'lum* or *little brain,* which is also double, consisting of two hemispheres. Each hemisphere, from its inner surface, sends out a multitude of fibres, which pass downward and forward toward the centre, unite into flattened bundles, emerge from the hemisphere, sweep across the base of the brain, pass up to the other hemisphere and spread out over its internal surface; thus originating in one hemisphere, and terminating in the other. The two sets of fibres cross in front of the Medulla Oblongata, in the middle line of the base of the cerebellum, forming *the bridge of the Cerebellum,* or the *Pons Varo'lii.*

At the pons, the medulla oblongata sends off from the

restiform bodies, bundles of fibres called the *Inferior Peduncles of the Cerebellum.* Passing under and among the fibres of the pons, and imbedding the *Ganglion of the Tuber Annula're,* are two bundles—one of fibres from the anterior pyramids and the front of the olivary bodies, the other from the posterior pyramids and the back of the olivary bodies; as they appear

FIG. 152.

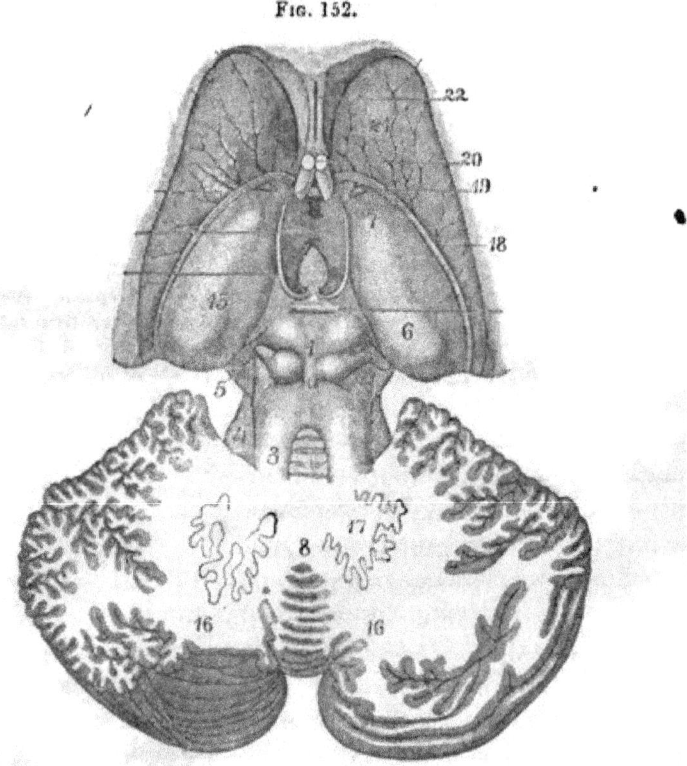

FIG. 152 (*Leidy*). STRIATED BODIES, THALAMI, QUADRIGEMINAL BODY AND CEREBELLUM.—1, Quadrigeminal body. 3, Superior peduncle of the cerebellum. 4, Superior portion of the middle peduncle. 5, Superior portion of the crus, or leg, of the cerebrum. 6, Posterior tubercle of the thalamus. 7, Anterior tubercle. 8, Fundamental portion of the cerebellum. 15, Thalamus. 16, Hemispheres of the cerebellum. 17, Dentated body. 18, Semicircular line. 19, Vein of the striated body. 20, Anterior crura of the fornix. 21, Striated body. 22, Fifth ventricle between the layers of the pellucid septum.

in front, they diverge, forming stalk-like bundles known as the *Peduncles of the Cerebrum,* as they seem to support the two hemispheres of the *cere'brum*, or *brain proper*, as the flower-stalk bears its flower. The *anterior* bundles pass upward to two large ganglia (one on each side the median line), called the *Cor'pora Stria'ta*, or *Striated Bodies;* the *posterior* bundles also pass upward to two ganglia situated a little in front of the striated bodies, and named the *Op'tici Thal'ami.* In these ganglia the fibres seem to terminate, while a new set connects the ganglia with the main surface of the cerebral hemispheres.

FIG. 153.

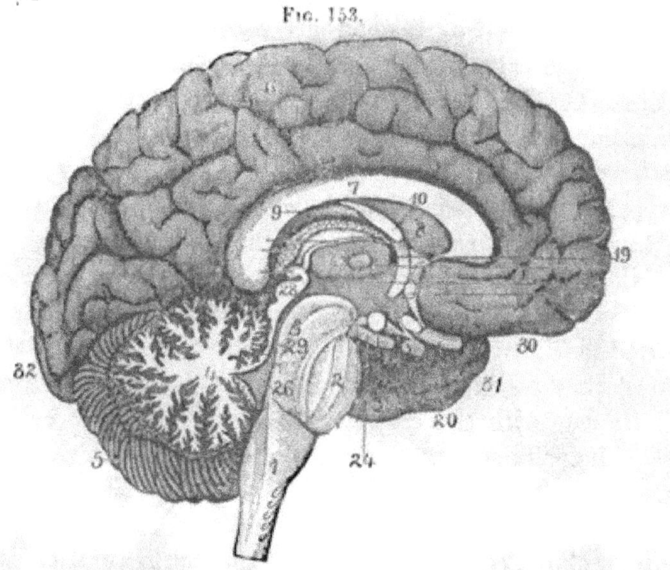

FIG. 153 (*Leidy*). SECTION OF THE BRAIN ALONG THE GREAT LONGITUDINAL FISSURE.—1, Medulla oblongata. 2, Pons. 3, Crus of the cerebrum. 4, Arborescent appearance in section of the fundamental portion of the cerebellum. 5, Left hemisphere of the cerebellum. 6, Inner surface of the left hemisphere of the cerebrum. 7, Corpus callosum. 8, Pellucid septum. 9, Fornix. 10, Anterior crus of the fornix. 19, Foramen of communication between the third and lateral ventricles. 20, Optic nerve. 24, Oculo-motor nerve. 26, Fourth ventricle. 28, Quadrigeminal body. 29, Entrance from the third to to the fourth ventricle. 30, 31, 32, Anterior, middle and posterior lobes of the cerebrum.

It will be noticed that these ganglia have an unbroken connection with the spinal cord through the peduncles of the cerebrum and the fibres of the medulla oblongata.

Extending backward from the optic thalamus, is a body divided on its upper surface into four eminences, hence called the *Corpora Quadrigemina* or the *Quadrigeminal body*. It consists of four small ganglia, sometimes named *Optic Ganglia* (as they send nerves to the eye), which are attached to the peduncles of the optic thalamus, to the cerebellum and cerebrum and to the medulla oblongata.

443. All the above-mentioned ganglia are variously connected with each other, with the peduncles of the cerebrum and cerebellum, and, through the medulla oblongata, with the spinal cord.

444. The hemispheres of the cerebrum are closely united in their central part by a transverse commissure, called the *Corpus Callosum*, formed by a dense band of transverse fibres radiating at each extremity to the inner surface of its corresponding hemispheres. The corpus callosum is arched in shape, and about four inches in length. It forms the roof of a large central cavity between the two ganglia, corpora striata, the cavity being divided by a thin, double membrane (the pellucid septum) into two communicating apartments called the Lateral Ventricles. Each of these has for its roof the corpus callosum; for its floor, the *Fornix*—a membrane continuous with the corpus callosum behind; for its inner wall, the pellucid septum; and for its outer wall, the corpus striatum. The floor of the lateral ventricles forms the roof of the *Third Ventricle*, which is a narrow cavity between the optic thalami, communicating with the fourth ventricle lying below and back of it, by a narrow passage-way. Hence it appears that the lateral ventricles, in the centre of the cerebrum, communicate with each other and with the third ventricle, the third with the fourth, and the fourth with the central canal of the spinal cord, making one unbroken communication through the whole extent.

445. Within the hemispheres are numerous other small ganglia, membranes and galleries, whose description our present limits will not allow.

446. The *hemispheres of the cerebrum* enclose all the other parts, in front, above and behind, like a great overshadowing dome. Their outer surface is of gray matter, hence they are

essentially two connected ganglia, and the largest in the system. Each hemisphere is marked off by fissures into three lobes, the frontal, middle and posterior lobe or ganglion, the frontal being the largest, and there is a little offshoot of the frontal lobe, called the *Olfactory*. Each of these lobes has its surface moulded into many tortuous and complicated elevations of the cerebral substance, termed *Convolutions*, which are marked off from each other by secondary winding fissures, named *Sulci;* thus there is formed "one unbroken but undulating sheet" over the whole surface of the brain.

FIG. 154.

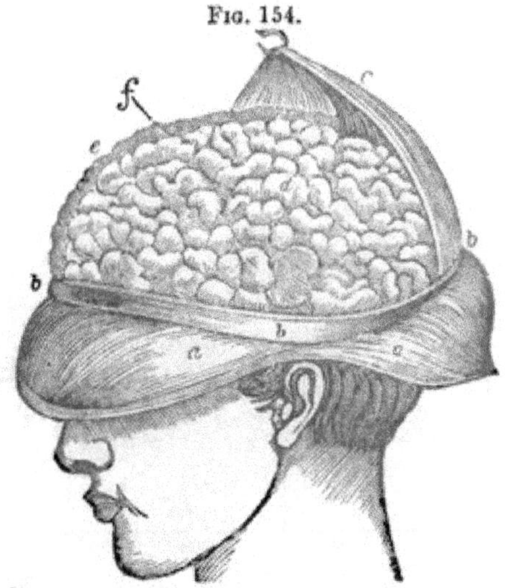

FIG. 154 REPRESENTS A CONVOLUTED CEREBRAL HEMISPHERE. *a*, *a*, The scalp turned down. *b*, *b*, *b*, The cut edge of the bones of the skull. *c*, The external strong membrane of the brain (dura mater), suspended by a hook. *d*, The left hemisphere of the brain.

447. The *general plan* of convolutions in the two hemispheres is the same, but in detail there is want of exact symmetry. It is a remarkable fact that the higher the mental development, the more unsymmetrical and complicated are the convolutions, and the deeper the depressions or Sulci. This is not only seen in comparing the lower animals with man, but in comparing different races of men. The brain of the "Hottentot Venus, who was no idiot," has been described as having the convolutions of the frontal lobe strikingly simple and regular, and as presenting an almost perfect

symmetry in the two hemispheres, such as is never found in the Caucasian race, and which much resembles that of the lower animals.

Fig. 155.

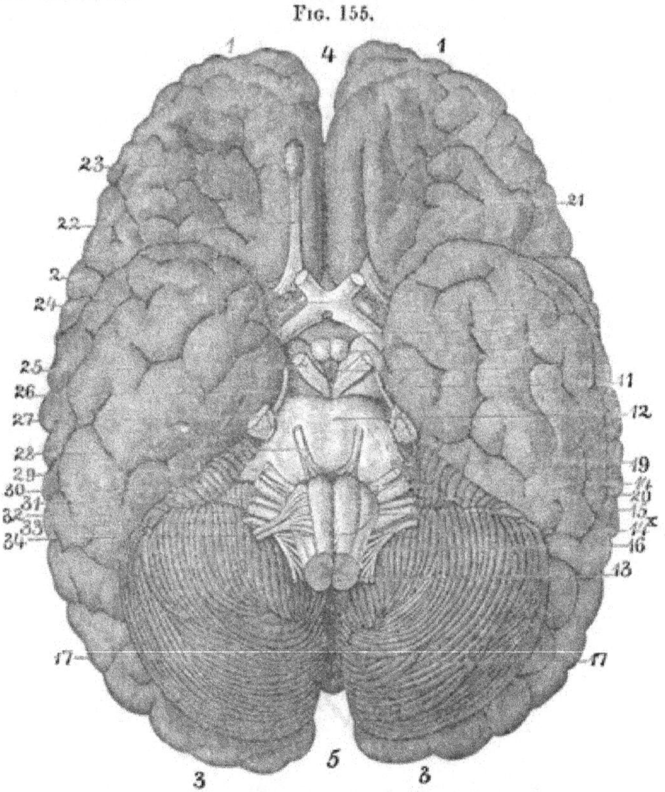

Fig. 155 (*Leidy*). Base of the Brain.—1, Anterior lobes of the cerebrum. 2, Middle lobes. 3, Posterior lobes. 4, 5, Anterior and posterior extremities of the great longitudinal fissure. 11, Crura of the cerebrum. 12, Pons. 13, Medulla oblongata. 14, Pyramidal bodies. 14*, Decussation of the pyramids. 15, Olivary body. 16, Restiform body. 17. Hemispheres of the cerebellum. 19, Crus of the cerebellum. 20, Pneumogastric lobule of the cerebellum. 21, Fissure which accommodates the olfactory nerve (22). 23, Bulb of the olfactory nerve. 24, Optic commissure. 25, Oculo-motor nerve. 26, Pathetic nerve. 27, Trifacial nerve. 28, Abducent nerve. 29, Facial nerve. 30, Auditory nerve. 31, Glosso-pharyngeal nerve. 32, Pneumogastric nerve. 33, Accessory nerve. 34, Hypoglossal nerve.

448. The cerebellum, like the cerebrum, has its hemispheres marked off into lobes. The lobes are highly subdivided on their sides and surface into thin plates or laminæ, by crescentic furrows or sulci. The white fibres within the cere-

bellum are so arranged that, when a vertical section is taken, it presents the appearance of the trunk and branches of a tree, and hence it bears the name of *Arbor Vitæ* (fig. 158).

449. The parts already described, viz., the brain and spinal cord, constitute the *Cerebro-Spinal Axis*, from which proceed

THE NERVES OF THE CEREBRO-SPINAL SYSTEM.

450. Certain of these nerves conduct nerve-force *from* the ganglia to their own distal ends in the tissues, chiefly muscular, where motion is produced; other nerves carry impressions from their extremities *to* the centres; the first are termed *Motory* from their function, and *Efferent* from the direction of conduction; the second are termed *Sensory* and *Afferent*. The anterior fibrous bundles of the medulla oblongata, passing upward to the corpora striata, form a *Motor Tract*, so distinguished by the endowments of the nerves that issue from it; the posterior bundles passing to the Thalami Optici form a *Sensory Tract*.

FIG. 156.

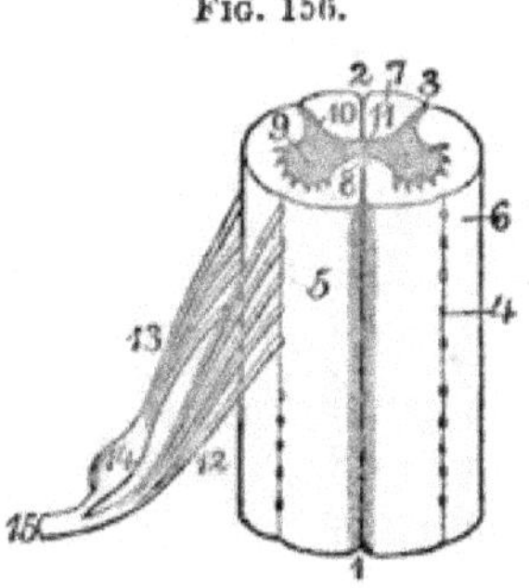

FIG. 156 (*Leidy*). SEGMENT OF THE SPINAL CORD.—1, Anterior median fissure. 2, Posterior median fissure. 3, Postero-lateral fissure. 4, Antero-lateral fissure. 5, Anterior column. 6, Lateral column. 7, Posterior column. 8, Anterior commissure. 9, Anterior horns of the gray substance. 10, Posterior horns. 11, Gray commissure. 12, Anterior root of a spinal nerve springing by a number of filaments from the antero-lateral fissure. 13, Posterior root from postero-lateral fissure. 14, Ganglion on the posterior root. 15, Spinal nerve formed by the union of the two roots.

451. The Cerebro-Spinal nerves are also distinguished as *Cranial* nerves when they pass directly from the brain, through openings in the cranium; and as *Spinal* when they issue from the vertebral openings of the spinal column.

The CRANIAL NERVES are arranged in twelve pairs, named numerically, counting from before backward, or from their function, destination or specific character. They may be arranged in three groups, according to their functions, as *Sensory*, *Motor* and *Mixed*.

452. Cranial Nerves.

Group	Pair	Name	Origin	Destination
1st Group. *Sensory.*	1st pair	Olfactory	Olfactory ganglia	Mucous membrane of nasal passages.
	2d "	Optic	Optic "	Retina of eye.
	8th "	Auditory	Medulla oblongata	Internal ear.
2d Group. *Motor.*	3d pair	Oculo-Motor	Cerebral peduncles	The muscles of eye, excepting external rectus and trochlear.
	4th "	Patheticus	" "	Trochlear muscle of eye.
	6th "	Abducent	Pons and medulla oblongata	External rectus " turning eye upward.
	7th "	Portio Dura	Medulla oblongata	Different muscles of face, giving expression.
	12th "	Hypo-glossal	" "	Muscles of tongue.
	11th "	Spinal Accessory	Spinal cord, low in the neck	Muscles of neck.
3d Group. *Mixed.*	5th pair	Trifacial	Pons Varolii	Motor branches to muscles used in mastication. Sensory " the teeth, tongue and different parts of the face. (See fig.)
	9th "	Glosso-pharyngeal	Medulla oblongata	Mucous membrane of the tongue and throat.
	10th "	Pneumogastric	" "	Motor branches to the pharynx, larynx, trachea, lungs, heart, œsophagus, stomach and intestines. Sensory to ditto. (See fig.)

FIG. 157.

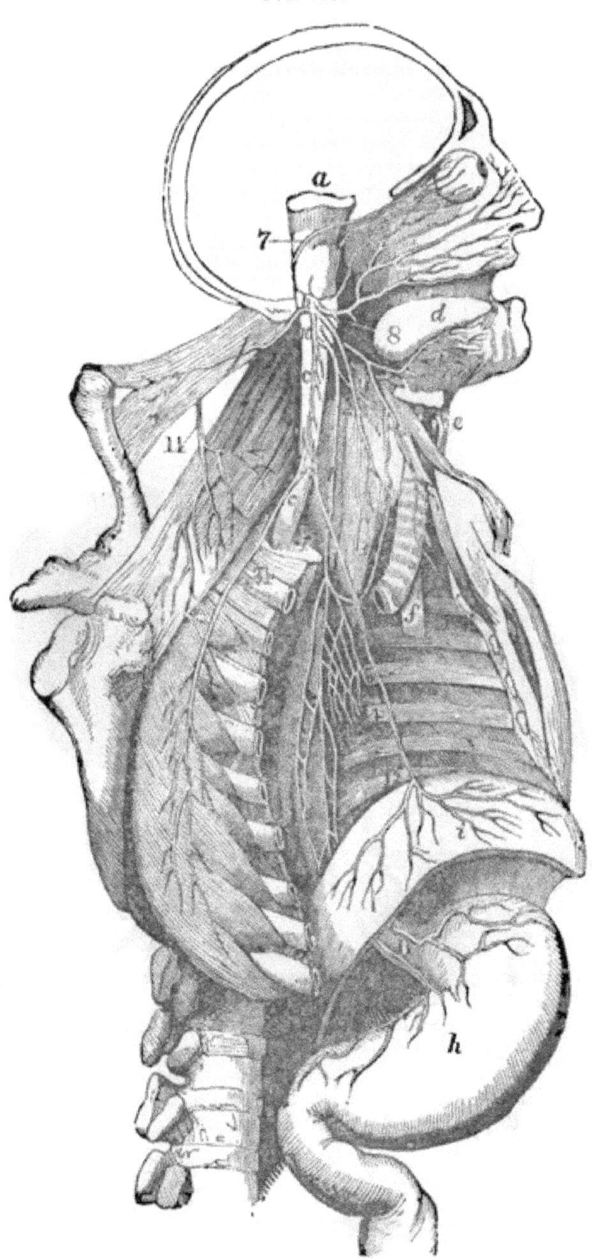

Fig. 157. Distribution of Pneumogastric Nerves.—*a*, Section of the brain and medulla oblongata. *b*, The lateral columns of the spinal cord. *c*, *c*, The respiratory tract of the spinal cord. *d*, The tongue. *e*, The larynx. *f*, The bronchia. *g*, The œsophagus. *h*, The stomach. *i*, The diaphragm. 1, The pneumogastric nerve. 2, The superior laryngeal nerve. 3, The recurrent laryngeal nerve. (These two ramify on the larynx.) 4, The pulmonary plexus of the tenth nerve. 5, The cardiac plexus of the tenth nerve. These two plexuses supply the heart and lungs with nervous filaments. 7, The origin of the fourth pair of nerves, that passes to the superior oblique muscle of the eye. 8, The origin of the facial nerve, that is spread out on the side of the face and nose. 9, The origin of the glosso-pharyngeal nerve, that passes to the tongue and pharynx. 10, The origin of the spinal accessory nerve. 11, This nerve penetrating the sterno-mastoideus muscle. 12, The origin of the internal respiratory or phrenic nerve, that is seen to ramify on the diaphragm. 13, The origin of the external respiratory nerve that ramifies on the pectoral and scaleni muscles.

Fig. 158.

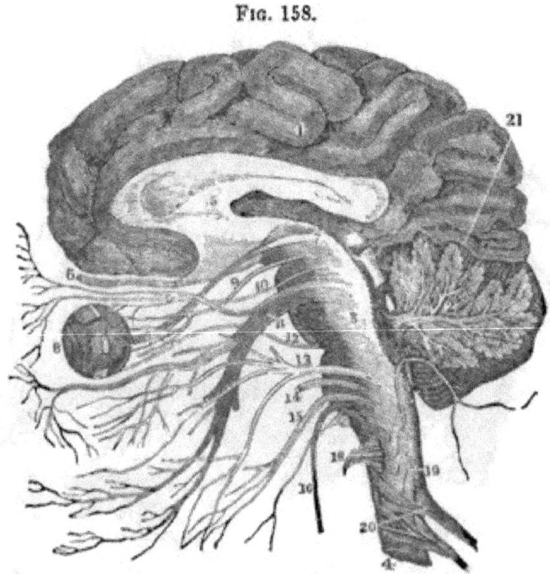

Fig. 158. A Vertical Section of the Cerebrum, Cerebellum and Medulla Oblongata, showing the relation of the cranial nerves at their origin. 1, The cerebrum. 2, The cerebellum, with its arbor vitæ represented. 3, The medulla oblongata. 4, The spinal cord. 5, The corpus callosum. 6, The first pair of nerves. 7, The second pair. 8, The eye. 9, The third pair of nerves. 10, The fourth pair. 11, The fifth pair. 12, The sixth pair. 13, The seventh pair. 14, The eighth pair. 15, The ninth pair. 16, The tenth pair. 19, The eleventh pair. 18, The twelfth pair. 20, Spinal nerves. 21 The tentorium.

453. The SPINAL NERVES are arranged in thirty-one pairs, and (unlike the cranial nerves, excepting the Trifacial) each arises by two roots: an anterior or *Motor* root springing from the anterior columns of the spinal cord, which are continuous with the *Motor tract* before mentioned; and a posterior or *Sensitive* root from the posterior columns of the spinal cord, and continuous with the *Sensory tract*. The Sensitive roots are larger than the Motor, and each has an imbedded ganglion, after the formation of which the two roots unite into one trunk, forming the spinal nerve, which passes out of the spinal column through the intervertebral openings.

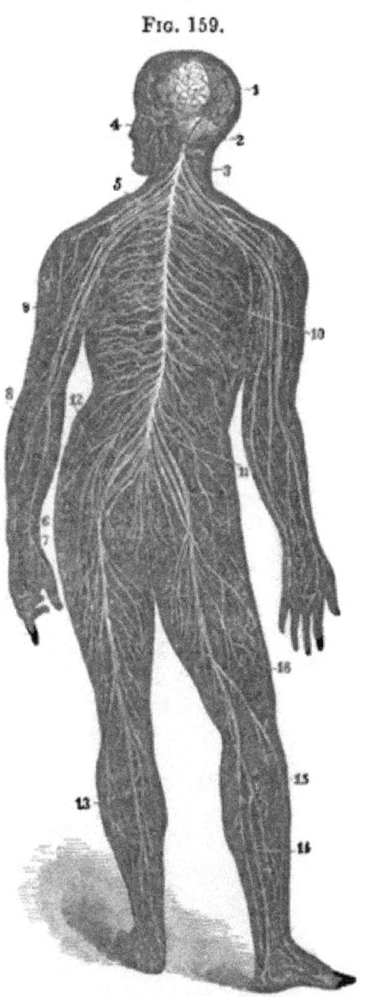

FIG. 159. A BACK VIEW OF THE BRAIN AND SPINAL CORD.—1, The cerebrum. 2, The cerebellum. 3, The spinal cord. 4, Nerves of the face. 5, The brachial plexus of nerves. 6, 7, 8, 9, Nerves of the arm. 10, Nerves that pass under the ribs. 11, The lumbar plexus of nerves. 12, The sacral plexus of nerves. 13, 14, 15, 16, Nerves of the lower limbs.

454. The Spinal Nerves are divided into—

Cervical........	8 pairs.
Dorsal..........	12 "
Lumbar........	5 "
Sacral...........	6 "

At some parts of their course certain branches of the nerves reunite, forming networks called *plexuses*. Thus the four upper cervical nerves anastomose, forming the *cervical plexus*, at the side of the neck; the four lower cervical, and the upper dorsal, form the *brachial* plexus (fig. 159), from which proceed six nerves which ramify upon the muscles and skin of the upper extremities; the last dorsal and four lumbar nerves form the *lumbar plexus*, which sends off six nerves to ramify upon the muscles and skin of the lower extremities; the last lumbar and four upper sacral form the *sacral plexus*, which distributes nerves to the muscles and skin of the hip and lower extremities.

Fig. 160.

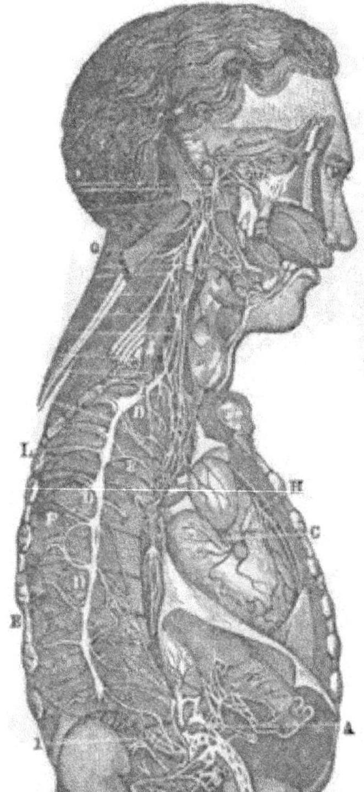

Fig. 160 represents the Sympathetic Ganglia, and their Connection with other Nerves, from the grand engraving of Manec, reduced in size. A, A, A, The semilunar ganglion and solar plexus, situated below the diaphragm and behind the stomach. This ganglion is situated in the region (pit of the stomach) where a blow gives severe suffering. D, D, D, The thoracic (chest) ganglia, ten or eleven in number. E, E, The external and internal branches of the thoracic ganglia. G, H, The right and left coronary plexus, situated upon the heart. I, N, Q, The inferior, middle and superior cervical (neck) ganglia. 1. The renal plexus of nerves that surrounds the kidneys. 2, The lumbar (loin) ganglion. 3. Their internal branches. 4, Their external branches. 5, The aortic plexus of nerves that lies upon the aorta. The other letters and figures represent nerves that connect important organs and nerves with the sympathetic ganglia.

THE SYMPATHETIC NERVOUS SYSTEM.

455. The Sympathetic system, like the Cerebro-Spinal, is double, consisting of two chains of ganglia, one on each side

of the spinal column, running through the deep parts of the neck, into the chest and abdomen. These ganglia communicate with each other, with the spinal cord and with the internal organs—as the heart, lungs, stomach, liver, pancreas, intestines and kidneys. In the neck and chest the ganglia are arranged in pairs; those of the neck are three in number and the largest of the system; those of the chest, twelve in number, a ganglion resting upon the head of each rib; in the abdomen the arrangement is irregular.

456. A peculiarity of the Sympathetic nerves is, that they *follow the distribution of the blood-vessels.* Starting from the heart, they envelop the large vessels with a close network, called the *Arterial plexus;* and in the abdomen, behind the stomach, the large blood-vessels are surrounded by many small ganglia, all united by networks of fibres called the *solar plexus,* because the other plexuses of the abdomen radiate from it, like the rays diverging from the sun. In all parts of the body, these nerves accompany the arteries which supply the different organs, and form networks around them which take the names of the organs, as the hepatic plexus, splenic plexus, mesenteric plexus, etc.

§ **45.** HISTOLOGY OF THE NERVOUS SYSTEM.—*Three Microscopic Elements of Nerve-Tissue. Nerve-Cells. Nerve-Fibres. Membranes of Cerebro-Spinal System.*

457. NERVOUS TISSUE is composed of three microscopic elements—*Nerve-Cells,* or *Ganglionic Corpuscles, White* or *Tubular Fibres,* and *Gray* or *Gelatinous Fibres.*

458. The NERVE-CELLS are *nucleated* cells; that is, vesicular matter containing, besides a pulpy substance, an eccentric, roundish body, or *nucleus,* enclosing one or more *nucleoli* surrounded by colored granules (32). These nerve-cells have various branches or offsets starting from any part of the cell-wall and completely continuous with it and with the contents of the cell itself. The branches connect the cells with each other, and also with the nerve-fibres. Their number varies from one to twenty, and the cells are accordingly dis-

K*

tinguished as *unipolar*, *bipolar* and *multipolar* (fig. 16). A collection of nerve-cells constitutes the essential part of a *Ganglion*. They are imbedded in a matrix of fine, soft, granular matter, and variously mingled and interwoven with multitudes of fibres. Composed of such masses, do we find the whole convoluted surface of the brain, the thalami optici, the corpora striata, the quadrigeminal body and some other minute bodies; from these, one unbroken, gray tract may be traced through the interior of the peduncles of the brain, the interior of the medulla oblongata and of the spinal cord (fig. 164). The various ganglia of the sympathetic system are also of the same substance.

FIG. 161.

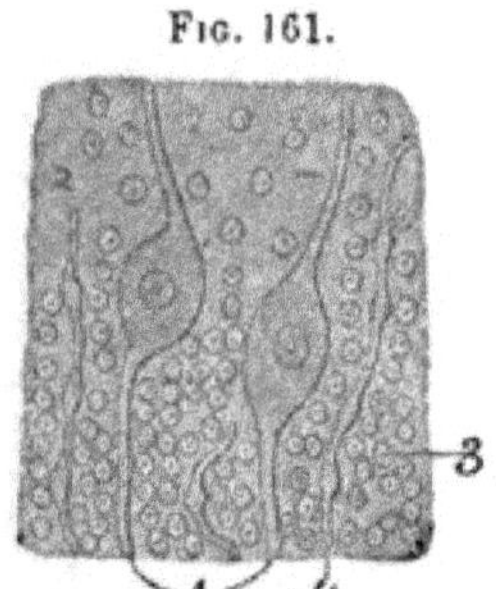

FIG. 161 (*Leidy*). PORTION OF GRAY SUBSTANCE, FROM THE EXTERIOR OF THE CEREBELLUM.—1, Two nerve-cells with bipolar prolongations. 2, Granular matter. 3, Nuclear bodies. 4, Nerve-fibres.

459. The WHITE or TUBULAR FIBRES, or the ultimate nerve-filaments, consist of an outer, structureless membrane enclosing a layer of transparent fluid fat, or medullary matter, within which is a firmer part —a gray, ribbon-like thread—called the *central band-axis*, or the *axis cylinder*. This is identical in structure with the processes of the nerve-cells with which it is continuous, and is very important, as it is sometimes the only part of the nerve-fibre left within the structureless sheath; thus constituting the so-called *pale, non-medullated* nerve-fibre. As the medullary matter encloses the band-axis, it is often, though improperly, called the *medullary sheath*.

FIG. 162.

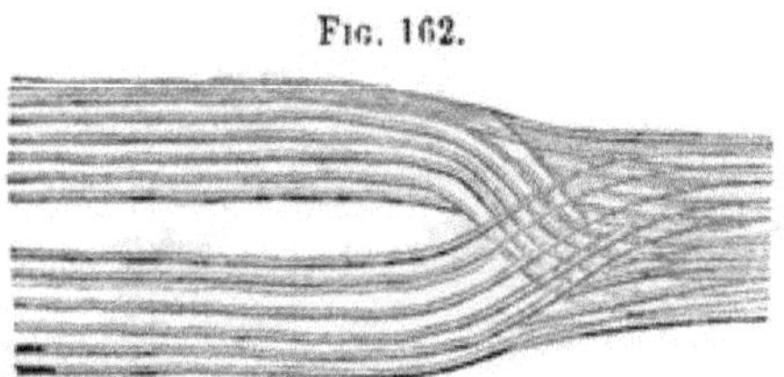

FIG. 162. NERVE-FILAMENTS, decussing with their sheath.

460. The nerve-filaments are distributed to the skin, muscles and glandular organs, in all parts of the body. From these points they approach each other, uniting into

little bundles or fibres, and then into larger bundles, till they are of sufficient size to be seen by the naked eye, when they constitute a *nerve*. The filaments do not blend with each other, but lie in simple juxtaposition, each retaining a complete individuality from its origin to its termination. Like the fibres of a muscle, they are bound together and protected by a covering of areolar tissue, called its *Neurilem'a*, or sheath, which also contains the blood-vessels for the nutrition of the nerve. The filaments become gradually finer toward their outer extremities, till at length the sheath, medullary portion and band-axis become undistinguishable. Their mode of termination is uncertain, though the sensory nerves, at least, seem to have free extremities.

FIG. 163.

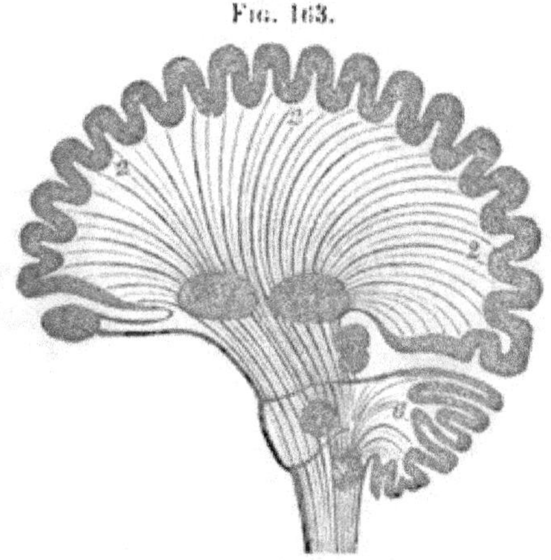

FIG. 163. DIAGRAM OF HUMAN BRAIN, IN VERTICAL SECTION, showing the situation of the different ganglia and the course of the fibres. 1, Olfactory ganglion. 2, Hemisphere. 3, Corpus striatum. 4, Optic thalamus. 5, Tubercula quadrigemina. 6, Cerebellum. 7, Ganglion of tuber annulare. 8, Ganglion of medulla oblongata.

461. The tubular fibres compose the white parts of the brain and the spinal cord; the chief substance of the nerves; and also pass into and mix with the gray substance of the brain, cord and all the ganglia. They vary in size, being finest of all in the superficial layers of the brain, fine in the nerves of special sense and in the ganglia, larger in the fore part of the spinal cord, and largest in the motor nerves.

462. Besides the White tubular fibres, there are found,

chiefly in the Sympathetic System, Gray or Gelatinous Fibres, which are flattened, more minute than the white fibres, and greatly resembling their band-axis. Some have considered these but a form of connective tissue, but whether they be so considered, or as true nervous elements, they seem to be produced by the coalescence of elongated nucleated cells, the contents of which, as the cells enlarge, become soft and finely granular, while the nuclei appear wider and wider apart.

463. The Membranes of the Cerebro-Spinal System are four in number—the *Dura Mater*, the *Pia Mater*, the *Arach'noid Membrane* and the *Epen'dyma.* The Dura Mater is a tough, fibrous membrane lining the bony walls of the skull and spinal column, forming their *periosteum.* The Pia Mater is another fibrous and very vascular membrane which closely invests the brain and spinal cord, and sends processes into all their fissures. The inner surface of the dura mater, and the outer surface of the pia mater, each become very delicate in structure and are lined with an epithelium: this gossamer membrane is named the Arachnoid Membrane. Its two layers unite at many points, thus forming closed sacs, which, like other serous membranes, secrete a fluid called the *arachnoid fluid.*

464. The dura mater not only firmly invests the brain and spinal cord, but sends off supporting partitions—that which descends between the hemispheres of the cerebrum being called the *Cerebral Falx;* that between the hemispheres of the cerebellum, the *Cerebellar Falx;* and that between the cerebrum and cerebellum, the *Tentorium.* Through separations in the layers of the dura mater, channels are formed, performing the office of veins: they are named *Sinuses of the Dura Mater,* and are lined with a continuation of the ordinary epithelium of blood-vessels. The dura mater also furnishes the areolar sheaths to the several cranial and spinal nerves; therefore it is continuous from the lining of the cranium to the extremity of the nerves in the different parts of the body.

465. The *Ependyma* is a delicate, transparent, serous membrane, lining the ventricles of the brain and the central canal of the spinal cord.

§ 46. PHYSIOLOGY OF THE NERVOUS SYSTEM.—*Man's Compound Nature. Relation of the Nervous System to this Nature. The Rank of the Nervous System. Relation of the Nervous Centres to the Sensitive and Motor Nerves. Classification of the Centres. System of Dependencies. General Function of the Organic Centres. Their Modes of Reflex Action. Peculiarity of Sympathetic Action. Functions of the Reflex or Spinal Centres. Their Acquired Action, and the Theory Explaining it. Practical Importance of the Automatic Tendency produced by Repetition and Association. Character of the Sensational Centres and their Action. Internal Stimuli to the Activity of these Centres. Functions of the Ideational Centres. Ideas Suggested by the Same Object different in Different Individuals. Various Manifestations of Reflex Action in the Ideational Centres. Emotional Character of these Centres. Volitional Character. Relation of the Emotions to the Will. Influence of the Physical Nature for Good or for Evil. The Language of the Muscles.*

466. At different periods of the world's history, many different opinions have prevailed concerning the respective existence of *body* and *soul*, and their relations to each other. The pagan Greek included all under the one word ψυχή and the Roman under that of *anima*, which was almost "equally applicable to the vegetative life of a cabbage, the animal life of a sheep, and the spiritual life of an apostle." During the fifth century before the Christian era, Anaxagoras advanced a shadowy idea of man's compound nature, which at the day-dawn of Christianity assumed a clear and definite outline. At length philosopher and Christian advocated the supremacy of the immaterial nature over the material, and eventually regarded their interests as antagonistic. The body was deemed the source of all evil—the work of the Prince of Darkness. At the present day, more than at any former period, efforts are being made to rightly balance the two natures, and yet many seem to regard the body as a gloomy prison-house in which God has shut us up, rather than as a beautiful "temple" in which the mind and soul may dwell as priest and priestess, using all its appointments in rendering service to the Lord of the temple.

467. The NERVOUS SYSTEM is the border-land where the material touches the immaterial. It possesses that highest

refinement of physical organization through which the mind may manifest itself, and by means of which it may control and bring into service, not only the various organs of the body, but other matter more external and remote.

468. The organisms heretofore described have no *inherent* active power, but are entirely dependent upon the nervous system: thus, the bones are dependent for movement upon the contractility of the muscles; this contractility, upon the stimulus of the nerves; this stimulus, upon the energetic action of the nerve-centres; and these centres are graded in rank and measurably dependent, the lowest upon the next higher, and so on to the highest, or convoluted centres of the hemispheres.

469. In their function, the *nervous centres* are intermediate between the sensitive and motor fibres; as the sensitive fibres, being acted upon at their distal extremities, convey impressions *inward to the centre;* and the motor fibres, being acted upon at the centres, convey nerve-force *outward*, and produce motion *at their distal extremities.* Let any part of the surface of the body be touched by a hot iron, and muscular contraction instantly follows; but there has been time enough for the sensation of pain to be conveyed to the nervous centre, and for an impulse to be sent from that centre to the muscles: such action is called the *Reflex Action of the Nervous System.* By this means a communication is established between the different organs. This communication is never direct, but from one organ inward to the nervous centre, then outward to another organ: so are the different functions associated and exercised for the common good of the whole.

470. In dealing with the functions of the Nervous System we adopt the following classification of the Nervous Centres: viz.—1st, The *Primary* or *Ideational* Centres, comprising the gray matter of the convolutions of the hemispheres (446); 2d, The *Secondary* or *Sensational* Centres, comprising the gray matter between the floors of the lateral ventricles and the decussation of the pyramids (442); 3d, The *Tertiary* Centres, or Centres of *Reflex Action,* comprising the gray

matter of the spinal cord (441); 4th, The *Quarternary* or *Organic* Centres, comprising the gray matter of the Sympathetic System (455).

471. The arrangement of this system of centres is like that of a well-ordered body politic. Each distinct department, or nerve-centre, acts independently within certain limits, but beyond these limits it is subordinate to the next higher: thus, the Organic Centres are subordinate to the Reflex or Spinal Centres; the Reflex, to the Sensorial; and all, to the Ideational or Supreme Centres. In each centre the individual cells probably differ in rank, some having a higher dignity, some a lower, but each its special appointment, its assigned duty. Such would be the inference from their varied form, color, size, and mode of branching. There are probably important differences of chemical constitution and action, but of this we have no means of proof. There is, then, reason for supposing that from the lowliest cell in the Organic centres, to that of highest rank in the Ideational, there is a long series of dependencies, and, so nice is the adjustment, that if one fails to conform to the laws of the organization the others must suffer. Slight disturbances may take place in the lower centres without the knowledge of the Supreme Authority, but any serious matter beyond their control is early reported; this is the meaning of pain, weariness, etc. If the warning is disregarded at headquarters, there is liability to an open rebellion that will shake the system to its foundations, and not unlikely result in its complete overthrow. "The well-being and power of the higher individuals are entirely dependent upon the well-being and contentment of the humbler workers, which do so great a part of the daily work of life."

472. The Organic or Sympathetic Centres are not well understood, but the distribution of their nerves would indicate that they exercise a controlling influence over the involuntary functions of digestion, absorption, circulation and assimilation. From the fact that these nerves reach their ultimate destination supported on the arterial vessels,

it is probable that their influence is exerted through a certain control over the muscular coat of the heart and arteries, thus hastening or retarding the course of the blood, and increasing or diminishing its quantity in various organs. Thus the functions of nutrition, secretion, etc., depending so much upon the state of the circulation, are made to sympathize with each other very closely; hence the name, "Sympathetic" System (456).

473. The organic centres being connected with the various organs by sensitive and motor nerves, are capable of an *independent reflex action.* They are also connected with the cerebro-spinal system, and are more or less assisted by and subordinate to it. In health the brain takes no cognizance of their action; when diseased, however, the centres report to the highest authority by means of cramps and other severely acute pains. In its normal action, a centre seems to expend only so much force as is disposed of by the motor nerves; in diseased action there is a surplus, which is conveyed to the next higher centre, to be disposed of by its motor nerves; if there is still a surplus, it passes on as before.

474. There are *three kinds of reflex action* taking place either wholly or partially through the Sympathetic System; viz.—1st, The reflex action from the internal organs to the voluntary muscles and sensitive surfaces: examples are seen in the convulsions of children, caused by the irritation of undigested food in the intestines; and in adults, in the attacks of temporary blindness or confused vision so often accompanying indigestion. 2d, The reflex action from the sensitive surfaces to the involuntary muscles and the internal organs; as mental and moral impressions received by the senses disturb the motions of the heart and affect the circulation, digestion and secretion; disagreeable sights or odors produce nausea and other functional derangements. 3d, The reflex action between the internal organs; as the associated action of the stomach, liver, etc. The variation in the capillary circulation of the abdominal viscera, according as they are active or inactive, is probably referable to a similar influence.

475. One marked peculiarity of the Sympathetic System is, that its nerves and ganglia act with much less rapidity than those of the Cerebro-Spinal System; hence, inflammation of the internal organs is not manifest for several hours after the application of the exciting cause; as the effects of a chill or cold do not usually follow immediately after the exposure. Because of this tardy action, the effect remains long after the cause is removed. A very beautiful example of the slow action of the sympathetic nerves is seen in the movements of the iris of the eye. The ciliary nerves controlling these movements originate in the brain, but pass through, and are affected by, a sympathetic ganglion. In passing from the dazzling sunshine into the house, we are scarcely able to distinguish objects about us, and some minutes are often required to adapt the iris to the less amount of light; and the same slow movement is evident in passing from a less degree of light to a greater. Were these nerves purely *cerebro-spinal*, the action would take place instantly.

476. The Tertiary, Reflex or Spinal Centres. The white, tubular substance of the spinal cord connects the muscles and integuments below with the brain above, and thus assists in the production of conscious sensation and voluntary motion. The gray matter forms nerve-centres, which *exert a general protective influence over the whole body.* They *preside over the involuntary movements of the limbs and trunk;* if a finger touch a heated surface, it is suddenly withdrawn, and that without effort of the will, and often in opposition to it. The same movement takes place upon tickling the foot of a person asleep. They *regulate the action of the sphincter muscles,* as in the rectum and bladder. They *exercise a certain control over the changes of secretion, nutrition, etc.,* as is manifest in cases of disease. Thus we see that many human activities are performed by the *reflex action* of the *spinal centres,* inherent in their *natural* constitution.

477. They are, however, capable of an *acquired* reflex action, which is matured through experience. An act or an association of acts becomes easier to them by repetition.

This acquired power of reflex action has been accounted for by a theory* which is at least beautifully *illustrative* of the *facts* in the case. Every display of energy in the nerve-cells causes a change or waste of nervous element which is repaired by nutrition. This theory assumes that the *character of the waste determines the character of the deposit;* that the particle deposited is necessarily *endowed* according to the particular kind of activity manifested, and that this endowment *inclines* the particle to the *same kind* of activity again. By each repetition, the *tendency* becomes stronger and more definite, till, after a longer or shorter series of repetitions, the action becomes *automatic.*

478. When a certain class of movements have, after many voluntary efforts, become *associated*, they become perceptibly more and more easy. Walking is at first a very conscious and voluntary act; but it may become so far reflex and automatic that one in a profound abstraction may continue to walk without being at all conscious where he is going, and when he wakes from his revery may find himself in some other place than that which he intended to visit. Multitudes of our daily acts are the result of this acquired reflex action of the spinal centres. The wisdom of such an arrangement is very evident, for but little could be accomplished if acts became no easier by repetition and association. Conscious efforts of the will soon produce exhaustion, while the automatic acts of which we are speaking occasion comparatively little weariness. We often say of certain rounds of duties that they do not weary us, for we are *accustomed* to them. In speaking of this acquired power of which the spinal centres are capable, Dr. Maudsley says, "Like the brain, the spinal cord has its *memory.* A spinal cord without memory would be an *idiotic* spinal cord, incapable of culture —a degenerate nervous centre in which the organization of special faculties could not take place. It is the lesson of a good education so consciously to exercise it in reference to

---

* Dr. Maudsley.

its surroundings that it shall act automatically in accordance with the relations of the individual in his particular walk of life."

479. The SENSATIONAL CENTRES, including the gray matter of the medulla oblongata, and of the base of the brain as far as the lateral ventricles, consist chiefly of the nervous centres of the higher or *special senses*, as sight, hearing (442), etc. Any one of these senses is quickly destroyed by destroying its ganglion: the loss of the quadrigeminal body destroys the sight as effectually as putting out the eyes. That these centres have an independent reflex action may be seen by the involuntary closure of the eyelid when a strong light falls upon the eye, or by the involuntary contortions of the face when an article is sour or bitter to the taste. These are examples of *natural* reflex action, but, like the spinal cord, these centres are capable of an *acquired* reflex action; as in the articulation of words upon seeing their signs; adapting the movements of the body to the rhythm of music, in dancing, marching, etc. Most of the sensations of the special senses become clear and definite only after a long course of training; for instance, the visual sensation of the adult is a very different matter from that of the child whose eyes have recently opened upon the world. "The sensation of the cultivated sense thus sums up, as it were, a thousand experiences, as one word often contains the accumulated acquisitions of generations."

480. In speaking of the acquired reflex action of the spinal centres, we referred to the theory that a *relic*, or *residuum*, of every activity remained in the nerve-cell as a special endowment; that perhaps the character of the activity determined the character of the nutritive deposit. This theory is equally applicable to the sensational centres, and equally illustrative of the certain fact that acts of this class are rendered easier by repetition.

481. The sensational centres are excited to activity not only by impressions from the organs of the special senses, but by sensations from *within the body*, both from the organic

and ideational centres. Of the former, examples are afforded when the higher nervous centres are weakened by disease, or when the organic stimuli have an unnatural activity, as is the case with the intemperate man.

482. The IDEATIONAL CENTRES seem to have the power of fashioning into ideas the impressions received by the sensational centres. When the various properties of an object are presented by the different senses, these centres reject the unessential, and, selecting the essential, mould them into an organic unity, or idea. By means of the sensorial centres and nerves, we may gain perceptions or impressions of the qualities of a rose, but these would be isolated, and we should have no clear and definite idea of the rose, without the moulding and vivifying influences of the ideational centres.

483. Different persons obtain very different ideas of the same object; the character of the idea being dependent upon the character of the organization both of the sensational and ideational centres, and the character of the organization upon natural endowment, or *inherited organization*, and also upon the education received.

484. The ideational centres, like those already described, are capable of an independent, reflex action, which may be manifested in different ways: 1st, This may take place *through the motor tract*, thus giving rise to what have been named *ideomotor* movements. This energy may be exerted either upon the voluntary or involuntary muscles, and in the former case either with or without consciousness: the idea that vomiting must take place when a qualmish feeling exists will hasten or even produce vomiting, affording an example of the reflex action of an idea upon the involuntary muscles, conformable to what has been said of the subordination of the organic nervous centres to those of the cerebro-spinal system. Examples of the reflex action of ideas upon our voluntary muscles are seen every hour of our waking life; these may be unconscious, as is seen in most persons who talk to themselves, or they may be *conscious*, and yet without the intervention of the will, as when a quick-tempered indi-

vidual quickly resents an insult by a blow. 2d, The reflex action of an ideational nerve-cell may not only operate downward upon the muscular system, but downward upon the *sensory centres:* the idea of a nauseous taste may excite the sensation to such a degree as to produce vomiting. The action of ideas upon our sensory ganglia is indeed a regular part of our mental life, for the co-operation of sensory activity is necessary to clear conception and representation, and by it we may see our own ideas as actual images. Those great writers who delight us with their vivid descriptions of scenery or events possess this power in a high degree. 3d, Another very important reflex action of these centres is that which modifies the *secretions and nutrition:* a flow of saliva may be produced by the thought of food, or a flow of tears by a sympathetic idea. 4th, There may be in these centres a reflex action *among the cells themselves.* One cell reacts to a stimulus from a neighboring cell, then transfers or *reflects* this energy to another. This may be the condition of activity among these cells during that process of the mind which we call *Reflection.*

485. These ideational centres are also the seat of the *Emotions.* When an idea is attended with some feeling, either pleasant or unpleasant, it is so far *Emotional;* and when the feeling *preponderates,* the idea is obscured, and the state of mind is then called an *Emotion,* or, when rising above the ordinary degree and becoming impatient of restraint, a *Passion.* The capacity for emotion depends essentially upon the range and vigor of ideas. The man of great strength of mind, as a Milton or Napoleon, is capable of deeper emotion than the man of dwarfed and puny intellect. Indeed, just here lies, in no small degree, the secret of his superior power. The same stimulus may at one time produce simply an idea, and at another time an emotion, according to the condition of the nerve-cells.

486. Every centre of idea is also a centre of *Volitionary* reaction. When an idea acts *directly* downward, we call the effect *ideomotor;* but when there is deliberation or reflection

delaying the action, and it afterward takes place downward, we call the effect *volitional.* Volition is also exercised in preventing as well as in producing an action.

487. The exercise of the *Will* is the highest energy of which the supreme centres are capable. Within certain limits, the ideas and emotions are subject to its control. Suppose a being endowed with the intellectual and emotional natures, but not with the will: though possessing the intelligence of a man, his capacities for action would be inferior to those of the brutes, for, like them, his actions would be the result of mere sensational impulses, and yet he would be destitute of that natural guide of brutes which we call *instinct.* This represents the wretched condition of a man whose will is by any means so enfeebled that it fails to control the mental and physical powers.

488. The *power* of the will depends both upon the inherited organization and also upon the training it has received, for volitions, like sensations and ideas, become more easy and definite by repetition. A naturally weak will may be greatly strengthened by due care and training. According to the theory before mentioned, each volition leaves its relic, trace or residuum which inclines the portion of nerve-element exercised to a like activity again. If we accustom ourselves to decide promptly, to act energetically and to carry out our purposes in the many smaller and less important affairs of life, we gain a power of will which may be carried into higher departments of action and into circumstances of greater embarrassment and difficulty.

489. The Will bears very important relations to the *Emotions.* If they are allowed to react independently, as is their natural tendency, they weaken the will; if duly controlled and co-ordinated, they strengthen it. The passionate nature of the child may, by proper training, become a potent force for good in after years, "giving a white heat, as it were, to the expression of thought, an intensity to the will." Untrained, it will become a no less potent force for evil, and the individual under the mastery of his passions will

be tossed about as helplessly as a boat in the rapids of Niagara.

490. The free action of the will requires an unimpeded association of ideas, and the ease and completeness of such an association depends upon the condition of nervous element, very slight disorders of which declare themselves in the deterioration of the will-power. As in the spinal centres disturbance of the nerve-element weakens their control over movements, so in the ideational centres disordered nerve-element is quickly manifested in the loss of will-power; and as in great disorder of the spinal centres all co-ordinating power is lost and convulsions ensue, so in great disorders of the ideational centres all co-ordinating power over the thoughts and feelings is lost, convulsive reactions of the cells take place, and the individual becomes *insane.*

491. We have seen that the mind is closely *united* and yet distinct from the material organ through which it acts—dependent for its manifestations, but independent in essence. So intimate is the union, that the body exercises a powerful influence in leading us upward into a true and higher life, or downward into a low and sensual existence. What this influence shall be depends somewhat upon inherited organization, but more upon education. Accepting the theory already advanced as at least illustrative, we see that if the thoughts, feelings and volitions are pure and true and good, their impressions or residua remaining in the nerve-cells are of the same character, and tend to give a right direction to the future activities of these cells. If the thoughts, feelings and volitions are evil in nature, the impressions or residua will also be evil, inclining to evil activities in the future. When we resist a temptation to wrong action, then we not only avoid the particular evil, but lay up that which will render the next resistance easier and more natural. If we yield to the temptation, we are not only guilty of the particular wrong, but lay up that which will make resistance more difficult or yielding more easy and natural for the future. When a man sets his *heart* to do right, all his physical being struggles

to give him aid; and when he sets his heart to do wrong, its energies are expended in dragging him downward.

492. The *visible impress* which the workings of the mind leave upon the body is worthy our notice. The character of the man is declared by the lines of his muscles, which tell no lies. Especially is this true of the muscles of the face. Let him narrow his soul by penuriousness, become the victim of rasping jealousy, wear the nettles of envy against his heart, or be the slave of defiling lust, and in spite of any natural comeliness or studied concealment, his true character will be proclaimed to all who have learned aught of the language of the muscles. "Be sure your sin will find you out," says He who has made the fleshly lineaments to reveal the most hidden vice. The more secret the viciousness, the deeper is the impress. But if the spirit of evil thus leaves the traces of its blackened pen upon the face, the spirit of goodness writes thereon in no less legible characters of light. Purity of heart, nobleness of purpose, restfulness of soul, soften, irradiate, *spiritualize* the outer man, giving a higher beauty than that of form or complexion, even to him who is wrinkled by years, bowed by infirmity and scarred by the battles of life.

§ **47.** HYGIENE OF THE NERVOUS SYSTEM.—*Two Classes of Agencies Affecting the Health of the Nervous System. Natural Heritage. Importance of the Physical Agency—Air—Diet—Exercise and Sleep. The Effect of Mental Impressions on the Body. Mental Exercise. Recreation and Amusement. Harmonious Development of the Different Mental Powers.*

493. We have seen that different organs of the body are entirely dependent for functional action upon the stimulus afforded by the nervous system; and since this is the material organization through which the mind acts, we are led to the inevitable conclusion that the physical condition of this system must affect, more or less, the *mental* manifestations. It becomes, then, a matter of primary importance that we

understand the conditions essential to the health of this system, especially as suffering from nervous disease exceeds that of other diseases, as the delicacy of the organization exceeds that of other organizations of the body.

494. In considering the hygiene of the nervous system, it is necessary to have reference both to physical and mental agencies. The highest health and vigor of the nervous system doubtless require—1st, A sound nervous organization by inheritance. 2d, A nutrition equal to the demands of repair and growth. 3d, The harmonious action of the various mental powers.

495. 1st, A Sound Organization by Inheritance. "Each of us is only the footing-up of a double column of figures that goes back to the first pair," is the striking expression of a great truth. Every-day observation shows that children inherit not only the features, but the physical, mental and moral constitution of their parents. Even those utterly ignorant of the laws of transmission are wont to estimate the child according to its family; favorably, if of a "good family" or "good blood;" unfavorably, if of a "bad family" or "bad blood."

Every formation of body, internal and external, all intellectual endowments and aptitudes, and all moral qualities, are or may be transmissible from parent to child. If one generation is missed, the qualities may appear in the next generation. It is important to notice that not only the *natural constitution* of the parents may be inherited, but their *acquired habits* of life, whether virtuous or vicious, but especially is this true of vice. Even when the identical vice does not appear, there is a morbid organization and a tendency to some vice akin to it. Not only is the evil tendency transmitted, but what was the simple practice, the *voluntarily* adopted and cherished vice of the parent, becomes the passion, the overpowering impulse of the child.

496. M. Morel sketches the history of four generations as follows: "First Generation.—The father was a habitual drunkard, and was killed in a public-house brawl. Second

Generation.—The son inherited his father's habits, which gave rise to attacks of mania, terminating in paralysis and death. Third Generation.—The grandson was strictly sober, but full of hypochondriacal and imaginary fears of persecutions, etc., and had homicidal tendencies. Fourth Generation.—The fourth in descent had very limited intelligence, and had an attack of madness when sixteen years old, terminating in stupidity nearly amounting to idiocy; with him the race probably becomes extinct."

497. Says a learned physician, after long and close observation of the evil effects of tobacco: "If the evil ended with the individual who, by the indulgence of a pernicious custom, injures his own health and impairs his faculties of mind and body, he might be left to his enjoyment, his *fool's paradise,* unmolested. This, however, is not the case. In no instance is the sin of the father more strikingly visited upon the children than the sin of tobacco-smoking. The enervation, the hysteria, the insanity, the dwarfish deformities, the consumption, the suffering lives and early deaths of the children of inveterate smokers bear ample testimony to the feebleness and unsoundness of the constitution transmitted by this pernicious habit."

498. Should we trace the effects of the whole list of vices, it would be with equally sad results; even of the great love of money-getting, the celebrated Dr. Maudsley says: "I cannot but think, after what I have seen, that the extreme passion for getting rich, absorbing the whole energies of a life, does predispose to mental degeneration in the offspring, either to moral defect, or to moral and intellectual deficiency, or to outbreaks of insanity."

499. Any kind of nervous disease in the parents, whether natural or acquired, seems to predispose to innate feebleness in the child. From this instability of nervous element, the slightest irritation often produces convulsions in the young child and loss of equilibrium in the adult. Such a natural constitution may be improved by a judicious education and strict obedience to physical and men-

tal laws; but the original defect can never be entirely removed.

500. 2d, A Nutrition equal to the Demands of Repair and Growth. The relation of the nervous centres to the blood is the same in kind as that between other parts of the body and their blood-supply. Great waste is produced by nervous action; hence, the centres are very largely supplied with blood-vessels, especially the Ideational centres. The activity of ideas is largely dependent upon the active flow of blood to the nerve-cells. Activity of thought invites the blood which, in turn, is so necessary to activity. The nerve-centres, then, must be supplied with the proper quality and quantity of blood; hence, whatever deteriorates the blood impairs the health of the nervous system. It is evident, then, that—

501. *The nervous system may be impaired by impure air.* Everybody knows that bad air injures the lungs, but few realize that, on the whole, it injures the brain still more. As the nerve-tissue is the most delicate part of the body, it soonest feels the evil effects of imperfectly oxygenated blood. (See Respiration.)

502. *The nervous system may be impaired by improper diet.* We are wont to believe that improper diet may affect the digestive organs, but seldom consider the *mental* and *moral* effects of such diet. Improper food poisons the blood, and thus the nerve-centres are cheated of their nutriment and also poisoned; hence, the ideas become confused, the emotions morbid and the will weakened. The whole man is crippled, physically, mentally and morally. It is an indisputable fact that *bad bread*, for instance, may thus have a very *immoral* influence. Those much engaged in mental labor suffer most from bad diet. No teacher can teach well, no lawyer can plead well, no physician can practice well, no minister can preach well, who habitually takes improper food. (See Digestion.)

503. If such be the effect of improper food, what shall we say of such poisons as alcohol, opium, haschish, tobacco, etc.,

which act so directly and powerfully upon the nervous system? The same poison does not equally affect all the nerve-centres; thus, strychnine acts upon the spinal centres, but not the cerebral; haschish, upon the sensory centres, giving rise to hallucinations; alcohol, upon the cerebral centres particularly. The alcoholic poison first produces an increased activity of the muscles, then alternate exaltation and depression, both physical and mental; finally, stupor, relaxation of the muscles and deep sleep. These symptoms are transitory; but let the poisoning process be continued, and true delirium, so well known as "delirium tremens," follows, and at length what is known as "chronic alcoholism;" and while intoxication lasts a few hours, and delirium tremens a few days or weeks, chronic alcoholism spreads its baneful influence over years, unless death prevents the full development of the tragedy. The victim of alcoholic poisoning is equally enfeebled in body and mind. The nervous system becomes exhausted, the moral sentiments perverted, the will-power broken, and he seems powerless to cease from the fatal habit which has produced the change.

504. With the *opium-eater* the diseases of the nervous system declare themselves even more rapidly than with the drunkard. Says M. Morel: "Given the period at which a person begins to smoke opium, and it is easy to predicate the time of his death; his days are numbered."

505. *Tobacco* is one of the most virulent poisons. It soothes the nerves temporarily, only to leave them more enfeebled and irritable.

Even excessive use of *tea* and *coffee* may prove disastrous to the health of the nervous system.

506. *The nervous system may be impaired by want of physical exercise.* Among other agencies that affect the nervous system, none exert a wider influence than bodily exercise. It seems to be required to complete the change which the blood undergoes while passing through the lungs and skin, without which the waste of nerve-element could not be repaired. In persons who are merely sedentary, having little occasion for

active thought, this want of exercise is sufficiently mischievous; but when there is great mental activity, the mischief is vastly increased. Thousands of ministers, lawyers, those who sit in the bank and counting-room, shorten their days because of this neglect; especially is this the case in America. The English nobility, notwithstanding their many indulgences, are a long-lived race, and this is doubtless owing to their spending so much time in open-air exercise (208).

507. *The nervous system may become impaired by taking too little sleep.* "Sleep knits up the raveled structure" of nervous element; for during sleep, organic assimilation is restoring what has been expended in functional energy. A periodical renewal of nervous energy as often as once a day is an institution of Nature. Among the wise arrangements of the Creator, none harmonizes with the wants of the system more perfectly than the alternation of day and night. The amount of sleep necessary depends upon the age, health, natural temperament and occupation of the individual. The more rapid the exhaustion of nervous energy from any cause, the more sleep will be required. The young and the aged need more sleep than the person of middle life; the sick, more than the well; those engaged in mental pursuits, more than those wearied by manual labor; persons of great sensibility, more than the sluggish natures whose normal condition is more nearly allied to sleep; woman, more than man. We may say in general that the time should not be less than from six to eight hours, and most persons require a longer period. The time, however, must be proportioned to the need.

Among the more affluent classes, the customs of the times are quite incompatible with those habits of sleep which are essential to mental vigor. Where amusements are pursued till late hours, night after night, the nervous system greatly suffers, and every department of the mind becomes unhealthy. The man who, eager to become rich, takes time from his sleep for business purposes, draws from his brain capital. The mother—alas! here we must stop. Mothers are the one class who hardly get any rest till the "blessed Father takes them in his arms and gives his beloved sleep."

508. 3d, HARMONIOUS ACTION OF THE VARIOUS MENTAL POWERS. That the bodily organs may be directly affected by impressions purely *mental* does not admit of doubt. Of this fact the skillful physician never loses sight; for a hopeful, healthful influence of the mind may be made a remedial agency quite as powerful as that of drugs and plasters.

509. *Regular and systematic mental exercise is essential to the health of nerve-tissue.* Exercise increases the flow of blood to the active part. We have seen this to be the case in the muscle, and that by use it is both enlarged and strengthened. In like manner the nerve-tissue needs exercise; and as the gymnast becomes expert, not by spasmodic muscular efforts, but by accurate, persistent drill, so must the mental athlete gain his power by the regular performance of such exercise as he is able to bear. The gymnast at first feels pain in his muscles, but he has only to persevere, with proper intervals of rest, and what was at first so difficult becomes easy, while power is gained for severer feats. So the person unaccustomed to mental gymnastics feels headache and confusion at first, but frequent repetition will make easy and natural the very thoughts which struggled so painfully into existence, and the nerve-tissue will gain the firmness which increases its capability of action. Under such a course of training, the change in the brain-tissue is often so great as to modify perceptibly the form of the head.

510. Says Dr. Ray: "I have no hesitation in saying that, of all the means for preserving health, there is nothing more sure, or better suited to a greater variety of persons, than habits of regular and systematic mental occupation of some dignity and worth. In this proposition I would embrace all those kinds of employment which pass under the general name of business, and which, little as we are disposed to recognize the fact, bear the same relation to the health of the mind that food, exercise, etc., do to the health of the

body. Work is the condition of our being as active and progressive creatures."

Employment which is steadily pursued as a part of the established routine of life, and felt to be, in some degree, a matter of necessity, has an effect on the mind far more salutary than that which depends on the impulse of the moment, and is determined by no sense of necessity, no force of habit.

511. *The saddest effects of the absence of stated useful employment are seen among women of easy circumstances.* "It is a poor view of woman's duties and capacities that confines her to a little busy idleness, because the chances of fortune have placed her beyond the necessity of earning a living; and they must have but a narrow view of the exigencies of social life who believe that any woman of tolerable health and strength may not find abundant opportunities of that kind of work which affords no other recompense than the consciousness of doing good, and therefore to be done, if done at all, by those who can dispense with every other compensation. A life of idleness and luxurious ease can be no more honorable to one sex than to the other, and we know very well that in a man it creates no claims upon the respect and confidence of the community."

The little accomplishments of needlework, so generally diffused, cannot be dignified with the name of work. Many a mind, liberally endowed, from want of mental exertion becomes dwarfed, or may end in mental depression, particularly if ill health or deep affliction throws its weight into the scale.

512. *The amount of exercise should be adapted to the health and age of the individual.* If from any cause the nervous system be weakened, an amount of exercise which would be quite harmless to one in health may prove disastrous. The nerve-tissue of children and youth needs the same care as has been shown requisite for other tissues, and overwork, that in the adult is followed by fatigue, easily removed by rest, in the child may result in irreparable injury. At this period,

the tissue is soft and yielding, and when the blood-vessels become long distended by great activity, they may become permanently enlarged, and permanent congestion produced.

Parents and teachers should not fail to remember that there are important differences in the *quality* of different brains. In some children the mental reaction to impressions is sluggish and incomplete; in others, the reaction, though slow, is quite complete; in others, again, the reaction is rapid and lively, but evanescent, so that, though quick at perception, they retain ideas with difficulty; while in others there is that just equilibrium between the internal and external in which the reaction is exactly adequate to the impression. These differences should be taken into the account, and the dull intellect roused, while that unduly active should be restrained. It is too often the case, however, that exactly the opposite course is pursued. The fond parents and ambitious teachers, misled by the early promise of genius, excite the child to new activity by unceasing cultivation and the never-failing stimulus of praise. For a time the progress of the child is all they could desire, but in exact proportion as the picture of the future brightens to their fancy, the probability of its realization lessens. The brain, worn out by premature exertion, loses its tone, leaving the mental powers weakened and depressed for the remainder of life. The expected prodigy is then outstripped by many whose dull outset promised him an easy victory.

513. We often hear the saying, "The valedictorian is never heard of after Commencement-day;" and it is too often true when the honors are gained at an early age. The present tendency is to treat the mind like a race-horse, goading it on to make a certain round in a given time, and that before the brain-tissue has gained the consolidation requisite for severe exertion. Mary Lyon, with her characteristic wisdom, refused to admit to the Mount Holyoke course of study girls under sixteen years of age, and from her long list of applicants usually selected those not less than eighteen.

Let the material organ of the mind be subjected to a

systematic, thorough, gymnastic training, taking for it the *necessary time*, and the firm, educated tissue will be fitted for enduring labor in later years; but let it be weakened in *youth*, and it must ever work under a burden, if indeed it work at all.

Moderation in mental exertion is also a necessity with the aged, as they have no vitality for recuperation after severe exhaustion.

514. *Intense activity too long continued impairs the strongest brain.* The nerve-cells in a state of rest are neutral in their chemical character, but after severe exercise they become acid. When in this condition it is hazardous to continue the exercise. Sufficient rest should be taken to restore them to their normal condition. Congestion, or an undue accumulation of blood, also attends excessive functional action. The effect of severe congestion in the spinal centres is to produce convulsions; in the sensory centres, roaring in the ears, flashes of light before the eyes and various hallucinations; in the ideational centres, stagnation of ideas, swimming in the head, and, if long continued, irregular and convulsive action of the cells, causing delirium. The co-ordination of function is destroyed, the will-power abolished. The delirious ideas are the expression of a condition of things in the supreme centres analogous to that which in the spinal cord utters itself in convulsions.

515. *The required rest is often afforded by recreation and amusement.* "Important as stated employment unquestionably is to the mental health, amusement or recreation is scarcely less so. Few persons, whatever their mental character or temperament, can safely dispense with these altogether. Even the most commanding intellects sometimes seek the recreation which their exhausting labors make necessary in forms of amusement which, to those who feel the necessity less, seem to be frivolous and puerile. Endowed as we are with the faculty of being amused, it seems to be a reflection on the Author of our being to regard amusements as something to be carefully shunned rather than sought and enjoyed.

"To those whose life is one of severe toil and harassing care, amusements constitute almost the only practicable means for repairing the constant waste of the nervous energy. Especially is this want felt by women in the humbler walks of life, whose daily round of care and toil not only draws more largely than that of the stronger sex on the physical and mental energies, but is lightened by none of that relief which is afforded by a greater variety of duties and more frequent periods of rest."

Observation.—The brain, when severely taxed, is often rested by some kind of mental exercise which, without being fatiguing, requires just enough effort to impart interest. Hence, a change from Mathematics to the Languages, or from these to music, poetry or painting, will give the needed relaxation.

516. To maintain the highest mental vigor, each faculty of the mind should receive its due share of cultivation. Our various faculties were not bestowed at random, to be used or not as inclination may prompt, but each has its appointed place in the mental economy. Each bears some relation to every other, making one harmonious whole. All cannot and need not receive the same amount of cultivation; but let any one power be so neglected that it might as well be wanting, or let it be applied to some other than its destined use, and an element of strength is lost, the mind becomes to a greater or less extent weakened and one-sided, and therefore jars in its working. One must form habits of attention, accustom the mind to continuous thought, cultivate the reasoning powers and beget a taste for exact knowledge, if he would be in any measure equal to the intellectual effort essential to true success in every calling of life. He must, however, also call into action the creative power of the mind, the *imagination*, to give vividness to his conception, to add force to his reasoning and to light up the whole horizon of his thought. Many cry out against this faculty, forgetting that it is God-given, and capable of a culture that shall make it of inestimable value. It is the *abuse* of it, not the use, which we are

to guard against. Its exercise must not be indulged to an extent incompatible with the claims of the other faculties. It must not be allowed to fashion with unbridled power our principles and motives, our aims and ends. Give it, however, the purest material to work with, and, within proper bounds, no faculty is of more real service or more worthy of our regard. Especially is it of value in presenting to the mind an ideal of excellence, a standard of attainment, practicable and desirable, but loftier than anything we have yet reached.

517. The *æsthetic* faculty, the *love of the beautiful,* should not be allowed to remain inactive. Its importance is recognized only as we understand its value. An object is beautiful to us just in proportion to our power to discover through the material form the *thought* of which this form is but the expression; for beauty is but the spirit looking out through the visible, the material. Is not he, then, a happier, a wiser and a better man, who so develops this faculty that he may not only read the thought and sentiment embodied in the works of art, but also the thoughts of the *Creator* in their varied forms of expression through all the kingdoms of Nature?

518. Man has also a *moral* faculty, the power of discriminating between right and wrong, which is quickly followed by the feeling of obligation to do the right and avoid the wrong. Upon the right use of these faculties depend the happiness and the destiny of man. The power of an approving conscience over the human mind, and consequently over the health of the Nervous System, cannot be over-estimated, while on the other hand, the torments of an accusing conscience not only "cut the sinews of the soul's inherent strength," but snap one by one the gossamer filaments of the brittle thread of life.

We have given only a glance at a few of the mental faculties, but should we take them one by one, through all the departments of mind, and note their uses, we should find none which could be unused or misused without detriment to our health, happiness or usefulness.

519. Concerning the hygienic influence OF A HARMONIOUS DEVELOPMENT OF THE MENTAL POWERS, Dr. Ray says: "A partial cultivation of the mental faculties is incompatible not only with the highest order of thought, but with the highest degree of health and efficiency. The result of professional experience fairly warrants the statement that in persons of a high grade of intellectual endowment and cultivation, other things being equal, the force of moral shocks is more easily broken, tedious and harassing exercise of particular powers more safely borne, than in those of an opposite description, and disease, when it comes, is more readily controlled and cured. The kind of management which consists in awakening a new order of emotion, in exciting new trains of thought, in turning attention to some new matter of study or speculation, must be far less efficacious, because less applicable, in one whose mind has always had a limited range than in one of larger resources and capacities. In endeavoring to restore the disordered mind of the clodhopper who has scarcely an idea beyond that of his manual employment, the great difficulty is to find some available point from which conservative influences may be projected. He dislikes reading, he never learned amusements, he feels no interest in the affairs of the world; and unless the circumstances allow of some kind of bodily labor, his mind must remain in a state of solitary isolation, brooding over its morbid fancies, and utterly incompetent to initiate any recuperative movement."

§ 18. COMPARATIVE NEUROLOGY.—*The Comparison of the Nervous System of other Mammals with that of Man—Of Birds—Of Reptiles—Of Fishes. The Arrangement of the Nervous System of Mollusks—Of Radiata.*

520. We have seen that in the Motory and Nutritive apparatuses, there is an arrangement and condition of tissues, organs and functions, in all classes of vertebrates, homologous to those in man. Analogy would induce the supposition, that in the arrangements and appointments of the nervous system a similar condition would be found. Here are found gang-

lia, commissures, and nerves afferent and efferent, but the highest development, the *convoluted* hemispherical ganglion, seems to be wanting in the lower orders of the mammalia, in birds, reptiles and fishes; with this general exception, all other homologous parts are more or less developed.

Fig. 164.

Fig. 165

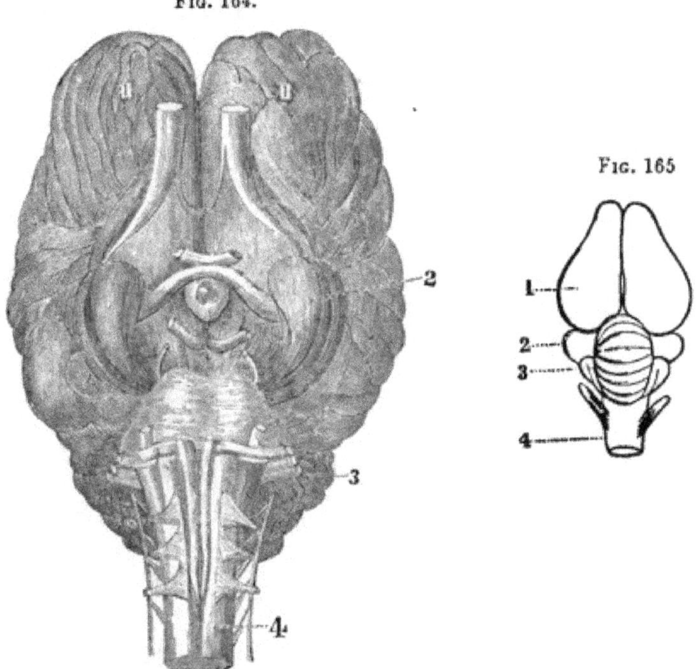

Fig. 164 (*Owen*). Base of Brain of a Horse.—1, Cerebrum. 2, Optic ganglion. 3, Cerebellum. 4, Medulla Oblongata and Spinal Cord.

Fig. 165. Brain of a Bird.—1, Cerebrum. 2, Optic ganglion. 3, Cerebellum. 4, Medulla Oblongata.

521. In the *Mammalia*, the relative size of the cerebrum and cerebellum, except in the lowest order (monotremata), as the ornithorynchus, is about the same as in man, but in birds the cerebellum is proportionately larger than the cerebrum; the sulci of the cerebrum and cerebellum of other mammals and birds are less developed than in man, and the same is true of the relative size of the brain, large and small, also certain ganglia are comparatively larger. In the horse.

22

ox, etc., the olfactory, optic and auditory ganglia are large, and the senses of smell, sight and hearing are acute. This is particularly apparent in birds, as the eagle, vulture and buzzard. In these, vision is not only far-reaching, but acute, and the same is true, to a certain extent, of smell and hearing. In some animals, as the mole, where vision is feeble, and in others where smell or hearing is obtuse, the ganglionic bulbs are very small and the nerves very delicate.

522. In *Birds* the hemispheres are not united by a corpus callosum, as in mammals; the cerebellum is proportionately larger than the medulla oblongata; and the comparative weight of the brain to the body is less than in mammals.

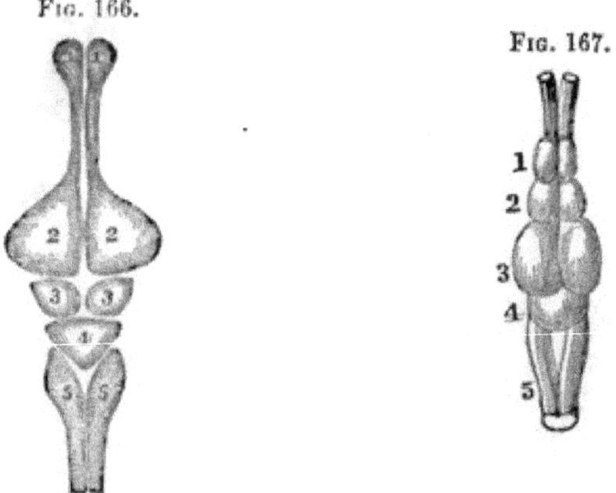

Fig. 166. Brain of an Alligator.—1, Olfactory ganglia. 2, Cerebrum. 3, Optic ganglia. 4, Cerebellum. 5, Medulla Oblongata and Spinal Cord.

Fig. 167. Brain of a Fish.—1, Olfactory ganglia. 2, Cerebrum. 3, Optic ganglia. 4, Cerebellum. 5, Medulla Oblongata and Spinal Cord.

523. The brain of *Reptiles* constitutes but a very small part of the body. It is smooth, and without convolutions. The hemispheres are hollow, and there is no striated body. The cerebellum sends no prolongations across the medulla oblongata, so as to form a kind of ring, as in mammals.

524. The relative size of the cerebrum in man, compared with that of the mammals, birds, reptiles and fishes, varies much. In some few, the relative weight between the brain

and body is about the same as in man, while in others it is less—seemingly an homologous appendage, as in the ornithorynchus. As we descend in the scale of animal life, the cerebellum, the medulla oblongata and the ganglia of some of the special senses, as of smell and vision, are larger relatively than in the corresponding parts in man, since these animals depend upon them for their subsistence and safety.

525. The spinal cord of all mammals, birds, reptiles and fishes varies most in length, but in structure, investment and function it resembles that of man. The number of pairs of spinal nerves correspond to the number of the vertebræ, but the size of the cord is relatively larger than the cerebrum, also the cerebellum and several of the ganglia.

526. The brain of the *Fish* is small; it does not fill the whole cranial cavity, there being found within it a spongy, fatty mass. The investment and protection of some of the organs of special sense are modified, as seen in the eye of some fishes, the deep-sea shark for instance, where the sclerotic tunic of the eye is bony, in order to protect this organ from the great pressure of the water. Perhaps the most wonderful arrangement is found in the electric fishes, as the common Torpedo, Malapterurus, and the Electric Eel of South America.

Fig. 168.

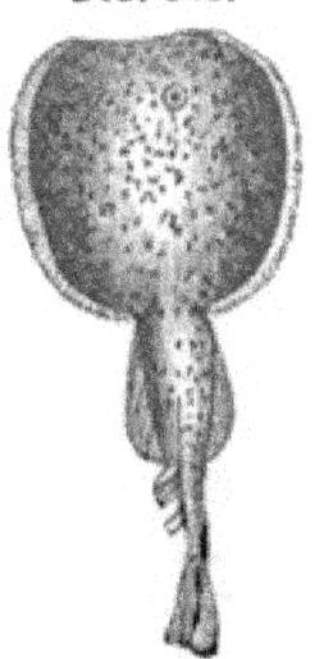

Torpedo.

527. "The torpedo is a cartilaginous fish. Its body is smooth, and represents a disc nearly circular, the anterior edge of which is formed by two prolongations of the muzzle, which on each side proceed to unite with the pectoral fins, and leave between these organs the head and the branchiæ, an oval space, in which is lodged the electric apparatus of the fish. This apparatus is composed of a number of vertical membranous tubes, closely packed like honeycomb, and subdivided by horizontal partitions filled with mucosites, and animated by several very large branches of the pneumogastric nerve, which, in this and other electric fishes, is larger

than the spinal cord. In these singular organs is produced the electricity which has now been proved to resemble in every respect common electricity. By experiment it has been ascertained that this property depends on the posterior lobe of the brain, and that by destroying this lobe, or cutting the nerves proceeding from it, the faculty is lost."

Fig. 169.

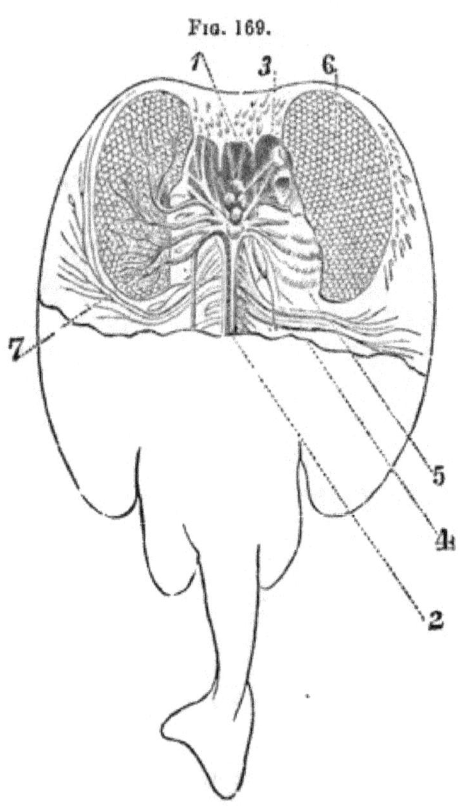

Fig. 169. Electric Organs of Torpedo.—1, Brain. 2, Spinal cord. 3, Eye and optic nerve. 4, Spinal nerve. 5, Branchiæ. 6, Electrical organ. 7, Pneumogastric nerve.

528. "The Gymnotus, or Electric Eel, possesses the power in the highest degree. It is met with in vast numbers in the rivulets and stagnant waters of the immense plains of South America. The electric shocks, which it discharges at will, are sufficiently strong to kill men and horses, and being transmissible through water, the gymnotus does not require to touch its prey. At first the electric discharges are feeble, but when roused they become terrible; but by this effort it becomes exhausted, and requires repose before it can renew the attack: this is the moment its captors avail themselves of to seize it. The electric organs are arranged along the back and tail."

In the Articulata, the body is different in its general

structure and the nervous system is correspondingly modified. The body is composed of several sections articulated with each other in a lineal series. The ganglia of the nerves of special sensation, as of sight and hearing, of motion, of respiration and nutrition, are larger than those of general sensation.

Fig. 170.

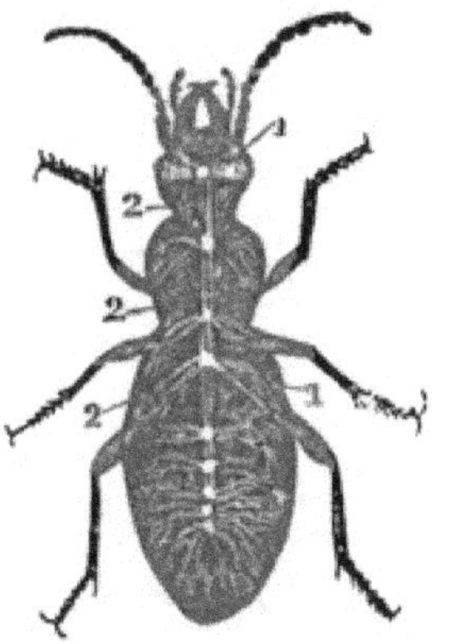

Fig. 170. Nervous System of an Insect.—1, 1, Central ganglia. 2, 2, 2, Nerves that connect the ganglia.

In the nervous system of the centipede, whose general structure is similar to that of other articulates, the ganglia are arranged in pairs of nearly equal size, except the ganglion that answers to the brain, which is larger, along the ventral surface of the alimentary canal. Each pair is connected

Fig. 171.

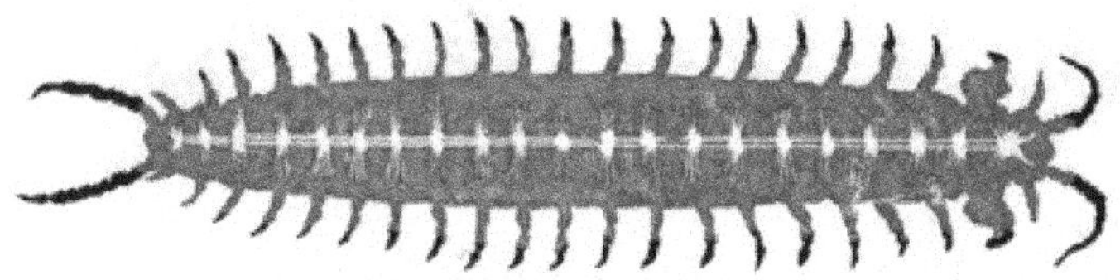

Fig. 171. Diagram of a Centipede.

with the preceding, with the integument or skin and with the muscles of its own segment, by sensitive and motor filaments of nerves.

529. In *Mollusks* are found the ganglia and commissure arrangement, with nerves sensitive and motor, afferent and efferent, and on a plan corresponding to the body. The structure of the organs of sense is less complete than in vertebrate animals. Some mollusks are gifted only with the sense of touch and taste; a great number have eyes, whose structure varies; none have yet been found possessing a special organ for smell.

530. In the Radiata, the star-fish manifests one of the

simplest forms of the nervous system. It consists of a central mass, with five arms radiating from it. In the centre is the mouth, and beneath it the stomach or gastric cavity, which sends prolongations to each limb. The nervous system consists of five similar ganglia situated in the central portion at the base of the arms. These ganglia are connected by commissures, and each sends off nerve-filaments to the corresponding limbs.

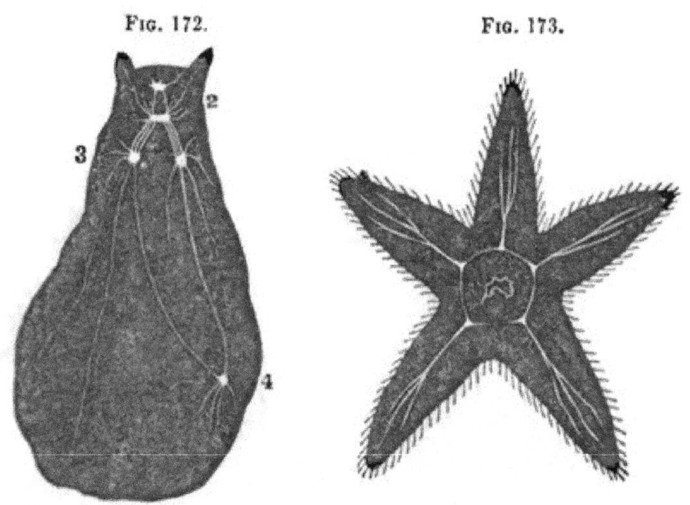

FIG. 172. DIAGRAM OF THE TYPE OF A MOLLUSK.—1, Œsophagal ganglia. 2, Cerebral ganglia. 3, Pedal or locomotive ganglia. 4, Respiratory ganglia.

FIG. 173. DIAGRAM OF A RADIATA—THE STAR-FISH.

531. We have seen that in all grades of the animal kingdom the cell-structure obtains, but in the lowest forms of animal life nerve does not exist. The stimulus which the little creature receives from without would seem to produce some change in the molecular relations of its almost homogeneous substance, and these insensible movements collectively to amount to the sensible movement which it makes; the molecular process in such case being perhaps not unlike that which ensues and issues in the coagulation of the blood when the fibrin is brought in contact, as some think, with a foreign substance. The perception of the

stimulus by the creature is the molecular change which ensues, the imperceptible motion passing, by reason of the homogeneity of its substance, with the greatest ease, from element to element of the *same kind*, as it were by an infection, or as happens in the folding of the leaves of the mimosa, or sensitive plant; and the sum of the molecular motions, as necessarily determined in direction by the form of the animal, results in the visible movement.

"With the differentiation of tissue and increasing complexity of organization which are met with as we ascend in the animal kingdom, the nervous tissue appears, but at first under a very simple form. Its simplest type may be represented as two fibres that are connected by a nerve-cell; the fibres are apparently simple conductors, and might be aptly compared to the conducting wires of a telegraph, while the cell, being the centre in which nerve-force is generated, may be compared to the telegraphic apparatus; in it the effect which the stimulus of the afferent nerve excites is transmitted along the efferent nerve, and therein is displayed the simplest form of that *reflex* action which plays so large a part in animal life."

The relations of the animal kingdom afford a striking evidence of divine unity, bound together in the closest harmony, and the work of Him who was the Beginning and will be the End.

CHAPTER XII.

THE ORGANS OF SPECIAL SENSE.

UNDER this head are classed the *Tongue*, the *Nose*, the *Eye*, the *Ear* and the *Tactile* portions of the Nervous System.

§ **49.** THE ANATOMY OF THE ORGANS OF SPECIAL SENSE.—*The Organ of Taste—Of Smell. The Coats of the Eye. The Humors of the Eye. The Muscles of the Eye. The Protecting Organs. Classification of the Organs of Hearing. The External Ear. The Labyrinth. The Internal Ear. The Organs of Touch. Two Layers of Skin. The Epidermis. The Dermis. The Hairs. The Sebaceous Glands. The Perspiratory Glands. The Nails.*

532. The organ of the SENSE OF TASTE is the mucous membrane which covers the Tongue, especially the back part of this organ, and the palate. Upon the upper surface of the tongue the mucous membrane has various little eminences, called *papillæ*, resembling the villi of the intestines. The

FIG. 174.

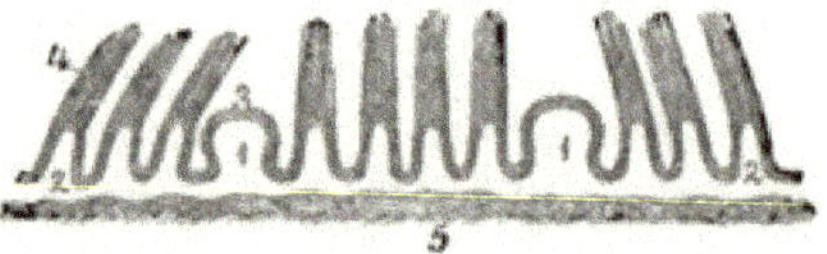

FIG. 174 (*Leidy*). DIAGRAM OF THE PAPILLÆ OF THE TONGUE, moderately magnified. 1, Capitate papillæ. 2, Conical papillæ. 3, Epithelium. 4, The same structure forming bunches of hair-like processes. 5, Connective tissue.

principal of these are of a composite character, and present three varieties—the *Circumvallate*, the *Fungiform* and the *Conical*. The CIRCUMVALLATE papillæ are shaped like the letter V with the point turned downward, and are surrounded by an annular wall-like elevation, whence their name. They are about a dozen in number, and are found upon the posterior part of the tongue. The FUNGIFORM papillæ are broad at the free extremity and narrow at the base, having something of the mushroom shape, whence their name. They are more numerous than the circumvallate, and are scattered over the surface of the tongue, but are especially numerous at and near the tip. The CONICAL papillæ are smaller and more numerous than the others, and are found in the intervals between them, arranged in rows diverging from the median line of the tongue. All the above-described papillæ and the spaces between are covered with *simple* papillæ, conical in form. From those surrounding the conical papillæ, the squamose epithelium rises in hair-like appendages, which give a brush-like arrangement, admirably adapted to the imbibition of liquids to be tasted. These hair-like appendages give the velvety character to the surface of the tongue, and upon them the furred condition of this organ depends. Minute blood-vessels and nerves pass up into these papillæ, thus giving a large extent of sensitive surface.

FIG. 175.

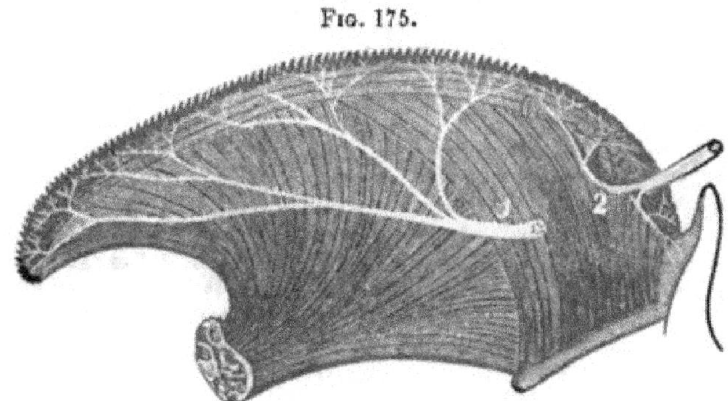

FIG. 175 (*Dalton*). DIAGRAM OF THE TONGUE, with its sensitive nerves and papillæ. 1, Lingual branch of fifth pair. 2, Glosso-pharyngeal nerve.

Nervous filaments are received from the fifth, ninth and twelfth pairs of nerves. The branch of the fifth, called the *Gust'a-to-ry*, is the nerve of taste and ordinary sensibility;

the twelfth, called the Hypo-glossal, of voluntary motion. By means of the ninth, or Glosso-pharyngeal, the tongue is brought into association with the fauces, œsophagus and larynx. It is of obvious importance that these parts should act in concert; and this is effected by the distribution of this nerve.

Fig. 176.

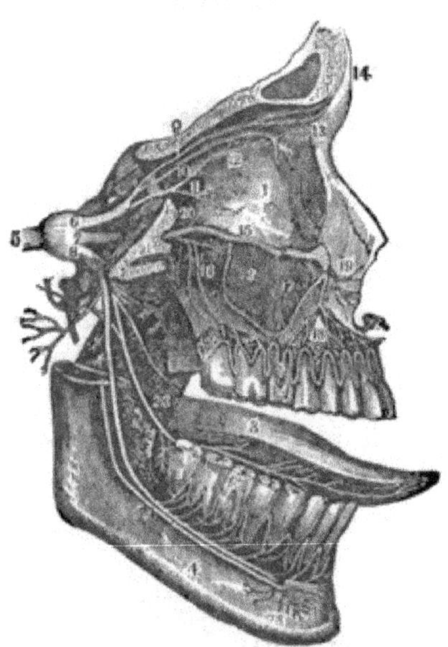

Fig. 176. The Distribution of the Fifth Pair of Nerves.—1, The orbit for the eye. 2, The upper jaw. 3, The tongue. 4, The lower jaw. 5, The fifth pair of nerves. 6, The first branch of this nerve, that passes to the eye. 9, 10, 11, 12, 13, 14, Divisions of this branch. 7, The second branch of the fifth pair of nerves is distributed to the teeth of the upper jaw. 15, 16, 17, 18, 19, 20, Divisions of this branch. 8, The third branch of the fifth pair, that passes to the tongue and teeth of the lower jaw. 23, The division of this branch that passes to the tongue, called the gustatory. 24, The division that is distributed to the teeth of the lower jaw.

533. The organ of the Sense of Smell is a part of the delicate mucous membrane lining the nasal passages. These passages extend from the opening of the nostrils in front, to the pharynx behind; they are high, vaulted and narrow, and separated from each other by a partition partly bony, and partly cartilaginous. This double cavity is separated

from the mouth by a bony floor (the hard palate), which is continued backward to the root of the tongue by a fleshy curtain, called the soft palate. In ordinary positions of the mouth, this palate and the root of the tongue effect a closure between the mouth and the pharynx. Each of the outer walls of the nasal chamber has three bony processes called *turbinated* bones, arranged one above another, like shelves. The front of the chamber is bounded by a thin plate of bone, filled with perforations, hence, named the *cribiform*, or sieve-like

FIG. 177.

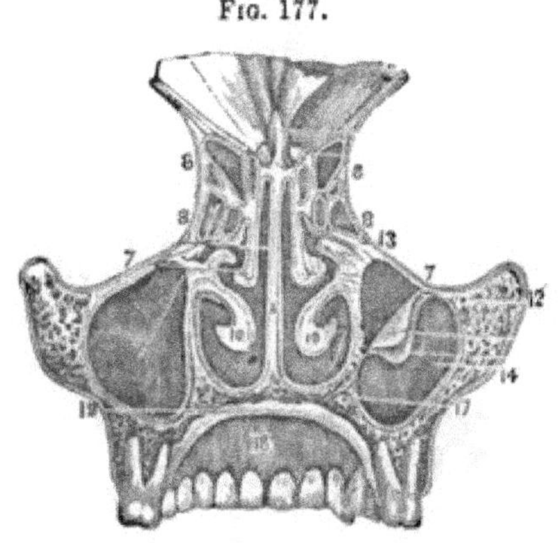

FIG. 177. A VERTICAL SECTION OF THE MIDDLE PART OF THE NASAL CAVITIES.—7, The middle spongy bones. 8, The superior part of the nasal cavities. 10, The inferior spongy bones. 11, The vomer. 12, The upper jaw. 13, The middle channel of the nose. 14, The lower channel of the nose. 17, The palatine process of the upper jaw-bone. 18, The roof of the mouth covered by mucous membrane. 19, A section of this membrane.

plate. Upon it, rest the olfactory lobes which send numerous filaments through the perforations to the mucous membrane of the two upper turbinated bones, affording the special sense of smell; the membrane of the lower bone receives a branch from the fifth nerve, which is endowed with common sensibility only; the odor of cologne, for example, is distinguished by the olfactory nerve, and the pungency, by the branch of the fifth nerve.

FIG. 178.

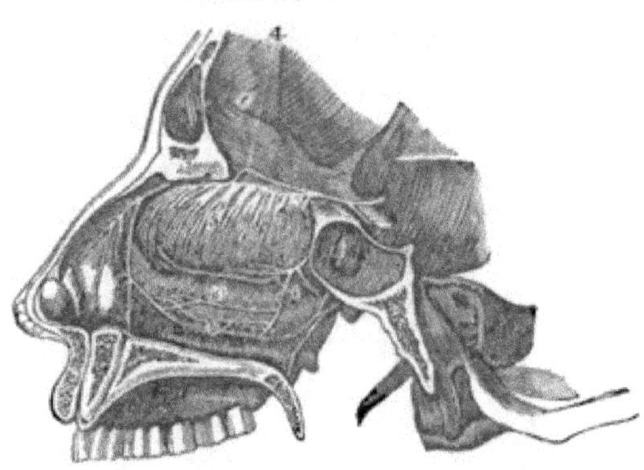

FIG. 178. A SIDE VIEW OF THE PASSAGE OF THE NOSTRILS, AND THE DISTRIBUTION OF THE OLFACTORY NERVE.—4, The olfactory nerve. 5, The fine divisions of this nerve on the membrane of the nose. 6, A branch of the fifth pair of nerves.

534. The chief organ of the SENSE OF SIGHT is the Eye. The globe of the eye, or eyeball, is composed of three concentric envelopes—viz., the *Sclerot'ica*, with the *Cornea* in front; the *Cho'roidea*, with the *Iris* in front; and the *Ret'ina*, which is internal. These make up most of the solid part of the eyeball, which is a hollow sphere filled with three fluid or semifluid substances—the *Aqueous Humor*, the *Crystalline Lens* and the *Vitreous Humor*.

FIG. 179.

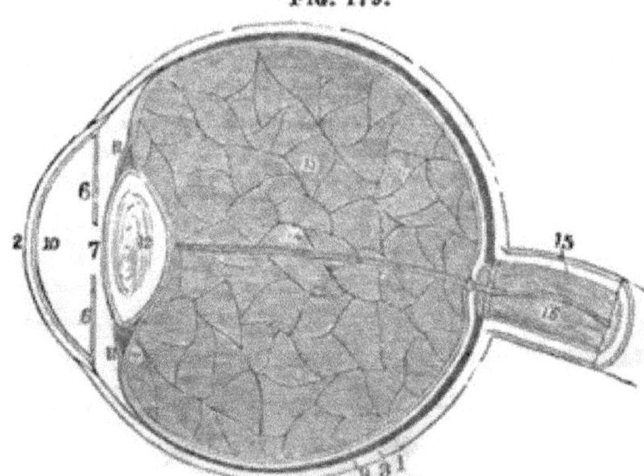

FIG. 179. A SECTION OF THE GLOBE OF THE EYE.—1, The sclerotic coat. 2, The cornea. (This connects with the sclerotic coat by a bevelled edge.) 3, The choroid coat. 6, 6, The iris. 7, The pupil. 8, The retina. 10, 11, 11, Chambers of the eye that contain the aqueous humor. 12, The crystalline lens. 13, The vitreous humor. 15, The optic nerve. 16, The central artery of the eye.

The SCLEROTICA, or Sclerotic coat, invests the globe of the eye, excepting the part covered with the cornea in front. It is composed of white fibrous tissue arranged in many layers, which cross each other at right angles, and form a tunic of great strength. It is white, glassy and opaque, and is commonly called "the white of the eye." It has few blood-vessels and seems destitute of nerves.

The CHOROIDEA, or Second Coat of the eye, has some fibrous tissue like the sclerotica, but is chiefly composed of blood-vessels and pigment-cells. These cells give the coat an intense black color on the inside, but externally it is brown. It lines the sclerotica, and is connected with it by a delicate areolar tissue. It is perforated behind, for the passage of the optic nerve, and terminates in front in the *cil'iary* ligament (composed chiefly of dense areolar tissue), in the anterior part of which the iris is inserted. This muscle also lies at the juncture of the sclerotica and cornea, being in connection with the first coat and cornea, and the second coat and iris.

The CILIARY PROCESSES consist of a number of minute, triangular folds, formed apparently by the plaiting of the internal layer of the choroid coat toward its front part. Their bases are toward the pupil, and the free portion rests against the circumference of the crystalline lens. These processes are covered with pigment-cells.

The IRIS occupies the opening of the choroidea in front, forms a partition between the anterior and the posterior chambers of the eye, and is pierced by a circular opening, which is called the *Pupil.* It is free, except at its peripheal attachments, and floats freely in the aqueous humor. The posterior surface of the iris, or uvea, is thickly covered with pigment; but the anterior surface gives the *color of the eye,* so remarkably and beautifully varied in different individuals, and presenting numerous blended tints of black, brown, blue and gray. The iris is generally regarded as a modification of muscular tissue. It has two layers of fibres—one layer of radiating fibres, converging from the circumference to the centre, the other of circular fibres.

The RETINA is the inner coat of the eye, formed by the expansion of the optic nerve upon the inner side of the choroid coat, but not extending so far forward. It ends at a short distance from the ciliary ligament, in a jagged edge, from which an exceedingly fine membrane extends to the ciliary processes. Its inner surface is bounded by an exceedingly delicate membrane, called the "membrana limitans," which separates it from the vitreous humor.

FIG. 180.

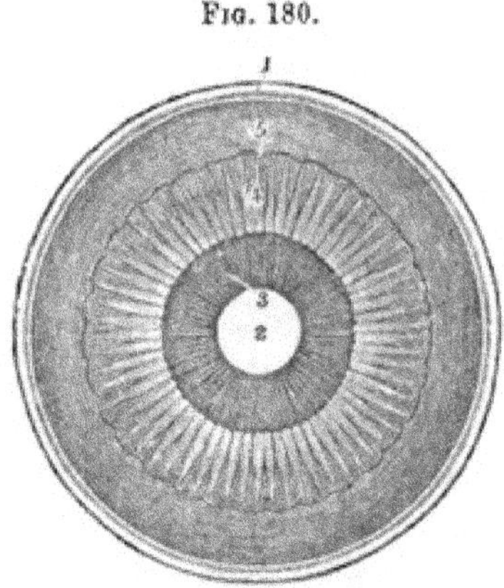

FIG. 180. A VIEW OF THE ANTERIOR SEGMENT OF A TRANSVERSE SECTION OF THE GLOBE OF THE EYE, seen from within. 1, The divided edge of the three coats—sclerotica, choroidea and retina. 2, The pupil. 3, The iris: the surface presented to view in this section being the uvea. 4, The ciliary processes. 5, The scalloped anterior border of the retina.

535. Of the three humors, or liquid substances of the eye, the AQUEOUS, or watery, is situated in the anterior portion of the organ behind the cornea. It is an albuminous fluid, with an alkaline reaction and liquid like water. The iris is placed vertically in the fluid, the space between it and the cornea being the *anterior chamber* of the eye, and that between the iris and crystalline lens behind, the posterior chamber. The two chambers are lined by a membrane secreting the aqueous humor.

The CRYSTALLINE humor, or lens, is situated immediately behind the pupil, and is surrounded by the ciliary processes. It is invested by a transparent, elastic membrane, called the capsule of the lens. The humor is more convex on the posterior than on the anterior surface. It is imbedded in the anterior part of the vitreous humor, from which it is separated by a thin membrane. The lens consist of thin layers, like the coats of an onion. The external layer is soft, but each successive one increases in firmness.

Observation.—When the crystalline lens or its investing membrane is changed in structure, preventing the rays of light from passing to the retina, the affection is called a *cataract.*

Fig. 181.

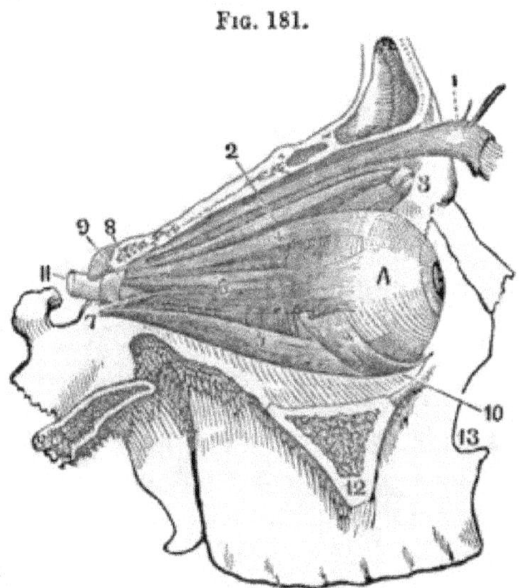

Fig. 181 (*Leidy*). Muscles of the Eye.—1, The palpebral elevator muscle. 2, The superior oblique. 3, The pulley through which the tendon of insertion plays. 4, Superior straight muscle. 5, Inferior straight muscle. 6, External straight muscle. 7, 8, Its two points of origin. 9, Interval through which pass the oculo-motor and abducent nerves. 10, Inferior oblique muscle. 11, Optic nerve. 12, Cut surface of the malar process of the superior maxillary bone. 13, The nasal orifice. A, The eyeball.

536. The Vitreous Humor forms the principal bulk of the globe of the eye. It is an albuminous fluid resembling the aqueous humor, but is more dense, and if once discharged by disease or accident, it is irrecoverably lost; while the aqueous humor may be lost and afterward restored. This humor is enclosed in a delicate membrane, called the *hy'a-loid*, which sends processes into the interior of the globe of the eye, forming the cells in which the humor is retained.

537. The MUSCLES of the eye are six in number. They are attached at one extremity to the orbit behind the eye; at the other extremity they are inserted by broad, thin tendons to the sclerotic coat, near the junction of the cornea. The white, pearly appearance of the eye is caused by these tendons.

Observation.—If the external muscle is too short, the eye is turned out, producing the "wall eye;" if the internal muscle is contracted, the eye is turned inward toward the nose, and is called a "cross eye."

538. The PROTECTING ORGANS are the *Orbits, Eyebrows, Eyelids* and *Lach'rymal Apparatus.*

The ORBITS are deep, bony sockets in which the globes of the eye are placed. The bottom of each orbit has a large perforation, giving passage to the optic nerve. These cavities are lined with a thick cushion of fat.

The EYEBROWS, forming the upper part of the boundary of the orbits, are two tegumentary prominences covered with coarse hair.

The EYELIDS are two movable curtains, having a delicate skin on the outside, muscular fibres beneath, and a narrow cartilage on their edges, which tends to preserve the shape of the lid. Internally, they are lined by a smooth mucous membrane, which is reflected on the front of the eye upon the sclerotica. This membrane is called the *Conjuncti'va.*

Observation.—When this membrane is inflamed, it sometimes deposits a whitish material called *lymph*, which accounts for the films, opacities and white spots seen upon the eye after the inflammation has subsided.

On the internal surface of the cartilage there are found several small glands, which have the appearance of parallel strings of pearls. They open by minute apertures upon the edges of the lids.

The edges of the eyelids are furnished with a triple row of hairs, called eyelashes, which curve upward from the upper lid, and downward from the lower.

The LACHRYMAL APPARATUS which secretes the tears

consists of the *Lachrymal Gland* with its ducts, *Lachrymal Canals* and the *Nasal Duct.*

The Lachrymal Gland is situated at the outer and upper angle of the orbit, occupying a depression in the orbital plate of the frontal bone. Ten or twelve small ducts pass from this gland and open upon the upper eyelid, where they pour upon the conjunctiva the lachrymal fluid, or tears.

Fig. 182.

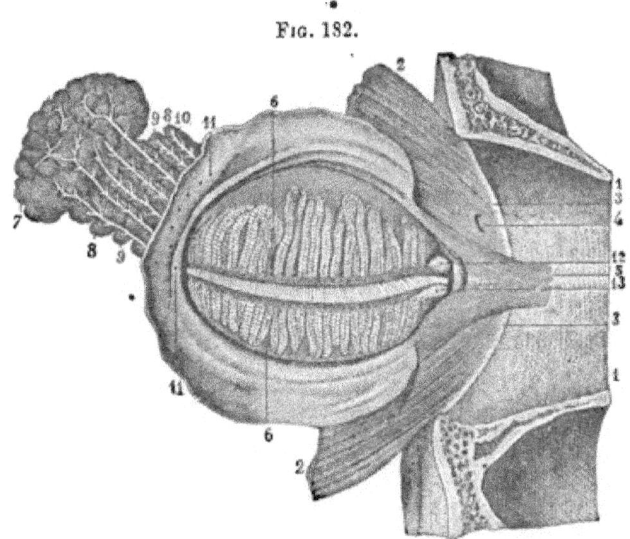

Fig. 182 (*Leidy*). The Left Eyelid and Lachrymal Gland, turned Forward and Inward, to show their Inner Surface.—1, Upper and lower part of the orbit. 2, Portion of the palpebral orbicular muscle. 3, Attachment of this muscle to the inner margin of the orbit. 4, Perforation for the passage of the external nasal nerve. 5, Offset described as the tensor muscle of the eyelids. 6, Palpebral glands. 7, Posterior, and 8, anterior portions of the lachrymal glands. 9, 10, Ducts. 11, Orifices opening on the inner surface of the upper eyelid. 12, 13, The lachrymal orifices at the summits of the lachrymal papillæ.

The Lachrymal Canals commence at the free borders of each eyelid, near the internal angle of the eye, by two minute orifices, called "punc'ta lach'rymalia" (tear points). Each of these ducts communicates with the sac at the upper part of the nasal duct.

The Nasal Duct is a short canal about three quarters of an inch in length, directed downward and backward to the inferior channel of the nose, where it terminates by an expanded orifice. The tears, secreted by the lachrymal gland, are conveyed to the eye by the small ducts before described. They are then taken up by the puncta lachrymalia and carried by the lachrymal canals into the lachrymal sac, from which they are passed to the nasal cavities by the nasal duct.

Fig. 183.

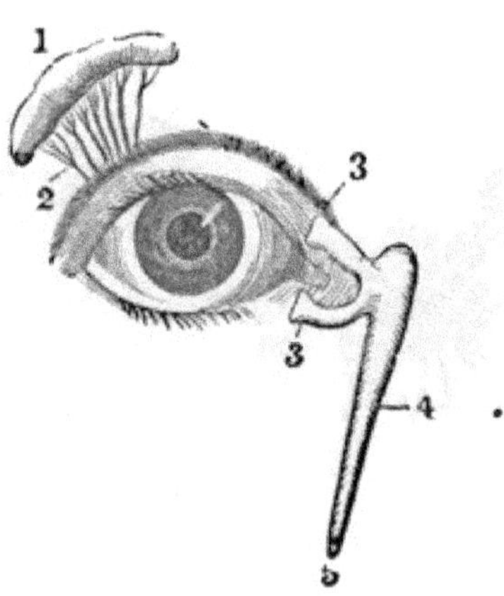

Fig. 183. View of Lachrymal Gland and Nasal Duct.—1, The lachrymal gland. 2, Ducts leading from the lachrymal gland to the upper eyelid. 3, 3, The puncta lachrymalia. 4, The nasal sac. 5, The termination of the nasal duct.

539. The Sense of Hearing does not strictly belong to one organ, but to several, which are grouped into three divisions—the *External Ear*, the *Tym'panum** and the *Labyrinth* or *Internal Ear.*

540. The Labyrinth is so called from its remarkable and varied configuration. It is divided into three portions—the *Vestibule*, the *Semicircular Canals* and the *Coch'lea.*†

541. The Vestibule is a small and somewhat triangular cavity about the size of a grain of wheat. It is placed almost vertically in the centre of the labyrinth, and is a kind of entrance-chamber or ante-room to the semicircular canals behind and the cochlea in front.

* Gr., *túmpanon*, a drum. † Gr., *kochlos*, to twist.

FIG. 184.

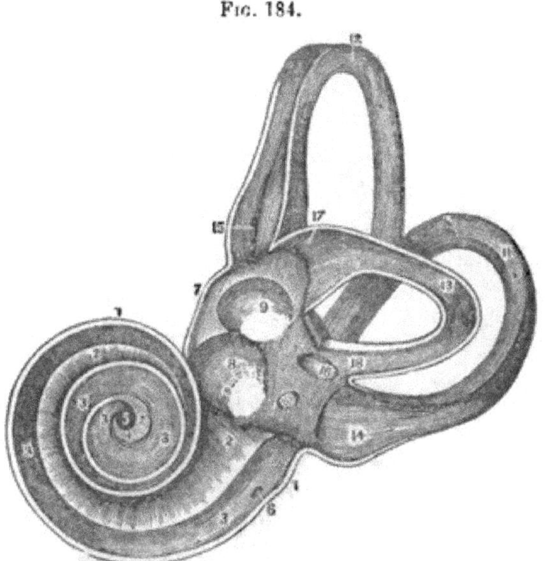

FIG. 184. A VIEW OF THE LABYRINTH LAID OPEN.—1, 1, Cochlea. 2, 3, Two canals, that wind two and a half turns around a hollow axis (5). 7, Central portion of the labyrinth (vestibule). 8, Fenestra rotunda. 9, Fenestra ovalis. 11, 12, 13, 14, 15, 16, 17, 18, The semicircular canals. Highly magnified.

542. The SEMICIRCULAR CANALS are three curved passages, describing more than half a circle, and are about the twentieth of an inch in diameter. Two of them open into the vestibule at both extremities, and the third at one extremity. Both the vestibule and the canals contain a transparent fluid like lymph, and in this fluid, without touching the walls of the cavity, floats a membranous labyrinth, corresponding in form to the osseous one, but considerably smaller. It is a sheath or bag enlarged at the vestibule, and sending out prolongations into the semicircular canals on the the one side and the cochlea on the other. It is filled with a lymph-like fluid of greater consistency than that in which it floats. The auditory nerve is distributed in the walls of this membranous labyrinth, and nervous filaments connect it with its osseous counterpart.

543. In front of the vestibule is the COCHLEA, so called from its resemblance to a snail-shell. It consists of a bony

canal which winds around a hollow axis nearly three times, gradually decreasing in diameter, and thus forming a spiral cone. The interior of the canal is divided into two passages by a membranous partition, upon which the remaining parts of the auditory nerve ramify. The passages are filled with lymph, and communicate with each other at the apex of the cone and at the apex of the base; one opens into the vestibule, the other into the *Tympanum* (the *Fenes'tra* Rotunda*).

Fig. 185.

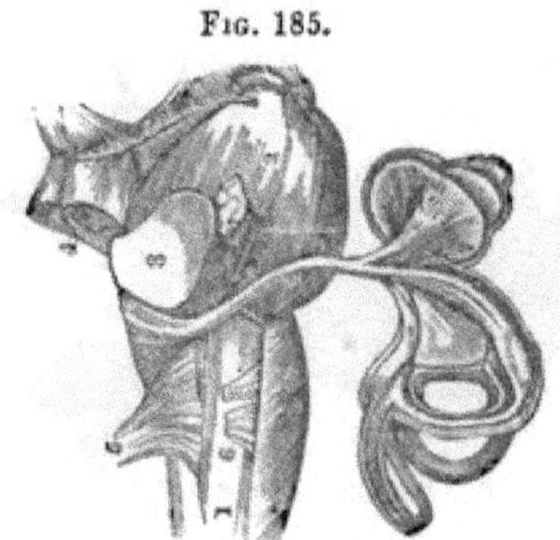

Fig. 185. A View of the Auditory Nerve.—1, Spinal cord. 2, Medulla oblongata. 3, Lower part of the brain. 4, Auditory nerve. 5, A branch to the semicircular canals. 6, A branch to the cochlea.

In the outer part of the bony wall of the vestibule is the *Fenestra Ovalis,* an oval-shaped perforation about one-eighth of an inch in length and one-sixteenth in width. This is closed by a thin fibrous membrane, which prevents the escape of the fluid from the vestibule, and through it the sonorous vibrations pass to the vestibule.

544. The Tympanum, or middle ear, is an irregular bony cavity larger than the vestibule and just outside of it. It is separated from the external ear by a thin, semi-transparent membrane of an oval shape. This is very closely fitted into a groove, between the tympanum and the auditory canal. The tympanum is often called the *Drum* of the ear, and very appropriately, for the membrane of the tympanum is in contact with the atmosphere whose sonorous vibrations beat upon

* Lat., a *window.*

it much like drumsticks upon the head of a drum. There are several openings into the tympanum, of which the largest is called the *Eustachian tube,* from the name of the first anatomist who described it. It is a trumpet-shaped canal somewhat over an inch and a half long, extending from the fore part of the tympanum obliquely inward, forward and downward to the pharynx. The tube is lined with a ciliated epithelium continuous with that of the pharynx and tympanum. In the tympanic cavity are three bones, or *ossicles,* the smallest in the body, weighing only a few grains. From their resemblance to the articles, they have been named the *Mallet, Anvil* (attached to this bone is a little tubercle, or *orbicular* bone, which is sometimes regarded as a separate ossicle) and *Stirrup.* The Mallet and Anvil articulate by a hinge-joint; the Anvil and Stirrup by a ball-and-socket joint.

Fig. 186.

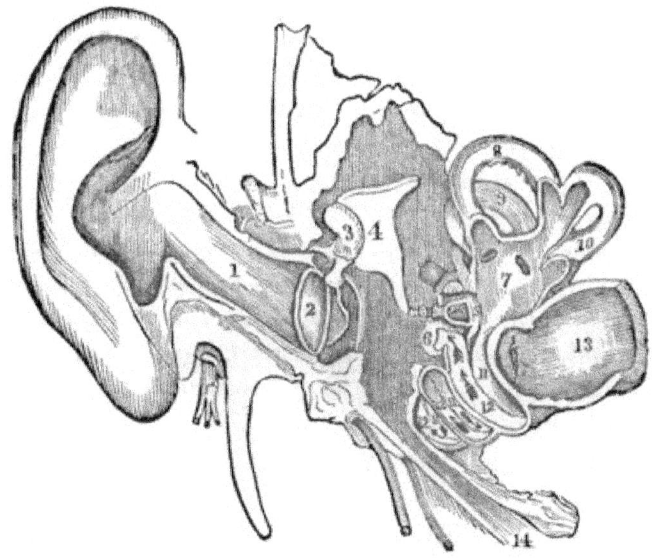

Fig. 186. A View of all the Parts of the Ear.—1, The canal that leads to the internal ear. 2, The membrana tympani. 3, 4, 5, The bones of the ear. 7, The central part of the labyrinth (vestibule). 8, 9, 10, The semicircular canals. 11, 12, The channels of the cochlea. 13, The auditory nerve. 14, The opening from the middle ear, or tympanum, to the throat (Eustachian tube).

545. The EXTERNAL EAR lies outside the membrane of the tympanum. It is composed of the auditory canal and the part which projects from the head. The canal, or *External Mea'tus* Audito'rius*, is partly bony and partly cartilaginous, about one inch in length, and narrower in the middle than at the extremities. Short, firm hairs are stretched across the tube, preventing the ingress of foreign bodies. Beneath the thin cuticle are small follicles which secrete the *Ceru'men*, or wax. The part of the external ear outside the cavity has numerous prominences and ridges.

(For Physiology of Hearing, see 565.)

546. The skin is the principal part of the body concerned in the SENSE OF TOUCH, but the tongue and lips also possess this sense. The skin consists of two layers. The external, or superficial layer, destitute of blood-vessels and nerves, is called the *Ep-i-derm'is*† (which consists of two layers, different in many respects, one being named the *Cuticle*, the other the *Soft Epidermis*); and an internal, or deeper layer, abundantly supplied with nerves and highly vascular, called the *Dermis*, *Cutis Vera* (or true skin). This layer presents two very different surfaces, of which the external is called the *Papillary* layer, the internal the *Co'ri-um.*‡

547. The skin covers the whole exterior of the body, and at the margins of the apertures is directly continuous with the mucous membrane, which last is an integument of greater delicacy, but has substantially the same composition—viz., a deep fibrous, sanguine, sensitive layer, a basement membrane, and an epithelium, or superficial, insensible and bloodless layer. Thus the whole body, externally and internally, has a complete epithelial investment.

548. The EPIDERMIS holds the same relation to the dermis that the epithelium does to the deeper layer of the mucous membrane. It varies in thickness, from the thin, delicate membrane upon the internal flexions of the joints, to the

---

* Lat., *meo*, to pass, a passage. † Gr., *epi*, upon, and *derma*, skin.
‡ Gr., *chorion*, skin.

thickened covering of the soles of the feet. This variation is perceptible in infants, before exercise can have had any influence.

FIG. 187.

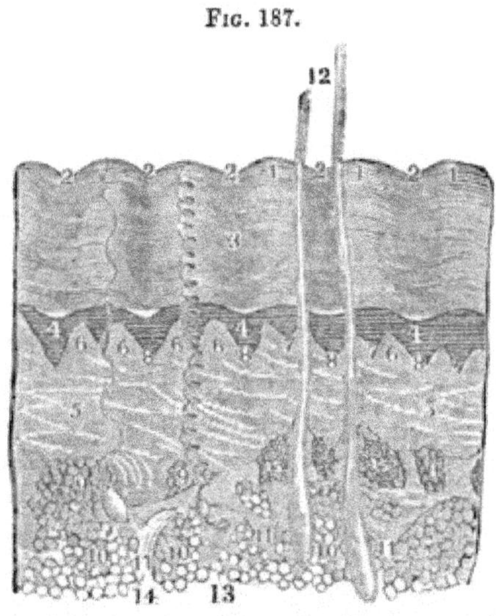

FIG. 187. A VERTICAL SECTION OF THE SKIN.—1, 1, The lines, or ridges of the cuticle, cut perpendicularly. 2, 2, 2, 2, 2, The furrows or wrinkles of the same. 3, The epidermis. 4, 4, 4, Colored layer. 5, 5, Dermis, or cutis vera. 6, 6, 6, 6, 6, Papillæ. 7, 7, Small furrows between the papillæ. 8, 8, 8, 8, Deeper furrows between each couple of the papillæ. 9, 9, Cells filled with fat. 10, 10, 10, Adipose layer, with numerous fat vesicles. 11, 11, 11, Cellular fibres of the adipose tissue. 12, Two hairs. 13, A perspiratory gland, with its spiral duct. 14, Another perspiratory gland, with a duct less spiral. 15, 15, Oil-glands with ducts opening into the sheath of the hair (12). A diagram.

During life the EPIDERMIS is constantly undergoing loss, throwing off the superficial epidermoid scales, which are constantly renewed by fresh cells, originating on the surface of the true skin. These gradually undergo transformation from the spherical to

FIG. 188.

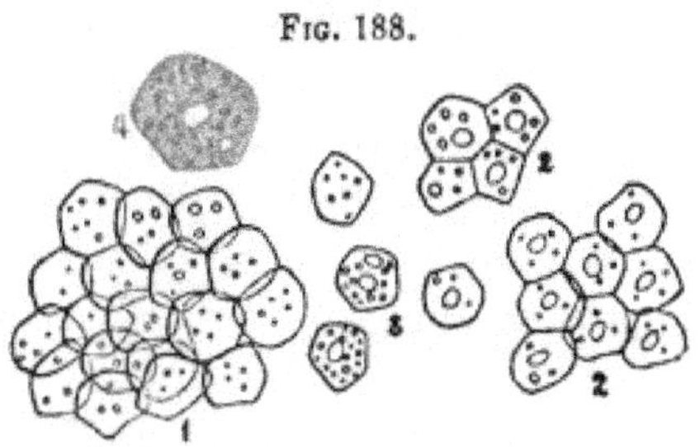

FIG. 188 (*Leidy*). FRAGMENT OF DANDRUFF FROM THE HEAD.—1, Portion of dandruff, consisting of non-nucleated cells. 2, Several fragments, consisting of nucleated cells. 3, Isolated cells, some with and some without nuclei. 4, A cell more highly magnified, exhibiting granular contents and a nucleus.

the flattened shape, as they approach the surface of the cuticle.

The soft epidermic layer is the seat of the color of the skin. The difference between the blonde and brunette, the European and the African, lies only in the deep, newly-formed layers of the epidermis. In the whitest skin, the cells of the epidermis always contain a slight amount of the pigmentary tint, which disappears from the cells as this soft layer is transformed into the cuticle.

549. The CUTICLE is a translucent, horn-like membrane. Its deeper surface is continuous with the soft epidermic layer from which it is constantly renewed. Its free surface is incessantly wearing away, or shed in small flakes, constituting scurf or dandruff.

FIG. 189.

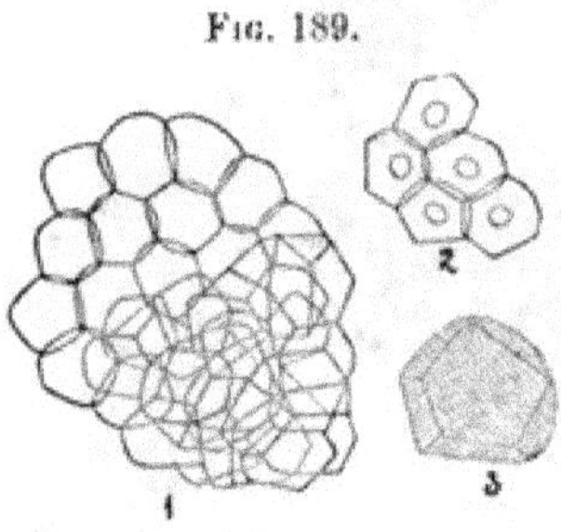

FIG. 189 (*Leidy.*) SCURF FROM THE LEG.—1, A fragment of scurf, consisting of dried, flattened, non-nucleated cells or scales. 2, A few cells with a nucleus. 3, A cell more highly magnified, to exhibit its polyhedral form.

550. The DERMIS, or TRUE SKIN, is made up of interlacing bundles of white areolar tissue, mixed with yellow elastic fibres. These are so interwoven as to constitute a firm, strong and flexible web. In the superficial part, the web is so close as to resemble felt cloth. In the deepest layers the network is loose, and encloses the hair-follicles with their sebaceous glands, and small masses of fat.

In most situations, plain muscular fibres are found mixed with the fibrous and elastic tissues; these are always present where hairs exist, to which parts they are often attached; but on the palms and soles, where these are absent, no muscular fibres are ever seen.

551. The outer surface of the dermis, as seen when denuded, is provided with little conical-shaped projections, called *Papillæ* (6, fig. 187). These are prolongations of the upper compact tissue of the corium into the newly-formed layer of the epidermis. The papillæ are very

numerous on the palm of the hand and on the free border of the lips.

552. The cutis vera is abundantly supplied with *blood-vessels, lymphatics* and *nerves.* Its general surface is covered with a close capillary network, from which looped vessels project and enter the papillæ. The lymphatics also form a close network on the surface. The nerves pass upward from the subcutaneous areolar tissue, and form, as they approach the surface, minute plexuses, from which the nerve-fibres are given off. Some of these fibres are lost in the compact tissue of the dermis; others end, perhaps, in loops; and many pass into certain of the papillæ, for it is said that some of these do not receive nerve-fibres. In the papillæ these fibres end in loops, or, as in the fingers, the sole of the foot, and perhaps on the red margin of the lips and the point of the tongue, they appear to terminate in small oval, condensed bodies, called *tactile corpuscles,* situated in the centre of the papillæ. In any case, it is supposed that the nerve-fibre turns back to rejoin some nerve-cell in the nervous centres.

FIG. 190.

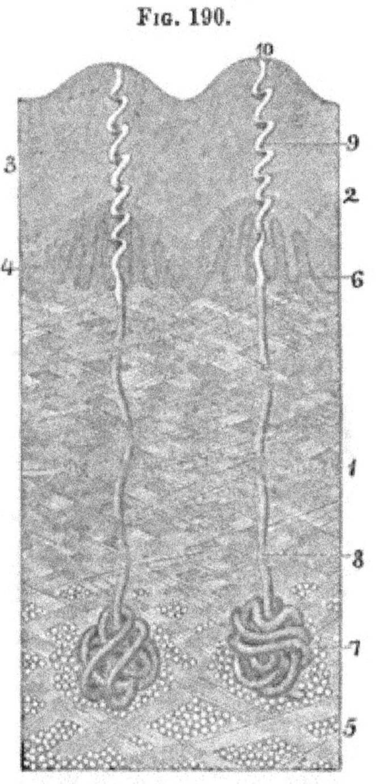

FIG. 190 (*Leidy*). VERTICAL SECTION OF THE SKIN OF THE FOREFINGER ACROSS TWO OF THE RIDGES OF THE SURFACE; highly magnified. 1, Dermis, composed of an intertexture of bundles of fibrous tissue. 2, Epidermis. 3, Its cuticle. 4, Its soft layer. 5, Subcutaneous connective and adipose tissue. 6, Tactile papillæ. 7, Sweat glands. 8, Duct. 9, Spiral passage from the latter through the epidermis. 10, Termination of the passage on the summit of ridge.

The network of nerves imbedded in the upper porous layer of the true skin is derived from nerves which take their winding course through the fat, distended openings of the corium.

553. The minute depressions from which the hairs of the skin emerge are called the *Hair-follicles*, or sacs. They are buried in the corium, or true skin. At the bottom of the follicle is a more or less elevated portion of the dermis, often forming a distinct papilla, which is destitute of cuticle. The root of the hair is composed of soft, pale and somewhat compressed nucleated cells; it is adherent to the lining of the follicle, or *root-sheath.* When a hair is plucked out, the sheath adheres to it, but the vascular papilla at the bottom of the follicle remains, and a new hair is generated upon it. If the papilla is destroyed, no new hair can be formed. All these papillæ, except those of the finest hairs, probably receive nervous fibrils. The part of the hair projecting above the surface is called the *Shaft.* The shaft is usually cylindrical, but sometimes flattened. It consists of an outer part, called the *Cortex*, composed of a single layer of imbricated scales whose edges are directed toward the point of the hair. Beneath the cortex is the so-called fibrous part of the hair, which constitutes its bulk, and consists of fusiform cells clustered into flattened fibres, running longitudinally and intermixed with pigment granules. Lastly, the very deepest cells, occupying the centre of the shaft and constituting the pith, are not elongated, but polyhedral and loosely connected together, and containing chiefly pigment or fat granules.

Fig. 191.

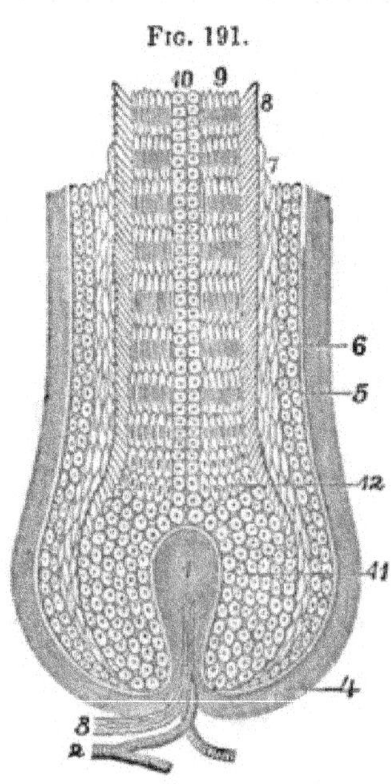

Fig. 191 (*Leidy*). Diagram of Structure of the Root of a Hair within its Follicle.—1, Hair papilla. 2, Capillary vessel. 3, Nerve-fibres. 4, Fibrous wall of the hair-follicle. 5, Basement membrane. 6, Soft epidermic lining of the follicle. 7, Its elastic cuticular layer. 8, Cuticle of the hair. 9, Cortical substance. 10, Medullary substance. 11, Bulb of the hair, composed of soft polyhedral cells. 12, Transition of the latter into the cortical substance, medullary substance and cuticle of the hair.

Many of the unstriated muscular fibres from the true skin pass obliquely down from the surface of the dermis to the under side of the slanting hair-follicles. The contraction of these fibres erects the hairs, and by drawing the follicles to the surface and pulling in a little point of the skin, produces that roughness of the integument called "goose-skin," or *Cutis Anserina.* The standing on end of the hair of the head, as the result of extreme fright, may be partly due to the contraction of such fibres, as well as to the action of the occipito-frontalis muscle.

FIG. 192.

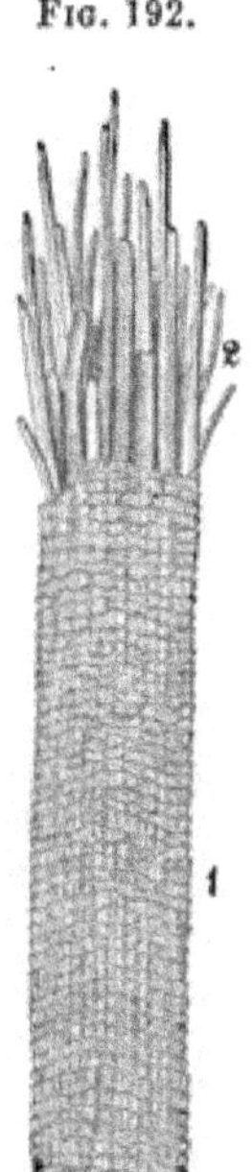

FIG. 192 (*Leidy*). PORTION OF A HAIR FROM THE OUTER PART OF THE THIGH, magnified. 1, Shaft of the hair covered with transverse markings indicating the projecting edges of the cuticular scales. 2. Cortical substance at the end of the hair, broken up into coarse fibres, as the result of friction of the clothing.

554. Each hair-follicle receives, in nearly all cases, the ducts of two *Sebaceous*, or *Oil-Glands*, which are situated in the dermis. They are found only where hairs exist. Each gland is a flask-shaped body, composed of from five to twenty little sacs, clustered around and leading into a common duct. These glands are lined by a fine epithelium, and the unctuous secretion first anoints the hair-bulb, and then oozes out upon the neighboring surface of the cuticle. The sebaceous glands are of considerable size.

555. Immediately beneath the skin, over the whole surface of the body, there are a multitude of little glandular bodies, called *Perspiratory*, or *Sweat Glands.* Each gland consists of a minute, cylindrical spiral duct, which passes inward through the epidermis, and terminates in a globular coil, in the deeper meshes of the cutis vera. The opening of the duct upon the cuticle is called the "pore." This aperture is oblique in

direction, and possesses all the advantage of a valvular opening, preventing the ingress of foreign injurious substances to

FIG. 193.

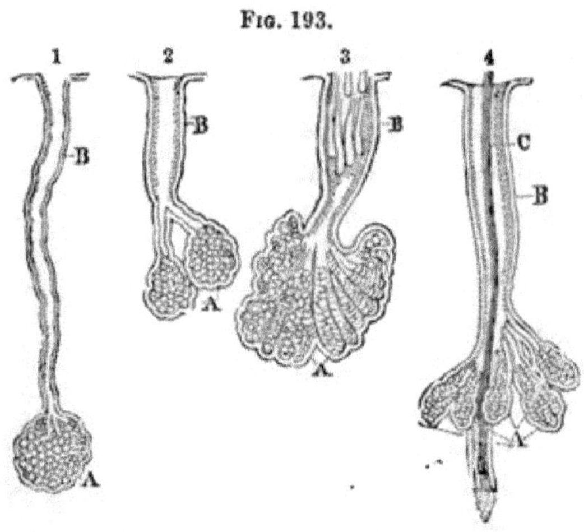

FIG. 193. OIL-GLANDS AND DUCTS, magnified thirty-eight diameters. 1, A, Oil-gland from the scalp; B, Its duct. 2, A, Two glands from the skin of the nose; B, Common duct. 3, A, Oil-gland from the nose; B, The duct filled with the peculiar animalculæ of the oily substances; the heads are directed inward. 4, A, Cluster of oil-glands around the shaft of the hair (C); B, Ducts.

the interior of the duct or gland. These glands, coming in contact with the capillary blood-vessels, receive a watery fluid (the perspiration) from the blood, having the following composition:

Water	995.00
Animal matters, with lime	.10
Chlorides of sodium and potassium and spirit extract	2.40
Acetic acid, acetates, lactates and alcoholic extracts	1.45
Sulphates and substances soluble in water	1.05
	1000.00

The formation of this watery fluid is constant, but usually evaporation takes place as fast as it reaches the surface. This is called the "insensible transpiration" of the skin.

556. The NAILS are horny appendages of the skin, and correspond with the hoofs and claws of animals. They are

flexible, translucent plates continuous with the epidermis, and rest on the depressed surface of the cutis vera, called the matrix, or bed. By maceration or severe scalding, the nail becomes detached with the epidermis, even in life.

Fig. 194.

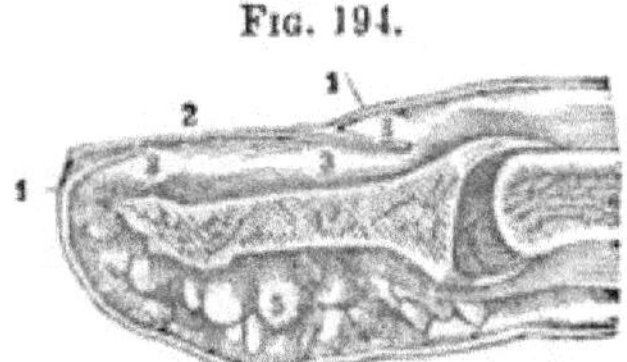

Fig. 194. A Section of the End of the Finger and Nail.—4, Section of the last bone of the finger. 5, Fat, forming the cushion at the end of the finger. 2, The nail. 1, 1, The cuticle continued under and around the root of the nail, at 3, 3, 3.

The horny layer of the nail answers to the cuticle; it is composed of numerous layers of flattened, nucleated cells, or scales, while the soft layer corresponds with the deep parts of the epidermis, and is made up of delicate polyhedral, nucleated cells. The nails increase in length by the constant addition of cells at the root; they grow in thickness by the formation of cells on the under surface. This double development explains why nails are thickest at their most convex portion.

For Physiology of the Skin, see 570.

§ 50. Physiology of the Organs of Special Sense.—*Primary Use of the Sense of Taste—Of Smell. Some of the Laws of Light. The Adaptation of the Eye to the Laws of Light. Cause of Short-Sightedness—Of Long-Sightedness—Defect remedied. Functions of the Different Coats of the Eye. The Accessory Parts of the Eye. Hearing. Function of the External Ear—Auditory Canal—Eustachian Tube—Cochlea and Semicircular Canals. Sounds reach the Fluid of the Labyrinth by Two Paths. Special Organ of the Sense of Touch. Functions of the Skin. Use of the Epidermis—Of the Cuticle—Of Cutaneous Papillæ. Vessels of the Corium. Function of the Oil-Glands. Uses of Perspiration.*

557. The primary use of the Sense of Taste is to guide animals in the selection of food, that noxious articles may not be introduced into the stomach. In man, this sense has been so abused and perverted by the introduction of stimulants and condiments, and the endless admixture of different articles of food, that the natural action seems to have been almost entirely superseded by acquired taste. This sense be-

comes very acute by cultivation, as may be seen in those persons whose business it is to judge of the quality of certain articles by the taste; as tasters of wine, tea, etc. The acuteness of taste, however, varies in different persons according to the sapid bodies themselves.

558. In man, the SENSE OF SMELL is one of inferior importance. It furnishes the mind with but few ideas, and these are mostly subservient to his physical well-being. This sense leads us to avoid disagreeable odors or putrescent food, and when acute, to escape the injurious effects of many vapors which endanger health.

559. THE STRUCTURE OF THE EYE is beautifully adapted to the laws of light, a few of which it is necessary for us to notice.

When light passes through a medium of unvarying density, the rays are in straight lines, but when it passes from a medium of one density into another of different density, they are refracted, or bent from a straight course, unless striking the medium perpendicularly, when they are unchanged.

When light passes from one medium to another having a convex or concave surface, instead of a flat surface, a great degree of refraction is produced, and the greater the curvature, the greater will be the amount of refraction. Fit a convex lens in an opening of the shutter of a darkened room; the rays of light will cross each other in the lens, and an inverted image of any object outside, as a tree or house, will be reflected upon a screen placed in the room, at a certain distance in front of the lens. The exact point where the image is most distinct, is called the focus of the lens, and the distance from the lens to the image, the focal distance. Now, in the eye, the pigment of the choroid coat gives the darkened room, the retina the screen, the pupil is the opening in the shutter, and the three humors are the curved lenses. The rays of light from any object cross each other, and an inverted image is formed on the retina.

560. The shape of the cornea and aqueous humors is convexo-concave; the vitreous humor is concavo-convex; while

the crystalline humor, or lens, is convexo-convex. It may at first seem that only one lens is necessary; but light is composed of three primary colors, which are not equally refracted by the same lens, hence, there would be upon the edges of any single lens prismatic colors which would interfere with the distinctness of the image. This is obviated, by the adaptation of the curvatures of the lenses to the different colors.

561. Suppose our object outside the darkened room to be at that distance from the lens which will give a distinct image upon the screen; now, if the object approach the lens, the image will be indistinct unless a more convex lens be substituted for the first, or the distance between the lens and screen be increased. If the object recede, the image will be indistinct unless a less convex lens be substituted for the first, or the distance be lessened between the lens and screen.

By a very nice adjustment, the eye is able to change the convexity of its lenses, and also to vary the focal distances, thereby adapting it to a wide range of vision. This is accomplished by the ciliary ligament and the muscular fibres connected with the ciliary processes, which change the curvature of the crystalline lens and the cornea by compression at the circumference, and at the same time throw the lens forward, increasing the distance between it and the retina. The iris also aids, in adapting the eye to different distances. It contracts when viewing a near object, and dilates when viewing one more remote.

562. When the cornea or crystalline lens is too convex, or the latter is too far from the retina, *short-sightedness* is produced, and the defect is measurably remedied by the use of concave glasses; when there is too little convexity, *long-sightedness* is the result, and convex glasses should be used. In old age, the humors being deficient in quantity, cause the flattening of the convex parts, hence the need of convex glasses. In the selection of glasses, the lens for each eye should be chosen separately, as the foci of the two eyes do not usually exactly correspond, therefore a lens that will suit one eye may strain the other.

563. *The Sclerotic Coat* gives form to the eye, and serves for the attachment of the muscles which move the eye in various directions. The movements of the two eyeballs are always simultaneous and harmonious, but frequently not symmetrical. The function of the pigment of the choroid coat is to absorb all the luminous rays not necessary for vision. "If the sclerotic and choroid coats be carefully dissected off from the posterior part of the eye of an ox or any other large quadruped, leaving only the retina, and the eye so prepared be placed in a hole in a window-shutter, in a darkened room, with the cornea on the outside, all the illuminated objects of the external scene will be beautifully depicted, in an inverted position, on the retina.

"Few spectacles are more calculated to raise our admiration than this delicate picture, which Nature has, with such exquisite art and with the finest touches of her pencil, spread over the smooth canvas of the expansion of the optic nerve—a picture which, though scarcely occupying a space of half an inch in diameter, contains the delineation of a boundless scene of earth and sky, full of all kinds of objects, some at rest and others in motion, yet all accurately represented as to their forms, colors and position, and followed in all their changes without the least interference, irregularity or confusion."

564. *The Accessory Parts of the Eye* are of two kinds; the one designed to protect the globe, or ball, the other to move it and give the required direction to fulfill its office. To enable the eye to move in all directions, without friction, it is placed on a cushion of fat which lines the bony orbit, thus protecting the globe on all sides except in front; here are the *Eyelids,* which by their alternate movement of depression and elevation spread over the front of the eyeball a watery secretion, by which its surface is constantly bathed, and its brilliancy and transparency kept unimpaired.

By the contraction of a small ring-like muscle (the *Orbicularis*), the eyelids quickly draw together, and as they instantly separate, the secretion from the lachrymal gland is diffused over the conjunctiva. During life, this muscle is ever active and watchful for the safety of the eye. When a cinder or dust gets under the eyelids, it irritates the conjunctiva, and the movements of winking are very rapid. A viscid fluid is spread along the margin of the lid, which prevents the tears running over the eyelid.

The *Cilia*, or *Eyelashes*, so interlace that protection is given the eye from light substances floating in the air. The *Eyebrows* assist in shading the eyes when exposed to strong light, and they lend expression to some emotions of the mind.

565. HEARING is that function by which we obtain a knowledge of the vibratory motions of bodies, which constitute sounds. Independent of the sense of hearing, sound, as sound, has no existence in nature.

566. The *External Ear* collects the waves of sound and reflects them on the membrane of the tympanum; this membrane facilitates their transmission to the chain of bones in the tympanum, to the walls of the cavity and to the air it contains; from the stirrup to the oval window; from this membrane the vibrations are communicated to the fluid of the labyrinth, until finally they are received by the expansion of the auditory nerve, by which the sensation is communicated to the brain.

567. The *function of the Auditory Canal* is to receive and conduct sonorous vibrations to the membrane of the tympanum. This membrane is admirably adapted for the reception of atmospheric sound-waves. In hearing, the air in the tympanic cavity plays an important part; the design of the *Eustachian tube* is evidently to allow of equal atmospheric pressure upon both sides of the membrana tympani. The complicated communications of the internal ear contain the highly important parts of the organ of hearing. The *Vestibule* is the part essential to the simplest exercise of this sense. The *Cochlea* and *Semicircular* canals, or rather their contained membranous canals, receive vibrations through the mixed membranous and bony tympanic apparatus. It is asserted by some physiologists, that sound is communicated through the

cranial bones; the transmission, however, through the solid bones of the head, if it exists, is effected with difficulty.

568. By this sense, therefore, we distinguish the quality, intensity, pitch, duration and direction of sonorous impulses. The delicacy with which these distinctions are appreciated varies in different individuals. The complication and finish of the auditory apparatus, and the perfection and delicacy of its action, are second only to those of vision.

569. The SENSE OF TOUCH, though common to all parts of the *Skin* and adjoining mucous membranes, has for its special organ, *the hand.* It is most admirably adapted to its office, by reason of the number, size, arrangement, structure and abundant nervous supply of its papillæ.

570. The functions of the SKIN are threefold: 1st, As a *Protecting* membrane; 2d, As a *Medium* for the distribution of the tactile nerve-filaments; and 3d, As an *Eliminating* organ. The extent of the skin, as it invests the entire surface of the body, following all its prominences and curves, its arrangement in layers, differing in structure, vitality and function, make it an external envelope of harmonious unity, in appointment and end.

571. The uses of the *Epidermis* are various. It serves to cover and protect the delicate sensitive parts behind it; to prevent the too rapid escape of heat; and to restrain the evaporation of the fluids of the skin and its appendages, at the same time that it furnishes a medium through which those secretions can reach the surface of the body.

572. The *Cuticle* is constantly destroyed and replaced, as is proved by the disappearance from the skin of such stains as those produced by nitrate of silver; or the scales thrown off after some acute diseases, as scarlatina. The restoration of the cuticle is observed after the process of vesication by blisters, and in consequence of burns and scalds. By these means large patches of cuticle are removed; but they are renewed in short time, under favorable circumstances. The pigmentary substance is also capable of rapid reproduction.

573. The cutaneous *Papillæ* serve to increase the nutritive

and formative surface for the generation of the constantly wasting epidermis.

574. In the *Corium*, or internal layer of the skin, resides vitality. Here the arteries of the skin penetrate from beneath and end in a capillary network; the veins emerging from the skin are more numerous and much larger than the arteries. The skin is abundantly supplied with nerves, but their mode of termination has not been accurately ascertained.

575. The surface of the skin possesses the power of *absorbing* both liquids and vapors. The principal, if not sole, agents of this function on the surface of the body are the cutaneous *Lymphatic* vessels, which are active in proportion to the tenuity or absence of the cuticle. To a slight extent, the skin is a respiratory membrane in man, giving off carbonic acid gas, and actually absorbing oxygen.

576. The *sebaceous matter* from the *Oil-glands* anoints the hairs with oil in their progress of growth from the skin, and also imbues the cuticle, by which it is rendered repellent of water. The oiliness of the surface of the skin, occasioned by this material, permits the ready adhesion of dust and dirt, and necessitates the use of soap for the easy removal of its excess. This oily product often becomes inspissated and distends the glands, most frequently in the face, and especially on the nose; and at the mouths of the ducts it becomes mixed with dust. When pressed out it assumes the spiral form of the duct; hence it is commonly taken for a worm. In the healthiest individuals, the sebaceous matter contains a curious parasite, called the "pimple mite."

577. The uses of the *perspiration*, or sweat, are twofold: 1st, To free the system of a certain quantity of water; and 2d, To eliminate from the body certain special products of chemical changes.

The quantity of perspiration exhaled by different parts of the body differs widely. Its general quantity is influenced both by intrinsic and extrinsic conditions; thus, it is augmented by increased vascularity of the skin, by a higher temperature of the body, by a quicker circulation, and therefore by exercise and effort generally. Perspiration may also be induced by additional covering of the body, and also by peculiar conditions of the nervous system.

578. Of the external conditions which modify the quantity of perspiration, the *condition of the atmosphere* is most important. Thus, in warm air the activity of the cutaneous circulation is increased, which increases the perspiration, whilst cold air has the opposite effect; again, dry air increases the perspiration, whilst damp air diminishes it. Simple warmth acts by increasing the vascular action through the skin, whilst dryness operates by maintaining a constant evaporation from this membrane; on the other hand, cold diminishes the vascularity of the skin, and dampness of the air impedes evaporation. The combination of moisture with heat, however, increases the exhalation by the skin, which then appears in large drops. Large quantities of warm drinks also increase perspiration.

Observation.—The skin is said to regulate the quantity of fluid given off by the kidneys, and the quantity of fluid left in reserve in the blood and soft tissues generally; but the kidneys should rather be regarded as the true regulators. Observation shows that in cold weather the skin exhales less and the kidneys excrete more fluid; while in warm weather the skin eliminates more and the kidneys less.

579. The use of the non-vascular and insensible outgrowth of the epidermis, the *hair*, is protection; and the function of the *nails* is not only protection, but support to the yielding softness of the flesh at the finger-tips. When they reach exactly to the extremities of the fingers, they then fulfill the intention for which they were made, by enabling the fingers to hold both small and hard substances, and to tear and peel off skins of vegetables or animals. They are called into action where nicety of execution is required in art.

§ 51. HYGIENE OF THE ORGANS OF SPECIAL SENSE.—*Perversion of the Sense of Taste—Of Smell. How the Eye should be Used. Cause of Amaurosis. The Effect of Continued Oblique Position of the Eye—Viewing Objects at Different Distances. Bathing the Eye—Removal of Dust. Causes of Defective Hearing. Parts Essential to Hearing. Clothing. Kind of Material for Clothing. Class of Persons that need more Clothing. Cleanliness of Clothing. Bathing—Modes of Bathing—Time for Baths—General Rules for Bathing—Water a Curative Agent. Air Beneficial to the Skin. Effect of Light on the Skin. Treatment of Burns and Scalds—Of Corns—Of Frost-Bites.*

580. The *Sense of Taste* becomes perverted by the immoderate use of stimulants and condiments and the endless admixture of different articles of food. These indulgences lessen the sensibility of the nerve. In children, this sense is usually acute, and their preference is for food of the mildest character.

This sense is varied more than any other by the refinements of social life; thus, the Indian's like or dislike regarding particular articles of food generally extends to every individual of the tribe, but among civilized men no two persons are alike in all their tastes.

581. The *Sense of Smell* may become impaired by being frequently and powerfully stimulated by pungent articles, as "smelling salts;" also catarrh, or any influence that thickens the mucous membrane or renders it dry, diminishes the sensibility of the nerve of smell. Hence, the sense becomes very obtuse in persons addicted to the pernicious habit of "snuff-taking."

582. *The Eye is a delicate organ, requiring care to preserve it in health;* like other organs of the body, it should be exercised and then rested. The observance of this rule is particularly needful to those whose eyes are predisposed to inflammation. If the eye be used too long at one time, it becomes wearied and the power of vision diminished. On the contrary, if not called into exercise, its functions are enfeebled.

583. *Sudden transitions of light should be avoided.* The iris

enlarges or contracts according to the degree of light, but the change is not instantaneous. Hence the imperfect vision in passing from a strong to a dim light; an overwhelming sensation is experienced when passing from a dimly-lighted apartment to one brilliantly illuminated. A common cause of *Amaurosis*, or paralysis of the retina, is using the eye for a long time in a very intense light.

584. *Long-continued oblique position of the eye should be avoided*, or it may produce an unnatural contraction of the muscles called into action, producing squinting or strabismus. The vision of a cross eye is always defective, as only one eye is used in viewing the object toward which the attention is directed. The defect is remedied by a surgical operation. Children should not be allowed to imitate the "cross eye," as what is intended to be but temporary, may become permanent.

585. *The eye of the child should be trained to view objects at different distances.* The ciliary muscles are as capable of education as any others, and may be made to act very efficiently in adapting the lenses to view near or remote objects. Care on the part of the instructor and parent regarding the distance from the eye at which the child should hold his book or work would save many cases of defective vision.

586. *Bathing the eye in tepid or cold water is beneficial;* provided the eye be gently wiped and usually toward the inner angle; also, to remove the secretion from the lachrymal gland that sometimes collects at this angle, as it contains saline matter.

Observation.—Particles of dust or cinders should be removed from the eye by means of soft linen or silk. If the substance is concealed beneath the upper lid, take a smooth rod, like a knitting needle, place it over the upper lid in contact with and just under the edge of the orbit; hold it firmly by means of the lashes, turn the lid gently back over the pencil or needle, and remove the intrusive substance. If unsuccessful, too many attempts should not be made, as inflammation may be induced, but consult a surgeon immediately.

587. The *Sense of Hearing*, like the other senses, is capable of great improvement. By cultivation, the blind are able to judge with great accuracy of the distance of bodies in motion, and even of the height of buildings. The Indian will distinguish sounds inaudible to the untrained ear.

588. *Hearing may be impaired* by the destruction of the membrane of the tympanum. The obstruction of the Eustachian tube is not unfrequently the cause of defective hearing. By its closure, the vibratory effect of the air within the tympanum is diminished in the same manner as in the closure of the side of a drum. Enlarged tonsils, inflammation of the fauces and nasal passages, often attend and follow colds and attacks of scarlet fever, etc. For such deafness, remedial means should be directed by a skillful physician.

Observation.—The *nostrums* for the cure of deafness are usually of an oily character, and may be useful in cases of defective hearing caused by an accumulation of wax in the external canal of the ear; but a few drops of any animal oil will serve the purpose as efficiently.

589. In hearing, the integrity of the drum of the ear is not absolutely essential for the due performance of the function. The loss of the small bones does not necessarily cause deafness unless the stirrup is diseased; but if the auditory nerve or membranous vestibule becomes diseased, there is no remedial agent for the loss of hearing.

590. The *Hygiene of the Skin*, the chief organ of the Sense of Touch, holds important relation to the general health of the body. To maintain its healthy action in every part, attention must be given to *Clothing*, *Bathing*, *Light* and *Air*.

591. CLOTHING is chiefly useful in preventing the escape of too much heat from the body, and in protecting the body from exposure to the evil effects of a varying temperature of the atmosphere. In selecting and applying clothing, the following should be observed:

592. *The material for clothing should be a bad conductor of heat.* As air is a non-conductor, material should be chosen which is capable of retaining much air in its meshes; and as

moisture increases the conducting power, the material should not be such as will absorb or retain moisture. *Furs* retain much air in their meshes and absorb scarcely any moisture, and consequently, are well adapted to those subject to the great exposures of very cold climates. *Woolen cloth,* next to furs and eider down, retains the most air and absorbs the least moisture, hence it is a good article of apparel for all persons, unless too irritable to an over-sensitive skin. In that case, the flannel may be lined with cotton, or *silk* may be substituted. When of sufficient body or thickness, silk is a good article for inner clothing, excepting when it produces too much disturbance of the electricity of the system. Next to these articles, *cotton* is well adapted for garments worn next the skin. *Linen* should never be worn by persons in any way enfeebled, even in warm weather or in hot climates. It is a good conductor of heat and readily absorbs moisture; hence, with such covering, the body is surrounded by a layer of moisture instead of air.

593. *The clothing should be both porous and loosely fitted.* The necessity of porous clothing is seen in the wearing of India-rubber overshoes. In a short time the hose and under-boot become damp from retained perspiration. The residual matter thus left in contact with the skin is reconveyed into the system by absorption, causing headache and other diseases. Unimpeded transpiration, and a layer of air secured by loose clothing, enable the skin to imbibe oxygen, which gives it tone and vigor.

Observation.—As the design of additional clothing is to enclose a series of strata of warm air, we should, in going from a warm room into cold air, put on our extra covering some time previous to going out, that the layers of air which we carry with us may be warmed by the heat of the room, and not borrowed from the heat of the body.

594. *The clothing must be suited to the state of the atmosphere and to the condition of the individual.* Sudden changes of temperature should be regarded; but it is usually unsafe to make changes from thick to thin clothing, excepting in the

morning, when the vital powers are in full play. The evening usually demands an extra garment, as the atmosphere is more cool and damp, and we have less vital energy than in the early part of the day.

Observation.—Many a young lady has laid the foundation of a fatal disease by exchanging the thick dress, warm hose and shoes, for the flimsy fabric, thin hose and shoes which are considered suitable for the ball-room or party. All sudden changes of this kind are attended with hazard, which is proportionate to the weakness or exhaustion of the system when the change is made.

595. *The child and the aged person require more clothing than the vigorous person of middle age.* Judging from observation, we should infer that children needed less clothing than adults. The exposure to which the vain and thoughtless mother subjects her child very frequently lays the foundation for future disease. The system of "hardening" children, of which we sometimes hear, is as inhuman as it is unprofitable. To make the child robust and active, he must have nutritious food at stated hours, free exercise in the open air, and be guarded from the cold by *proper apparel.* Those who have outlived the energies of adult life also need special care regarding a proper amount of clothing.

596. *When a vital organ is diseased, more clothing is needed.* In consumption, dyspepsia, and even headache, the skin usually is pale and the extremities cold, because less heat is generated. Persons suffering from these complaints need more clothing than those with healthy organs.

597. *Persons of active habits need less clothing than those of sedentary employment.* Exercise increases the circulation of the blood, consequently, the vital activities become more energetic, and more heat is produced. We need less clothing when walking than when riding.

598. *The clothing should be kept clean.* Some portion of the transpired fluids of the body must necessarily be absorbed by the clothing. Hence, warmth, cleanliness and health require that it should be frequently changed and

thoroughly washed. Under-garments worn through the day should not be worn through the night, nor the reverse. When taken from the body, such garments should not be hung in the closet or put into the drawer, but exposed to a current of fresh air.

The covering of beds should be thoroughly aired every morning, and frequently renewed.

599. *Damp clothing is injurious.* All articles from the laundry should be well aired before being worn. When the clothing is wet by accident or exposure, it should be changed immediately, unless the person is exercising so vigorously as to prevent the slightest chill. When the exercise ceases, the body should be rubbed with a dry crash towel till a thorough reaction takes place.

Beds and bedding that have not been used for some weeks become damp, and should be dried before use. A hostess cannot be guilty of a more inhospitable act than that of sending her guest to her fine guest-chamber, to occupy a bed which has been long unused.

600. BATHING is indispensable to sound health as well as to cleanliness. The skin soon becomes covered with a mixture of perspirable matter, oil and dust, which, if allowed to remain, interferes with the action of the skin as an excretory organ. This increases the action of the lungs, kidneys, liver, etc., which take upon themselves the excretory work which the skin fails to perform. By overwork they soon become diseased, and if it is continued, the result will be consumption and other diseases of the vital organs. Again, obstruction of the pores will prevent respiration through the skin, and deprive the blood of one source of its oxygen and one outlet of its carbonic acid.

601. *Bathing gives tone and vigor to the internal organs.* When cool water is applied to the body, the skin instantly shrinks and the whole of its tissue contracts. This contraction diminishes the capacity of the blood-vessels, and a portion of the blood is thrown upon the internal organs. The nervous system is stimulated and communicates its stimulus

to the whole system. This causes a more energetic action of the heart and blood-vessels, and a consequent rush of blood back to the skin. This is the state termed *reaction*, the first object and purpose of every form of bathing. By this reaction the internal organs are relieved, respiration is lightened, the heart is made to beat calm and free, the tone of the muscular system is increased, the appetite is sharpened, the mind more clear and strong, and the whole system seems to possess new power. Regularity in bathing is necessary to produce permanently good effects.

602. *The simplest modes of bathing are by means of the sponge or the shallow baths.* The body may be quickly sponged over, wiped dry and followed by friction. The water may be warm or cold. If cold, the bath should be taken in the early part of the day, and followed by exercise. If exercise cannot be taken, the individual should rest under covering. The warm bath should usually be taken just before retiring. If taken at other hours, it should be followed by rest from half an hour to one hour under proper covering.

603. *The shallow bath, in which the body is partly immersed in water,* is very pleasant and safe, provided the bather exercises in it by vigorous rubbing and does not remain too long. For a cold bath it is not often safe to exceed five minutes, and with delicate persons the time should rarely exceed two or three minutes. A bath is considered cold when below 75°; temperate, from 75° to 85°; tepid, 85° to 95°. This and every other form of bath should be followed by thorough friction with a coarse towel or flesh-brush.

604. *The frequency of bathing must depend upon the condition and occupation of the individual.* Daily bathing may be practiced with profit by most persons, but to the studious and sedentary it is in most cases absolutely indispensable.

605. *The hour for ablution is of importance.* It should neither immediately precede nor follow a meal. The same is true of severe mental and muscular exercise. The bath is less beneficial in the afternoon than the forenoon. The best time for cold baths is two or three hours after breakfast.

The system is then at "flood-tide," while from that time till the retiring hour the tide is ebbing; hence, the worst time for a cold bath is at bed-time. For those who cannot choose their time, the hour of rising will answer very well—that is, for many persons, especially if they become accustomed to the use of water by beginning at another and a better hour. If the mind and body are brightened by the early bath, and an exhilaration follows, the bath is beneficial; if on the contrary, languor follows, and the skin looks blue or too pale, it is injurious. That the bath is to be followed by exercise must not be forgotten.

606. *In diseases of the skin, and many chronic ailments of the internal organs, bathing is a remedial measure of great power.* In disease which has baffled the skill of physicians depending wholly upon internal remedies, the effect of a systematic course of baths is often surprising. Like other curative means, the baths should be directed by those who thoroughly understand the use of water as a remedial agency. Matters of diet, exercise, etc., require adaptation to the treatment of the particular case. Those who desire the *full benefit* of these means must avail themselves of the appliances of a well-conducted water-cure establishment.

607. *A few simple rules must be observed in bathing.* The face and head should be wet in cold water before the bath. Cool baths should not be taken when the person is chilly, perspiring or greatly fatigued. All general baths should be taken briskly, the skin well rubbed and quickly dried, followed by a healthy glow over the whole body. Exercise should immediately follow all baths. Warm baths at night should be taken just before retiring; at other hours they should be followed immediately by rest, under coverings, after which exercise should be taken.

Soap is admirably adapted to the removal of dirt from the skin, but if it is too freely used on the general surface of the body, it dissolves the oily exudation of the sebaceous glands, leaving the skin dry or wrinkled. The external epithelial cells may be removed too rapidly when soap is used in excess, consequently the skin is not properly protected.

608. Pure AIR is an agent of great importance in the functions of the skin. It imparts to this membrane some oxygen, and receives from it carbonic acid gas. It likewise removes perspiration and portions of the oily secretion.

609. LIGHT exercises a very salutary influence upon the skin. It is no less essential to the vigor of animal than of vegetable life. Dwelling-houses should be built with reference to the free admission of sunlight and air into all occupied rooms. The dark, damp rooms so much used by indigent families and domestics in cities and large villages are fruitful causes of vice, poverty and suffering. Ladies often suffer seriously from too much exclusion of sunlight. Excepting in very warm weather, they should practice sitting or exercising in the full sunshine of the out-door world.

610. BURNS AND SCALDS. When blisters are formed, the epidermis is separated from the other layer of the skin by the effusion of serum; this fluid should be let free by puncturing the cuticle, care being taken not to remove the thin raised skin, as it makes the best possible protection to the sensitive, inflamed tissues beneath. When this thin outside layer of skin is removed, immediately cover the denuded parts with wheat flour, or a plaster made of lard and bees'-wax or the white of an egg; in a word, substitute a cuticle to protect the exposed nerves from the air. When dressings are applied, they should not be removed until they become dry and irritating.

To prevent vesication, when only a small patch of the skin is scalded or burned, apply *steadily* cold water until the smarting pain ceases; then put on a simple dressing, "not to take out the fire or heal it," but to protect the injured membrane.

611. When the epidermis, in particular spots, is exposed to excessive pressure or friction, it becomes too much thickened, producing *Corns*. They are not necessarily confined to the feet, but are produced in front of the clavicle of the soldier

from the pressure of his musket, or on the knee of the cobbler The pain of the callosity is due to their exciting inflammation in the sensitive dermis upon which they press. Remove the pressure, and the affected part is restored to its normal state.

612. FROST-BITE is usually manifested first upon parts unprotected by covering, as the face or ears, and especially the nose. In such case, the skin first becomes red, from congestion of the dilated capillary vessels; next it becomes bluish, from arrest of the circulation; and afterward of a dead white hue. To restore circulation and sensibility, rub the frozen part with snow or apply iced water. Keep the sufferer at first in a cold room, and let the return to a higher temperature be *gradual* and cautious, or *gangrene* may supervene. The *Chilblain* is not produced by the action of cold, but by the effect of heat on the chilled extremity.

APPENDIX.

CHAPTER XIII.

CARE OF THE SICK.

§ **1.** In every home, however humble or dignified, woman is usually the Nurse. Nature seems to have endowed her in an especial manner to minister at the couch of disease and suffering. To be a good nurse requires a high type of womanhood; she should have both mental and physical power, blended with integrity and Christian trust.

If "good nursing is half the cure," how important that the daughter be early taught how to prepare drinks and nourishments; to administer medicine; and to perform the varied and important duties of the faithful nurse!

The physician well knows that his attentions upon the sick are quite unavailing unless the nurse *obeys* his directions. For a nurse, or immediate relatives or friends of the sick, to put their judgment in opposition to that of the physician, is not only arrogant, but endangers the patient. The *room* for the sick should be selected where sunlight may enter, and as far from external noise as possible. It is poor economy, not to say unkind, to keep a sick man in a small, ill-arranged bed-room, when a more spacious and airy room is kept for only occasional "callers." All superfluous furniture should be removed from the sick room.

In the first stages of disease, it is always proper treatment to *rest* both body and mind. It is wrong to tempt the appetite of a sick person; the disinclination for food is the warning of Nature that the system cannot well digest it.

The beneficial effects of *bathing* can hardly be over-estimated, but the mode of the bath should be directed by the medical adviser. The best time, however, for bathing is when the patient feels most vigorous and freest from exhaustion. Care is necessary to wipe dry the skin, particularly between the fingers and toes, and also the flexions of the joints. Friction from a brush, moreen mitten or a dry flannel that has been saturated with salted water tends to relieve restlessness in patients. Air-baths have a tranquillizing influence.

Quiet should reign in the sick room. No more persons should enter or remain in it than the welfare of the patient demands. It is the duty of the physician to direct when visitors should be admitted or excluded, and the nurse should enforce the directions. The movements of the attendants should be gentle: no bustling to "clear up the room" at a fixed time; this should be done quietly and when it will give the least annoyance to the sick. (It may be necessary to use a *damp* cloth in dusting the furniture, also the carpet, especially if the patient has disease of the lungs.) Creaking hinges should be oiled; shutting doors violently and heavy walking avoided. All unnecessary conversation should be deferred. If a colloquy must be carried on, let the tone be so high that the patient, if interested, can thoroughly comprehend it.

The *making of the bed* is often badly conducted. All bunches should be removed, the material of the bed laid even and a thin quilt spread smoothly over a mattress. When convenient, have the head of the bed northerly (182), and so situated, at least, that the sick man may look on something more pleasurable than a table of glasses and phials. A nurse should never manifest impatience in arranging the pillows, but try to adapt them to the comfort of the weary patient.

All utensils employed in the sick room should be kept clean. Water designed for the patient to drink should not stand long in an open glass or pitcher, but be given fresh from a *spring* or well. A very sick person is fatigued by being raised to receive drinks, hence, a bent tube or a cup with a spout should be used.

Both the *apparel and the bed-linen* should be changed more frequently in sickness than in health, and oftener in acute than in chronic diseases. All clothing, whether from the laundry or bureau, should be well dried and warmed by a fire previous to being put on the bed or the patient.

No agent is of more importance to the sick room than *pure air;* hence, the nurse, with all convenient speed, should remove everything that can emit an unpleasant odor. She should be chary of keeping ripe fruit or bouquets of flowers any length of time in the sick chamber. When a disinfectant is needed, procure at the druggist's, chloride of lime. To change quickly and effectively the air of the sick room, cover the patient's bed with an extra blanket and closely envelop his head and neck, except the mouth and nose; the door and windows can then be safely opened for a short time without detriment. After the windows are closed, retain the extra coverings on the patient until the room is of proper warmth. Unless duly protected, the patient should never feel *currents of air*, although *fresh air* should be constantly admitted into the sick room.

A well-adjusted thermometer is indispensable, as the feelings of the patient or nurse are not to be relied on as a true index of the *temperature* of the room. Regulating the warmth of the patient is one of the many duties of the nurse. There is a "sweating temperature;" when this is exceeded, perspiration will cease if it has been present; or that it will not take place during a high temperature. The patient should no more be allowed to complain of *too much heat*, without an attempt at its reduction, than he should be permitted to remain chilly when it is possible to remove it.

The nurse should not confine herself to the sick room longer than six hours at a time. She should exercise daily in the open air, also eat and sleep as regularly as possible. No doubts or fears of the patient's recovery, either by a look or by a word, should be communicated by the nurse in the chamber of the sick; this duty devolves upon the physician.

Medicines assist the natural powers of the system to remove disease. They should be given regularly, judiciously and with a cheerful manner. Life itself is often at the mercy of the nurse, and depends on the faithful discharge of her duty.

Drinks have a more decided influence upon the system than is generally admitted; hence, the nurse should never depart from the quality of the drink, nor even exceed the due or prescribed quantity. Giving "herb teas" without the sanction of the physician may cause serious evil.

The *food* of the sick should be prepared in the neatest and most careful manner, and the nurse ought to obey implicitly the physician's directions about diet. When a patient is convalescent, the desire for food is generally strong; great care, firmness and patience is required, that the food be prepared suitably and given at the proper time.

We append a few modes of preparing nourishment for the sick.

CRUST COFFEE.—Take light, sweet bread or crackers, and brown them *thoroughly* as you would coffee berry; when wanted for use, pour over boiling water (the crusts will admit of several replenishings of boiling water); add sugar and cream to suit the condition of the patient.

GRUELS.—*Corn* meal requires to be boiled several hours to be suitable nourishment for the sick. The mode of preparing gruel should be suited to the case and directed by the physician. Wheat, or oat-meal, farina and sago, can be prepared in less time, though they must be well cooked. Add salt while cooking.

Egg Gruel.—Take the *yolks* of two eggs, boiled hard, and with a knife reduce them to a fine powder; beat this into a flour gruel made of new milk; salt and spices may be added if the condition of the patient admits.

BEEF TEA.—Meat contains principles that may be extracted, some by *cold*, others by *warm*, and others, again, by *boiling*, water; it should be cut very fine, and submitted for three hours each time, in succession, to half its weight of cold, of warm and of boiling water; the fluids strained from the first and second macerations are to be mixed with that strained from the boiling process, and the mixture should be brought to a boiling heat to cook it—the fat skimmed off; add a few drops of some acid, with salt, for a flavor.

§ **2.** The duty of the WATCHER is scarcely less responsible than that of the nurse; and, like the nurse, she should ever be cheerful, kind, firm and attentive in the presence of the patient.

The watcher should be prompt, and reach the house of the sick at an early hour; before entering the sick room, she should eat a simple, nutritious supper, and also during the night take some plain food. She should be furnished with an extra garment, as a heavy shawl, to wear toward morning, when the system becomes exhausted.

The directions about the sick, especially the administration of medicine, should be *written* for the temporary watcher. Whatever may be wanted during the night should be brought into the sick chamber or the adjoining room before the family retires to sleep, that the slumbers of the patient be not disturbed by haste or searching for needed articles.

Sperm candles are preferable for the sick room. Kerosene, in burning, emits a disagreeable odor, often annoying to the patient. All lights ought to be so arranged as not to be reflected in the part of the room where the sick lie.

It is not necessary that watchers make themselves acceptable to the patient by exhausting conversation. If two watchers are needed, it is more imperative that they refrain from talking, and particularly *whispering*.

Most sick persons have special need of nourishment about four or five o'clock in the morning.

When taking care of the sick, light-colored clothing should be worn in preference to dark apparel, especially if the disease is of a contagious character. It is always safe for the watcher to change her apparel worn in the sick chamber before entering upon her family duties. Disease is often communicated by the clothing.

It can hardly be expected that the farmer who has been laboring hard in the field, or the mechanic who has toiled during the day, is qualified to render all those little attentions that a sick person requires. Hence, would it not be more benevolent and economical to employ and *pay* watchers who are qualified by knowledge and *training* to perform this duty in a faithful manner, while the kindness and sympathy of friends may be *practically* manifested by assisting to defray the expenses of these qualified and useful assistants?

POISONS AND THEIR ANTIDOTES.

§ 3. POISONING, either from accident or design, is of such frequency, that every household should keep some available remedy, and every person should know *what to do* in such alarming contingencies. Nearly every poison has its antidote, which, *if used at once*, may prevent much suffering and even death.

When known that poison has been taken into the stomach, the first thing is to evacuate it by the use of the stomach-pump or an emetic, unless vomiting takes place spontaneously.

As an emetic, ground *Mustard* mixed in warm water is always safe. Take one tablespoonful to one pint of warm water. Give the patient one-half in the first instance, and the remainder in fifteen minutes, if vomiting has not commenced. In the interval, drink copious draughts of warm water. Irritate the throat with a feather or the finger, to induce vomiting. After vomiting has begun, give mucilaginous drinks; such as flaxseed tea, gum-arabic water, or slippery elm.

If the patient is drowsy, give a strong infusion of cold coffee, keep him walking, slap smartly on the back, use electricity; it may be well to dash cold water on the head, to keep the patient awake

After the poison is evacuated from the stomach, to sustain vital action, give warm water and wine or brandy. If the limbs are cold, apply warmth and friction.

In ALL cases of poisoning, call immediately a physician, as the after-treatment is of great importance.

POISONS.	ANTIDOTES OR REMEDIES FOR POISONS.
Aconite (Monkshood). Belladona (Deadly Night-Shade). Bryony. Camphor. Conium / Cicuta (Water Hemlock). Croton Oil. Digitalis (Foxglove). Dulcamara (Bitter-Sweet). Gamboge. Hyoscyamus (Henbane). Laudanum. Lobelia. Morphine. Opium. Paregoric. Sanguinara (Blood-Root). Savin Oil. Spigelia (Carolina Pink). Stramonium (Thorn Apple). Strychnine (Nux Vomica). Tobacco.	For *Vegetable* poisons give an emetic of *Mustard;* drink freely of warm water; irritate the throat with a feather to induce vomiting. Keep the patient awake until a physician arrives.
Arnica.	*Vinegar* and water.
Prussic Acid. Bitter Almonds (Oil of). Laurel Water.	Drink, at once, one teaspoonful of *Water of Hartshorn* (ammonia) in one pint of water.
Ammonia (Hartshorn). Potash. Soda.	Antidote is *Vinegar* or *Lemon Juice;* followed with sweet, castor or linseed oil. Thick cream is a substitute for oil. No emetic.
Iodine.	Starch or wheat flour beat in water. Take a Mustard emetic.

Poisons.	Antidotes or Remedies for Poisons.
Saltpetre (Nitrate of Potassa). Chili Saltpetre (Nitrate of Soda).	Take at once, a *Mustard* emetic; drink copious draughts of warm water; followed with oil or cream.
Lunar Caustic (Nitrate of Silver).	Two teaspoonfuls of table salt (chloride of sodium) mixed in one pint of water.
Corrosive Sublimate (bug poison). White Precipitate. Red Precipitate. Vermilion.	Beat the *Whites of six Eggs* in one quart of cold water; give a cupful every two minutes, to induce vomiting. A substitute for white of eggs is *soap-suds* slightly thickened with wheat flour. Emetics should not be given.
Arsenic. Cobalt (fly powder). King's Yellow. Ratsbane. Scheele's Green.	Use a *stomach-pump* as quickly as possible, or give a Mustard emetic until one is obtained. After free vomiting, give large quantities of *Calcined Magnesia.* The antidote for Arsenic is *Hydrated Peroxide of Iron.*
Acetate of Lead (Sugar of Lead). White Lead. Litharge.	Use a *Mustard* emetic; followed by Epsom or Glauber Salts. The antidote is diluted *Sulphuric Acid.*
Antimony (Wine of). Tartar Emetic.	The antidote is ground *Nutgall.* A substitute, oak or Peruvian bark; followed by a teaspoonful of paregoric.
Pearl-ash. Ley (from wood-ashes). Salts of Tartar.	Drink freely of *Vinegar* and water; followed with a mucilage, as flax-seed tea.
Sulphuric Acid (Oil of Vitriol). Nitric " (Aquafortis). Muriatic " (Marine). Oxalic Acid.	Drink largely of water or a mucilage. It is important that something *be given quickly,* to neutralize the acid. The antidote is *Calcined Magnesia.* Chalk, lime, strong soap-suds are substitutes for magnesia.

Poisons.	Antidotes or Remedies for Poisons.
Matches (Phosphorus). Rat Exterminator.	Give two tablespoonfuls of *Calcined Magnesia;* followed by mucilaginous drinks.
Verdigris. Blue Vitriol.	The antidote is *Cooking Soda,* or *White of Eggs.* Drink milk freely.
Sting of Insects.	*Ammonia,* or cooking soda moistened with water, applied in the form of a paste. The wound may be sucked, followed by applications of water.
Charcoal Fumes. Gas or Burning Fluid.	*Fresh* air and Artificial Respiration.

For the Treatment of Wounds and Arrest of Hemorrhage (363).
For the Recovery of Asphyxiated Persons (430).
For Burns and Scalds (610).

GLOSSARY.

AB-DO'MEN. [L. *abdo*, to hide.] That part of the body which lies between the thorax and the bottom of the pelvis.

AB-DOM'IN-IS. Pertaining to the abdomen.

AB-DUC'TOR. [L. *abduco*, to lead away.] A muscle which moves certain parts, by separating them from the axis of the body.

A-CE-TAB'U-LUM. [L. *acetum*, vinegar.] The socket for the head of the thigh-bone; an ancient vessel for holding vinegar.

A-CE'TIC. [L. *acetum*, vinegar.] Relating to acetic acid. This is always composed of oxygen, hydrogen and carbon in the same proportion.

A-CHIL'LIS. A term applied to the tendon of the two large muscles of the leg.

A-CRO'MI-ON. Gr. ακρος, *akros*, highest, and ωμος, *omos*, shoulder.] A process of the scapula that joins to the clavicle.

AD-DUC'TOR. [L. *adduco*, to lead to.] A muscle which draws one part of the body toward another.

AL-BU'MEN. [L. *albus*, white.] An animal substance of the same nature as the white of an egg.

A-LU'MIN-UM. [L.] The name given to the metallic base of alumina.

AL'VE-O-LAR. [L. *alveolus*, a socket.] Pertaining to the sockets of the teeth.

AM-MO'NI-A. An alkali. It is composed of three equivalents of hydrogen and one of nitrogen.

AM-PHI-AR-THRO'SIS. [Gr. αμφι, *amphi*, both, and αρθρωδια, *arthrodia*, well articulated.] A mixed articulation.

A-NAS'TO-MOSE. [Gr. ανα, *ana*, through, and στομα, *stoma*, mouth.] The communication of arteries and veins with each other.

AN-A-TOM'I-CAL. Relating to the parts of the body when dissected or separated.

A-NAT'O-MY. [Gr. ανα, *ana*, through, and τομη, *tomē*, a cutting.] The description of the structure of animals. The word *anatomy* properly signifies dissection.

AN-GI-OL'O-GY. [Gr. αγγειον, *angeion*, a vessel, and λογος, *logos*, discourse.] A description of the vessels of the body; as the veins and arteries.

AN'GU-LI. [L. *angulus*, a corner.] A term applied to certain muscles on account of their form.

AN-I-MAL'CU-LÆ. [L. *animalcula*, a little animal.] Animals that are only perceptible by means of a microscope.

AN'NU-LAR. [L. *annulus*, a ring.] Having the form of a ring.

AN-TI'CUS. [L.] A term applied to certain muscles.

A-ORT'A. [Gr. αορτη, *aortē*; from αηρ, *aēr*, air, and τηρεω, *tereo*, to keep.] The great artery that arises from the left ventricle of the heart.

AP-O-NEU-RO'SIS. [Gr. απο, *apo*, from, and νευρον, *neuron*, a nerve.] The membranous expansions of muscles and tendons. The ancients called every white tendon *neuron*, a nerve.

AP-PA-RA'TUS. [L. *apparo*, to prepare.] An assemblage of organs designed to produce certain results.

AP-PEND'IX. [L. *ad* and *pendeo*, to hang from.] Something appended or added.

A'QUE-OUS. [L. *aqua*, water.] Partaking of the nature of water.

A-RACH'NOID. [Gr. αραχνη, *arachnē*, a spider, and ειδος, *eidos*, form.] Resembling a spider's web. A thin membrane that covers the brain.

AR'BOR. [L.] A tree. *Arbor vitæ*. The tree of life. A term applied to a part of the cerebellum.

AR'TE-RY. [Gr. αηρ, *aēr*, air, and τηρεω,

tĕreo, to keep; because the ancients thought that the arteries contained only air.] A tube through which blood flows from the heart.

A-RYT-E′NOID. [Gr. αρυταινα, *arutaina*, a ewer, and ειδος, *eidos*, form.] The name of a cartilage of the larynx.

AS-CEND′ENS. [L.] Ascending; rising.

AS-PHYX′I-A. [Gr. α, *a*, not, and σφυξις, *sphyxis*, pulse.] Originally, want of pulse; now used for suspended respiration, or apparent death.

AS-TRAG′A-LUS. [Gr.] The name of a bone of the foot. One of the tarsal bones.

AUD-I′TION. [L. *audio*, to hear.] Hearing.

AUD-IT-O′RI-US. [L.] Pertaining to the organ of hearing.

AU′RI-CLE. [L. *auricula*, the external ear; from *auris*, the ear.] A cavity of the heart.

AX-IL′LA. [L.] The armpit.

AX′IL-LA-RY. Belonging or relating to the armpit.

A-ZOTE′. [Gr. α, *a*, not, and ζωη, *zoē*, life.] Nitrogen. One of the constituent elements of the atmosphere. So named because it will not sustain life.

BEN-ZO′IC. *Benzoic acid*. A peculiar vegetable acid obtained from benzoin and some other balsams.

BI′CEPS. [L. *bis*, twice, and *caput*, a head.] A name applied to muscles with two heads at one extremity.

BI-CUS′PIDS. [L. *bis*, two, and *cuspis*, a point.] Teeth that have two points upon their crown.

BILE. [L. *bilis*.] A yellow, viscid fluid secreted by the liver.

BI-PEN′NI-FORM. [L. *bis*, two, and *penna*, a feather.] Having fibres on each side of a common tendon.

BRACH′I-AL. [L. *brachium*.] Belonging to the arm.

BRE′VIS. [L.] *Brevis*, short; *brevior*, shorter.

BRONCH′I-A, -Æ. [L.] A division of the trachea that passes to the lungs.

BRONCH-I′TIS. [L.] An inflammation of the brenchia.

BUC-CI-NA′TOR. [L. *buccinum*, a trumpet.] The name of a muscle of the cheek, so named because used in blowing wind instruments.

BUR′SÆ MU-CO′SA [L. *bursa*, a purse, and *mucosa*, viscous.] Small sacs, containing a viscid fluid, situated about the joints, under tendons.

CÆ′CUM. [L.] Blind; the name given to the commencement of the colon.

CAL′CI-UM. [L.] The metallic basis of lime

CALX, CAL′CIS. [L.] The heel-bone.

CAP′IL-LA-RY. [L. *capillus*, a hair.] Resembling a hair; a small tube.

CAP′SULE. [L. *capsula*, a little chest.] A membranous bag, enclosing a part.

CA′PUT. [L.] The head. *Caput coli*, the head of the colon.

CAR′BON. [L. *carbo*, a coal.] Pure charcoal. An elementary combustible substance.

CAR-BON′IC. Pertaining to carbon.

CAR′DI-AC. [Gr. καρδια, *kardia*, heart.] Relating to the heart, or upper orifice of the stomach.

CAR′NE-A, -Æ. [L. *caro*, *carnis*, flesh.] Fleshy.

CA-ROT′ID. [Gr. καρος, *karos*, lethargy.] The great arteries of the neck that convey blood to the heart. The ancients supposed drowsiness to be seated in these arteries.

CAR′PUS, -I. [L.] The wrist.

CAR′TI-LAGE. [L. *cartilago*.] Gristle. A smooth, elastic substance, softer than bone.

CAU-CA′SIAN. One of the races of men.

CA′VA. [L.] Hollow. *Vena Cava*. A name given to the two great veins of the body.

CEL′LU-LAR. [L. *cellula*, a little cell.] Composed of cells.

CER-E-BEL′LUM. [L.] The hinder and lower part of the brain, or the little brain.

CER′E-BRO-SPI′NAL. Relating to the brain and spine.

CER′E-BRUM. [L.] The front and large part of the brain. The term is sometimes applied to the whole contents of the cranium.

CER′VI-CAL. Relating to the neck.

CER′VIX. [L.] The neck.

CHEST. [Sax.] The thorax; the trunk of the body from the neck to the abdomen.

CHLO′RINE. [Gr. χλωρος, *chloros*, green.] *Chlorine gas*, so named from its color.

CHOR′DA, -Æ. [L.] A cord. An assemblage of fibres.

CHO'ROID. [Gr. χοριον, *chorion.*] A term applied to several parts of the body that resemble the skin.

CHYLE. [Gr. χυλος, *chulos,* juice.] A nutritive fluid, of a whitish appearance, which is extracted from food by the action of the digestive organs.

CHYL-I-FI-CA'TION. [L. *chylus,* chyle, and *facio,* to make.] The process by which chyle is formed.

CHYME. [Gr. χυμος, *chumos,* juice.] A kind of grayish pulp formed from the food in the stomach.

CHYM-I-F-ICA'TION. [L. *chumos,* chyme, and *facio,* to make.] The process by which chyme is formed.

CIL'IA-RY. [L. *cilia,* eyelashes.] Belonging to the eyelids.

CIN-E-RI'TIOUS. [L. *cinus,* ashes.] Having the color of ashes.

CLAV'I-CLE. [L. *clavicula,* from *clavis,* a key.] The collar-bone; so called from its resemblance in shape to an ancient key.

CLEI'DO. A term applied to some muscles that are attached to the clavicle.

CO-AG'U-LUM. [L.] A coagulated mass; a clot of blood.

COC'CYX. [Gr.] An assemblage of bones joined to the sacrum.

COCH'LE-A. [Gr. κοχλω, *kochlo,* to twist; or L. *cochlea,* a screw.] A cavity of the ear resembling in form a snail-shell.

CO'LON. [Gr. κωλον, *kolon,* I arrest.] A portion of the large intestine.

CO-LUM'NA, -Æ. [L.] A column or pillar.

COM'MIS-SURE. [L. *committo,* I join together.] A point of union between two parts.

COM-MU'NIS. [L.] A name applied to certain muscles.

COM-PLEX'US. [L. *complector,* to embrace.] The name of a muscle that embraces many attachments.

COM-PRESS'OR. [L. *con,* together, and *premo, pressus,* to press.] A term applied to some muscles that compress the parts to which they are attached.

CON'DYLE. [Gr. κονδυλος, *kondulos,* a knuckle, a protuberance.] A prominence on the end of a bone.

CON-JUNC-TI'VA. [L. *con,* together, and *jungo,* to join.] The membrane that covers the anterior part of the globe of the eye.

COP'PER. A metal of a pale red color tinged with yellow.

COR-A'COID. [Gr. κοραξ, *korax,* a crow, and ειδος, *eidos,* form.] A process of the scapula shaped like the beak of a crow.

CO'RI-UM. [Gr. χοριον, *chorion,* skin.] The true skin.

CORN'E-A. [L. *cornu,* a horn.] The transparent membrane in the fore part of the eye.

COR'PO-RA. [L. *corpus,* a body.] The name given to eminences or projections found in the brain and some other parts of the body.

COS'TA. [L. *costa,* a coast, side or rib.] A rib.

CRIB'RI-FORM. [L. *cribrum,* a sieve, and *forma,* form.] A plate of the ethmoid bone, through which the olfactory nerve passes to the nose.

CRI'COID. [Gr. κρικος, *krikos,* a ring, and ειδος, *eidos,* form.] A name given to a cartilage of the larynx, from its form.

CRYS'TAL-LINE. [L. *crystallinus,* consisting of crystal.] *Crystalline lens,* one of the humors of the eye. It is convex, white, firm and transparent.

CU'BI-TUS, -I. [L. *cubitus,* the elbow.] One of the bones of the forearm, also called the *ulna.*

CU'BOID. [Gr. κυβος, *kubos,* a cube, and ειδος, *eidos,* form.] Having nearly the form of a cube.

CU-NE'I-FORM. [L. *cuneus,* a wedge.] The name of bones in the wrist and foot.

CUS'PID. [L. *cuspis,* a point.] Having one point.

CU-TA'NE-OUS. [L. *cutis,* skin.] Belonging to the skin.

CU'TI-CLE. [L. *cutis.*] The external layer of the skin.

CU'TIS VE'RA. [L. *cutis,* skin, and *vera,* true.] The internal layer of the skin; the true skin.

DE-CUS-SA'TION. [L. *decutio,* I divide.] A union in the shape of an X or cross.

DEL'TOID. [Gr. δελτα, *delta,* the Greek letter Δ, and ειδος, *eidos,* form.] The name of a muscle that resembles in form the Greek letter Δ.

DENT'AL. [L. *dens,* tooth.] Pertaining to the teeth.

DE-PRESS'OR. [L.] The name of a muscle

that draws down the part to which it is attached.

DERM'OID. [Gr. δερμα, *derma*, the skin, and ειδος, *eidos*, form.] Resembling skin.

DE-SCEND'ENS. [L. *de* and *scando*, to climb.] Descending, falling.

DI'A-PHRAGM. [Gr. διαφραγμα, *diaphragma*, a partition.] The midriff; a muscle separating the chest from the abdomen.

DI-AR-RHŒ'A. [Gr. διαρρεω, *diarrheo*, to flow through.] A morbidly frequent evacuation of the intestines.

DI-AR-THRO'SIS. [Gr. δια, *dia*, through, and αρθρουν, *arthroun*, to fasten by a joint.] An articulation which permits the bones to move freely on each other in every direction.

DI-AS'TO-LE. [Gr. διαστελλω, *diastello*, to put asunder.] The dilatation of the heart and arteries when the blood enters them.

DI-GES'TION. [L. *digestio.*] The process of dissolving food in the stomach and preparing it for circulation and nourishment.

DIG-I-TO'RUM. [L. *digitus*, a finger.] A term applied to certain muscles of the extremities.

DOR'SAL. [L. *dorsum*, the back.] Pertaining to the back.

DU-O-DE'NUM. [L, *duodenus*, of twelve fingers' breadth.] The first portion of the small intestine.

DU'RA MA'TER. [L. *durus*, hard, and *mater*, mother.] The outermost membrane of the brain.

DYS'EN-TER-Y. [Gr. δυς, *dūs*, bad, and εντερια, *enteria*, intestines.] A discharge of blood and mucus from the intestines attended with tenesmus.

DYS-PEP'SI-A. [Gr. δυς, *dūs*, bad, and πεπτω, *pepto*, to digest.] Indigestion, or difficulty of digestion.

EN-AM'EL. [Fr.] The smooth, hard substance which covers the crown or visible part of a tooth.

EN-DOS-MO'SIS. [Gr. ενδον, *endon*, within, and ωσμος, *osmos*, to push.] The transmission of fluids through membranes, inward.

E-PEN'DY-MA. [Gr.] The membrane which lines the ventricles of the brain.

EP-I-DERM'IS. [Gr. επι, *epi*, upon, and δερμα, *derma*, the skin.] The superficial layer of the skin.

EP-I-GLOT'TIS. [Gr. επι, *epi*, upon, and γλωττα, *glōtta*, the tongue.] One of the cartilages of the glottis.

ETH'MOID. [Gr. ηθμος, *ethmos*, a sieve, and ειδος, *eidos*, a form.] A bone of the skull.

EU-STA'CHI-AN TUBE. A channel from the fauces to the middle ear, named from Eustachius, who first described it.

EX'CRE-MENT. [L. *excerno*, to separate.] Matter excreted and ejected; alvine discharges.

EX'CRE-TO-RY. A little duct or vessel, destined to receive secreted fluids and to excrete or discharge them; also a secretory vessel.

EX-HA'LANT. [L. *exhalo*, to send forth vapor.] Having the quality of exhaling or evaporating.

EX-TENS'OR. [L.] A name applied to a muscle that serves to extend any part of the body; opposed to *Flexor*.

FA'CIAL. [L. *facies*, face.] Pertaining to the face.

FALX. [L. *falx*, a scythe.] A process of the dura mater shaped like a scythe.

FAS'CI-A. [L. *fascia*, a band.] A tendinous expansion or aponeurosis.

FAS-CIC'U-LUS, -LI. [L. *fascis*, a bundle.] A little bundle.

FAUX, -CES. [L.] The top of the throat.

FEM'O-RAL. Pertaining to the femur.

FE'MUR. [L.] The thigh-bone.

FE-NES'TRA, -UM. [L. *fenestra*, a window.] A term applied to some openings into the internal ear.

FI'BRE. [L. *fibra.*] An organic filament or thread which enters into the composition of every animal and vegetable texture.

FI'BRIN. A peculiar organic substance found in animals and vegetables; it is a solid substance, tough, elastic and composed of thready fibres.

FI'BRO-CAR'TI-LAGE. An organic tissue, partaking of the nature of fibrous tissue and that of cartilage.

FIB'U-LA. [L., a clasp.] The outer and lesser bone of the leg.

FIL'A-MENT. [L. *filamenta*, threads.] A fine thread, of which flesh, nerves, skin, etc., are composed.

FLEX'ION. [L. *flectio.*] The act of bending.

FOL'LI-CLE. [L. *folliculus*, a small bag.] A gland; a little bag in animal bodies.

FORE'ARM. The part of the upper extremity between the elbow and hand.

FOS'SA. [L., a ditch.] A cavity in a bone, with a large aperture.

FRÆ'NUM. [L., a bridle.] *Frænum linguæ*, the bridle of the tongue.

FUNC'TION. [L. *fungor*, to perform.] The action of an organ or system of organs.

FUN'GI-FORM. [L. *fungus* and *forma*.] Having terminations like the head of a fungus, or a mushroom.

GAN'GLI-ON, -A. [Gr.] An enlargement in the course of a nerve.

GAS'TRIC. [Gr. γαστηρ, *gastēr*, the stomach.] Belonging to the stomach.

GAS-TROC-NE'MI-US. [Gr. γαστηρ, *gastēr*, the stomach, and κνημη, *knēmē*, the leg.] The name of large muscles of the leg.

GEL'A-TIN. [L. *gelo*, to congeal.] A concrete animal substance, transparent and soluble in water.

GING'LY-FORM. [Gr. γιγγλυμος, *ginglymos*, a knife-like joint, and ειδος, *eidos*, a form.] An articulation that only admits of motion in two directions.

GLE'NOID. [Gr. γληνη, *glēnē*, a cavity.] A term applied to some articulate cavities of bones.

GLOS'SA. [Gr.] The tongue. Names compounded with this word are applied to muscles of the tongue.

GLOS'SO-PHA-RYN'GI-AL. Relating to the tongue and pharynx.

GLOT'TIS. [Gr.] The narrow opening at the upper part of the larynx.

GLU'TE-US. [Gr.] A name given to muscles of the hip.

GOM-PHO'SIS. [Gr. γομφουν, *gomphoun*, a nail.] The immovable articulation of the teeth with the jaw-bone, like a nail in a board.

HEM'OR-RHAGE. [Gr. αιμα, *haima*, blood, and ρηγνυω, *rēgnuo*, to burst.] A discharge of blood from an artery or brain.

HIS-TOL'O-GY. [Gr. ιστος, *histos*, tissue, and λογος, *logos*, discourse.] A description of the minute structure of the body.

HU'MER-US. [L.] The bone of the arm.

HY'A-LOID. [Gr.] A transparent membrane of the eye.

HY'DRO-GEN. [Gr. ὑδωρ, *hydor*, water, and γενναω, *gennao*, to generate.] A gas which constitutes one of the elements of water.

HY'GI-ENE. [Gr. ὑγιεινον, *hugieinon*, health.] The part of medicine which treats of the preservation of health.

HY'OID. [Gr. υ and ειδος, *eidos*, shape.] A bone of the tongue resembling the Greek letter Upsilon in shape.

HY'PO-GLOS'SAL. Under the tongue. The name of a nerve of the tongue.

IL'E-UM. [Gr. ειλεω, *eilō*, to wind.] A portion of the small intestines.

IL'I-UM. The haunch-bone.

IN-CI'SOR. [L. *incido*, to cut.] A front tooth that cuts or divides.

IN'DEX. [L. *indico*, to show.] The fore-finger; the pointing finger.

IN-NOM-I-NA'TA. [L. *in*, not, and *nomen*, name.] Parts which have no proper name.

IN-OS'CU-LATE. [L. *in*, and *osculatus*, from *osculor*, to kiss.] To unite, as two vessels at their extremities.

IN'TER. [L.] Between.

IN-TER-COST'AL. [L. *inter*, between, and *costa*, a rib.] Between the ribs.

IN-TER-NO'DI-I. [L. *inter*, between, and *nodus*, knot.] A term applied to some muscles of the forearm.

IN-TER-STI'TIAL. [L. *inter*, between, and *sto*, to stand.] Pertaining to or containing interstices.

IN-TES'TINES. [L. *intus*, within.] The canal that extends from the stomach to the anus.

I'RIS. [L., the rainbow.] The colored circle that surrounds the pupil of the eye.

I'VO-RY. A hard, solid, fine-grained substance of a fine white color; the tusk of an elephant.

JE-JU'NUM. [L., empty.] A portion of the small intestine.

JU'GU-LAR. [L. *jugulum*, the neck.] Relating to the throat. The great veins of the neck.

LA'BI-UM, LA'BI-I. [L.] The lips.

LAB'Y-RINTH. [Gr.] The internal ear, so named from its many windings.

LACH'RY-MAL. [L. *lachryma*, a tear.] Pertaining to tears.

LAC'TE-AL. [L. *lac*, milk.] A small vessel or tube of animal bodies for conveying chyle from the intestine to the thoracic duct.

LAM'I-NA, -Æ. [L.] A plate or thin coat lying over another.

LAR'YNX. [Gr. λαρυγξ, *larunx*.] The upper part of the windpipe.

LAR-YN-GI'TIS. Inflammation of the larynx.

LA-TIS'SI-MUS, -MI. [L., superlative of *latus*, broad.] A term applied to some muscles.

LE-VA'TOR. [L. *levo*, to raise.] A name applied to a muscle that raises some part.

LIG'A-MENT. [L. *ligo*, to bind.] A strong, compact substance serving to bind one bone to another.

LIN'E-A, -Æ. [L.] A line.

LIN'GUA, -Æ. [L.] A tongue.

LIV'ER. The name of one of the abdominal organs, the largest gland in the system. It is situated below the diaphragm, and secretes the bile.

LOBE. A round projecting part of an organ.

LON'GUS, LON'GI-OR. [L., long, longer.] A term applied to several muscles.

LUM'BAR. [L. *lumbus*, the loins.] Pertaining to the loins.

LYMPH. [L. *lympha*, water.] A colorless fluid in animal bodies, and contained in vessels called lymphatics.

LYM-PHAT'IC. A vessel of animal bodies that contains or conveys lymph.

MAG-NE'SI-UM. The metallic base of magnesia.

MAG'NUS, -NA, -NUM. [L. great.] A term applied to certain muscles.

MA'JOR. [L., greater.] Greater in extent or quantity.

MAN'GA-NESE. A metal of a whitish gray color.

MAR'ROW. [Sax.] A soft, oleaginous substance contained in the cavities of bones.

MAS-SE'TER. [Gr. μασσαομαι, *massaomai*, to chew.] The name of a muscle of the face.

MAS'TI-CATE, MAS-TI-CA'TION. [L. *mastico*.] To chew; the act of chewing.

MAS'TOID. [Gr. μαστος, *mastos*, breast, and ειδος, *eidos*, form.] The name of a process of the temporal bone behind the ear.

MAS-TOID'E-US. A name applied to muscles that are attached to the mastoid process.

MAX-IL'LA. [L.] The jaw-bone.

MAX'I-MUS, -UM. [L., superlative of *magnus*, great.] A term applied to several muscles.

ME-A'TUS. [L. *meo*, to go.] A passage or channel.

ME-DI-AS-TI'NUM. A membrane that separates the chest into two parts.

ME'DI-UM, -A. [L.] The space or substance through which a body passes to any point.

MED'UL-LA-RY. [L. *medulla*, marrow.] Pertaining to marrow.

ME-DUL'LA OB-LON-GA'TA. Commencement of the spinal cord.

ME-DUL'LA SPI-NA'LIS. The spinal cord.

MEM'BRA-NA. A membrane; a thin, white, flexible skin formed by fibres interwoven like network.

MES'EN-TER-Y. [Gr. μεσος, *mesos*, the middle, and εντερον, *enteron*, the intestine.] The membrane in the middle of the intestines by which they are attached to the spine.

MET-A-CAR'PUS. [Gr. μετα, *meta*, after, and καρπος, *karpos*, wrist.] The part of the hand between the wrist and fingers.

MET-A-TAR'SUS. [Gr. μετα, *meta*, after, and ταρσος, *tarsos*, the tarsus.] The instep. A term applied to seven bones of the foot.

MID'RIFF. [Sax. *mid*, and *hrife*, the belly.] See DIAPHRAGM.

MIN'I-MUS. [L.] The smallest. A term applied to several muscles.

MI'NOR. [L.] Less, smaller. A term applied to several muscles.

MI'TRAL. [L. *mitra*, a mitre.] The name of the valves on the left side of the heart.

MO-DI'O-LUS. [L. *modus*, a measure.] A cone in the cochlea around which the membranes wind.

MO'LAR. [L. *mola*, a mill.] The name of some of the large teeth.

MOL'LIS. [L.] Soft.

MO'TOR, -ES. [L. *moveo*, to move.] A mover. A term applied to certain nerves.

MU'COUS. Pertaining to mucus.

MU'CUS. A viscid fluid secreted by the mucous membrane, which it serves to moisten and defend.

MUS'CLE. A bundle of fibres enclosed in a sheath.

MY-O'DES. A term applied to certain muscles of the neck.

MY-O-LEM'MA. [Gr. μυς, *mus*, a muscle, and λεμμα, *lemma*, to receive.] The investing membrane of a fibre.

MY-OL'O-GY. [Gr. μυς, *mus*, a muscle, and λογος, *logos*, a discourse.] A description of the muscles.

NA'SAL. Relating to the nose.

NERVE. An organ of sensation and motion in animals.

NERV'OUS CEN'TRE. A collection of gray nervous matter, which receives impressions and originates the nervous impulses.

NEU-RI-LEM'A. [Gr. νευρον, *neuron*, a nerve, and λεμμα, *lemma*, a sheath.] The sheath or covering of a nerve.

NEU-ROL'O-GY. [Gr. νευρον, *neuron*, a nerve, and λογος, *logos*, a discourse.] A description of the nerves of the body.

NI'TRO-GEN. That element of the air which is called azote.

NU'CLE-US. [L. *nux*, a nut.] The central part of any body, or that about which matter is collected.

NU-TRI'TION. The art or process of promoting the growth or repairing the waste of the system.

OC'CI-PUT. [L. *ob*, and *caput*, the head.] The hinder part of the head.

OC'U-LUS, -I. [L.] The eye.

Œ-SOPH'A-GUS. [Gr. οιω, *oiō*, to carry, and φαγω, *phago*, to eat.] The name of the passage through which the food passes from the mouth to the stomach.

O-LEC'RA-NON. [Gr. ωλενε, *ōlene*, the cubit, and κρανον, *kranon*, the head.] The elbow; the head of the ulna.

O'LE-IN. An oily substance which is fluid at ordinary temperatures.

OL-FACT'O-RY. [L. *oleo*, to smell, and *facio*, to make.] Pertaining to smelling.

O-MEN'TUM. [L.] The caul.

O'MO. [Gr. ωμος, *ōmos*, the shoulder.] The name of muscles attached to the shoulder.

OPH-THAL'MIC. [Gr. οφθαλμος, *ophthalmos*, the eye.] Belonging to the eye.

OP-PO'NENS. That which acts in opposition to something. The name of two muscles of the hand.

OP'TI-CUS, OP'TIC. [Gr. οπτομαι, *optomai*, to see.] Relating to the eye.

OR-BIC'U-LAR. [L. *orbis*, a circle.] Circular.

OR'GAN. A part of the system destined to exercise some particular function.

OR'I-GIN. Commencement; source.

OS. [L.] A bone; the mouth of anything.

OS'MA-ZOME. [Gr. οσμη, *osmē*, smell, and ζωμος, *zōmos*, broth.] A principle obtained from animal fibre which gives the peculiar taste to broth.

OS'SE-OUS. Pertaining to bones.

OS'SI-FY. [L. *ossa*, bones, and *facio*, to make.] To convert into bone.

OS'TE-INE. [Gr. οστεον, *osteon*, a bone.] The albuminous ingredient of the bones.

OS-TE-OL'O-GY. [Gr. οστεον, *osteon*, a bone, and λογος, *logos*, a discourse.] The part of anatomy which treats of bones.

O-VA'LE. [L.] The shape of an egg.

OX-AL'IC. Pertaining to sorrel. *Oxalic acid* is the acid of sorrel. It is composed of two equivalents of carbon and three of oxygen.

OX'Y-GEN. A permanently elastic fluid, invisible and inodorous. One of the components of atmospheric air.

PA-LA'TUM. [L.] The palate; the roof of the mouth.

PAL-PE-BRA'RUM. [L. *palpebra*, the eyelid.] Of the eyelids.

PAL'MAR. [L. *palma*, the palm.] Belonging to the hand.

PAL-MA'RIS. A term applied to some muscles attached to the palm of the hand.

PAN'CRE-AS. [Gr. παν, *pan*, all, and κρεας, *kreas*, flesh.] The name of one of the digestive organs.

PAN-CRE-A'TIN. The albuminous ingredient of the pancreas.

PA-PIL'LA, -Æ. [L.] Small conical prominences.

PA-RAL'Y-SIS. Abolition of function, whether of intellect, sensation or motion.

PA-REN'CHY-MA. [Gr. παρεγχεω, *parengcheō*, to pour through.] The substance contained between the blood-vessels of an organ.

PA-RI'E-TAL. [L. *paries*, a wall.] A bone of the skull.

PA-ROT'ID. [Gr. παρα, *para*, near, and ωτος, *ōtos*, the gen. of ους, *ous*, ear, the ear.] The name of the largest salivary gland.

PA-TEL'LA, -Æ. [L.] The knee-pan.

PA-THET'I-CUS, -CI. [Gr. παθος, *pathos*, passion.] The name of the fourth pair of nerves.

PEC'TO-RAL. [L.] Pertaining to the chest.

PE'DIS. [L., gen. of *pes*, the foot.] Of the foot.

PEL'I-TONGS. A term applied to masses of fat.

PEL'LI-CLE. [L., dim. of *pellus*, the skin.] A thin skin or film.

PEL'VIS. [L.] The basin formed by the large bones at the lower part of the abdomen.

PEN'NI-FORM. [L. *penna*, a feather.] Having the form of a feather or quill.

PEP'SIN. [Gr. πεπτω, *peptō*, to cook.] An ingredient of the gastric juice, which acts as a ferment in the digestion of the food.

PER-I-CAR'DI-UM. [Gr. περι, *peri*, around, and καρδια, *kardia*, the heart.] A membrane that encloses the heart.

PER-I-CHON'DRI-UM. [Gr. περι, *peri*, around, and χονδρος, *chondros*, cartilage.] A membrane that invests cartilage.

PER-I-CRA'NI-UM. [Gr. περι, *peri*, around, and κρανιον, *kranion*, the cranium.] A membrane that invests the skull.

PER-I-MYS'I-UM. [Gr. περι, *peri*, around, and μυς, *mus*, a muscle.] The investing membrane of a muscle.

PER-I-STAL'TIC. [Gr. περιστελλω, *peristellō*, to involve.] A movement like the crawling of a worm.

PER-I-TO-NE'UM. [Gr. περι, *peri*, around, and τεινειν, *teinein*, to stretch.] A thin, serous membrane investing the internal surface of the abdomen.

PER'MA-NENT. Durable; lasting.

PER-SPI-RA'TION. [L. *per*, through, and *spiro*, to breathe.] The excretion from the skin.

PHAL'ANX, -GES. [Gr. φαλαγξ, *phalanx*, an army.] Three rows of small bones forming the fingers or toes.

PHA-LAN'GI-AL. Belonging to the fingers or toes.

PHA-RYN'GE-AL. Relating to the pharynx.

PHAR'YNX. [Gr. φαρυγξ, *pharunx*.] The upper part of the œsophagus.

PHOS'PHOR-US. [Gr. φως, *phōs*, the light, and φερω, *pherō*, to bear.] A combustible substance, of a yellowish color, semi-transparent, resembling wax.

PHREN'IC. [Gr. φρην, *phrēn*, the mind.] Belonging to the diaphragm.

PHYS-I-OL'O-GY. [Gr. φυσις, *phusis*, nature, and λογος, *logos*, a discourse.] The science of the functions of the organs of animals and plants.

PI'A MA'TER. [L., good mother.] The name of one of the membranes of the brain.

PIG-MEN'TUM NI'GRUM. [L.] Black paint; a preparation of colors.

PIN'NA. [L., a wing.] A part of the external ear.

PLA-TYS'MA. [Gr. πλατυς, *platūs*, broad.] A muscle of the neck.

PLEU'RA, -Æ. [Gr. πλευρα, *pleura*, the side.] A thin membrane that covers the inside of the thorax, and also forms the exterior coat of the lungs.

PLEU'RAL. Relating to the pleura.

PLEX'US. [L. *plecto*, to weave together.] Any union of nerves, vessels or fibres, in the form of network.

PNEU-MO-GAS'TRIC. [Gr. πνευμων, *pneumōn*, the lungs, and γαστηρ, *gastēr*, the stomach.] Belonging to both the stomach and lungs.

PNEU-MO-NOL'O-GY. [Gr. πνευμων, *pneumōn*, the lungs, and λογος, *logos*, a discourse.] A description of the lungs.

POL'LI-CIS. [L.] A term applied to muscles attached to the fingers and toes.

PONS. [L.] A bridge. *Pons varolii*, a part of the brain formed by the union of the crura cerebri and cerebelli.

POP-LIT-E'AL. [L. *poples*, the ham.] Pertaining to the ham or knee-joint. A name given to various parts.

POS'TI-CUS. [L.] Behind; posterior. A term applied to certain muscles.

POR'TI-O DU'RA. [L., hard portion.] The facial nerve; eighth pair.

POR'TI-O MOL'LIS. [L., soft portion.] The auditory nerve; seventh pair.

PO-TAS'SI-UM. [L.] The metallic basis of pure potash.

PRO-BOS'CIS. [Gr. προ, *pro*, before, and βοσκω, *boskō*, to feed.] The snout or trunk of an elephant or other animal.

PROC'ESS. A prominence or projection.

PRO-NA'TOR. [L. *pronus*, turned downward.] The muscle of the forearm that moves the palm of the hand downward.

PRO-TO′PLASM. [Gr. πρωτος, prôtos, first, and πλασμα, plasma, formed.] The formal basis of all living bodies.

PSO′AS. [Gr. ψοαι, psoai, the loins.] The name of two muscles of the leg.

PUL-MON′IC, PUL′MO-NA-RY, PUL-MO-NA′LIS. } [L. pulmo, the lungs.] Belonging or relating to the lungs.

PU′PIL. A little aperture in the centre of the iris, through which the rays of light pass to the retina.

PY-LOR′IC. Pertaining to the pylorus.

PY-LO′RUS. [Gr. πυλωρος, pulôros, a gate-keeper.] The lower orifice of the stomach, with which the duodenum connects.

RA′DI-US. [L., a ray, a spoke of a wheel.] The name of one of the bones of the forearm.

RA′DI-ATE. Having lines or fibres that diverge from a point.

RA′MUS. [L.] A branch. A term applied to the projections of bones.

REC-RE-MEN-TI′TIAL. [L. re, again, and cerno, to secrete.] Consisting of superfluous matter separated from that which is valuable.

REC′TUM. The third and last portion of the intestines.

REC′TUS, -I. [L.] Straight; erect. A term applied to several muscles.

REG′I-MEN. [L. rego, to govern.] The systematic regulation of the food and drink.

RE-SID′U-UM. [L.] Waste matter. The fæces.

RES-PI-RA′TION. [L. re, again, and spiro, to breathe.] The act of breathing. Inspiring air into the lungs and expelling it again.

RE-SPI′RA-TO-RY. Pertaining to respiration; serving for respiration.

RET′I-NA. [L. rete, a net.] The essential organ of sight. One of the coats of the eye, formed by the expansion of the optic nerve.

RO-TUN′DUM, -A. [L.] Round; circular.

RU′GA, -Æ. [L.] A wrinkle; a fold.

SAC′CU-LUS. [L., dim. of saccus, a bag.] A little sac.

SA′CRAL. Pertaining to the sacrum.

SA′CRUM. [L., sacred.] The bone which forms the posterior part of the pelvis, and is a continuation of the spinal column.

SA-LI′VA. [L.] The fluid which is secreted by the salivary glands, which moistens the food and mouth.

SAN′GUIN-E-OUS. [L. sanguis, the blood.] Bloody; abounding with blood; plethoric.

SAR-TO′RI-US. [L. sartor, a tailor.] A term applied to a muscle of the thigh.

SCA′LA, -Æ. [L., a ladder.] Cavities of the cochlea.

SCA-LE′NUS. [Gr. σκαληνος, skalênos, unequal.] A term applied to some muscles of the neck.

SCAPH′OID. [Gr. σκαφη, skaphê, a little boat.] The name applied to one of the wrist-bones.

SCAP′U-LA. [L.] The shoulder-blade.

SCAP′U-LAR. Relating to the scapula.

SCI-AT′IC. [Gr., pertaining to the loins.] The name of the large nerve of the loins and leg.

SCLE-ROT′IC. [Gr. σκληρος, sklêros, hard.] A membrane of the eye.

SE-BA′CEOUS. [L. sebum, tallow.] Pertaining to fat; unctuous matter.

SE-CRE′TION. The act of producing from the blood substances different from the blood itself, as bile, saliva; the matter secreted, as mucus, bile, etc.

SE-CRE′TO-RY. Performing the office of secretion.

SE-CUN′DUS. Second. A term applied to certain muscles.

SEM-I-CIR′CU-LAR. Having the form of a half circle. The name of a part of the ear.

SEM-I-LU′NAR VALVES. [L. semi, half, and luna, the moon.] Name of the three festooned valves of the heart, at the entrance of the great arteries.

SEM-I-TEN-DI-NO′SUS. [L. semi, half, and tendo, a tendon.] The name of a muscle.

SEP′TUM. [L.] A membrane that divides two cavities from each other.

SE′ROUS. Thin; watery. Pertaining to serum.

SE′RUM. [L.] The thin, transparent part of blood.

SER-RA′TUS. [L. serro, to saw.] A term applied to some muscles of the trunk.

SIG′MOID. [Gr.] Resembling the Greek ς, Sigma.

Si-li'ci-um. A term applied to one of the earths.

Si'nus. [L., a bay.] A cavity, the interior of which is more expanded than the entrance.

Skel'e-ton. [Gr. σκελλω, *skellō*, to dry.] The aggregate of the hard parts of the body; the bones.

So'di-um. The metallic base of soda.

Sphe'noid. [Gr. σφην, *sphēn*, a wedge, and ειδος, *eidos*, likeness.] A bone at the base of the skull.

Sphinc'ter. [Gr. σφιγγω, *sphingō*, to restrict.] A muscle that contracts or shuts an orifice.

Spi'nal Cord. A prolongation of the brain.

Spine. A thorn. The vertebral column; back-bone.

Splanch-nol'o-gy. [Gr. σπλαγχνον, *splanchnon*, the bowels, and λογος, *logos*, a discourse.] A description of the internal parts of the body.

Spleen. The milt. It is situated in the abdomen and attached to the stomach.

Sple'ni-us. The name of a muscle of the neck.

Squa'mose. [L.] Scaly.

Sta'pes. The name of one of the small bones of the ear.

Ste'ar-in. [Gr. στεαρ, *stear*, suet.] One of the proximate principles of animal fat, which is solid at ordinary temperatures.

Ster'num. The breast-bone.

Stom'ach. The principal organ of the digestive apparatus.

Stra'tum. [L. *sterno*, to spread.] A bed; a layer.

Sty'loid. [L. *stylus*, a pencil.] An epithet applied to processes that resemble a style, a pen.

Sub-cla'vi-an. [L. *sub*, under, and *clavis*, a key.] Situated under the clavicle.

Sub-lin'gual. [L. *sub*, under, and *lingua*, the tongue.] Situated under the tongue.

Sub-max'il-la-ry. [L. *sub*, under, and *maxilla*, the jaw-bone.] Located under the jaw.

Sul'phur. A simple mineral substance, of a yellow color, brittle, insoluble in water, but fusible by heat.

Su-pi-na'tor. [L.] A muscle that turns the palm of the hand upward.

Sut'ure. [L. *suo*, to sew.] The seam or joint that unites the bones of the skull.

Syn-ar-thro'sis. [Gr. συν, *sūn*, with, and αρθρον, *arthron*, a joint.] An immovable articulation.

Syn-o'vi-a. [Gr. συν, *sūn*, with, and ωον, *ōon*, an egg.] The fluid secreted into the cavities of joints for the purpose of lubricating them.

Sys'tem. An assemblage of organs, composed of the same tissues and intended for the same functions.

Sys-tem'ic. Belonging to the general system.

Sys'to-le. [Gr. συστελλω, *sūstellō*, to contract.] The contraction of the heart and arteries for expelling the blood and carrying on the circulation.

Tar'sus. [L.] The posterior part of the foot.

Ten'don. [Gr. τεινω, *teinō*, to stretch.] A hard, insensible cord, or bundle of fibres, by which a muscle is attached to a bone.

Tens'or. A muscle that extends a part.

Ten-tac'u-la, -æ. [L. *tenio*, to seize.] A filiform process or organ on the bodies of various animals.

Ten-to'ri-um. [L. *tendo*, to stretch.] A process of the dura mater which lies between the cerebrum and cerebellum.

Te'res. [L. *teres*, round.] An epithet given to many organs, the fibres of which are collected in small bundles.

Tho'rax. [Gr.] That part of the skeleton that composes the bones of the chest. The cavity of the chest.

Tho-rac'ic. Relating to the chest.

Thy'roid. [Gr. θυρεος, *thureos*, a shield.] Resembling a shield. A cartilage of the larynx.

Tib'i-a. [L., a flute.] The large bone of the leg.

Tis'sue. The texture or organization of parts.

Ton'sil. [L.] A glandular body in the throat or fauces.

Tra'che-a. [Gr. τραχυς, *trachus*, rough.] The windpipe.

Trans-verse', Trans-ver-sa'lis. Lying in a cross direction.

Tra-pe'zi-us. The name of a muscle, so called from its form.

Tri'ceps. [L. *tres*, three, and *caput*, head.] Three. A name given to muscles that have three attachments at one extremity.

TRI-CUS′PID. [L. *tres*, three, and *cuspis*, point.] The triangular valves in the right side of the heart.

TRIT′U-RAT-ING. Grinding to a powder.

TROCH′LE-A. [Gr. τροχαλια, *trochalia*, a pulley.] A pulley-like cartilage, over which the tendon of a muscle of the eye passes.

TRUNK. The principal part of the body, to which the limbs are articulated.

TU′BER-CLE. [L. *tuber*, a bunch.] A pimple, swelling or tumor on animal bodies.

TYM′PAN-UM. [L.] The middle ear.

UL′NA. [L.] A bone of the fore-arm.

UL′NAR, UL-NA′RIS. Relating to the ulna.

U-RE′TER. [Gr. ουρειν, *ourein*, to conduct water.] The excretory duct of the kidneys.

U′RIC. [Gr. ουρον, *ouron*, urine.] An acid contained in urine and in gouty concretions.

U-VE′A. [L. *uva*, a grape.] Resembling grapes. A thin membrane of the eye.

U′VU-LA. A soft body suspended from the palate, near the aperture of the nostrils, over the glottis.

VAC′CINE VI′RUS. [L. *vacca*, a cow, *virus*, poison.] Pertaining to cows; derived from cows.

VALVE. Any membrane, or doubling of any membrane, which prevents fluids from flowing back in the vessels and canals of the animal body.

VAS′CU-LAR. [L. *vasculum*, a vessel.] Pertaining to vessels; abounding in vessels.

VAS′TUS. [L.] Great, vast. Applied to some large muscles.

VEINS. Vessels that convey blood to the heart.

VE′NOUS. Pertaining to veins.

VEN′TRI-CLE. [L. *venter*, the stomach.] A small cavity of the animal body.

VER-MIC′U-LAR. [L. *vermiculus*, a little worm.] Resembling the motions of a worm.

VERM-I-FORM′IS. [L. *vermis*, a worm, and *forma*, form.] Having the form and shape of a worm.

VERT′E-BRA, -Æ. [L. *verto*, to turn.] A joint of the spinal column.

VERT′E-BRAL. Pertaining to the joints of the spinal column.

VES′I-CLE. [L. *vesica*, a bladder.] A little bladder, or a portion of the cuticle separated from the cutis vera and filled with serum.

VES′TI-BULE. [L.] A porch of a house. A cavity belonging to the ear.

VIL′LI. [L.] Fine, small fibres.

VI′RUS. [L., poison.] Foul matter of an ulcer; poison.

VI′TAL. [L. *vita*, life.] Pertaining to life.

VIT′RE-OUS. [L. *vitrum*, glass.] Belonging to glass. A humor of the eye.

VO′LAR. [L. *vola*, the hollow of the hand or foot.] Belonging to the palm of the hand.

VO′MER. [L., a ploughshare.] One of the bones of the nose.

ZYG-O-MAT′I-CUS. [Gr. ζυγος, *zugos*, a yoke.] A term applied to some muscles of the face, from their attachment.

27*

INDEX.

THE END.

KEY TO CUTTER'S

NEW OUTLINE ZOOLOGICAL CHARTS,

OR

HUMAN AND COMPARATIVE ANATOMICAL PLATES.

SUGGESTIONS TO TEACHERS.

In using these charts, we would suggest that the pupil carefully examine the illustrating cuts interspersed with the text in connection with the lesson to be recited. The similarity between these and the charts will enable the pupil to recite, and the teacher to conduct his recitation from the latter.

Let a pupil show the situation of an organ, or part, on an anatomical outline chart, and also give its structure, while other members of the class note all omissions and misstatements. Another pupil may give the use of that organ, and, if necessary, others may give an extended explanation. The third may explain the laws on which the health of the part depends, while other members of the class may supply what has been omitted. After thus presenting the subject in the form of topics, questions may be proposed promiscuously from each paragraph, and where examples occur in the text let other analogous ones be given.

If the physiology and hygiene of a given subject have not been studied, confine the recitation to those parts only on which the pupil is prepared. When practicable, the three departments should be united; but this can only be done when the chapter on the hygiene has been learned, while the physiology can be united with the anatomy in all chapters upon physiology.

CHART No. 1.

OSSEOUS SYSTEM—HUMAN AND COMPARATIVE.

A. *Bones of the Human Body.*—1, The frontal bone. 2, The superior maxillary (upper jaw-bone). 3, The inferior maxillary (lower jaw-bone). 4, The cervical vertebræ (bones of the neck). 5, The dorsal vertebræ (bones of the back). 6, The lumbar vertebræ (bones of the loins). 7, The sacrum (the basis of the spinal column). 8, The temporal bone. 9, The scapula (shoulder-blade). 10, 10, 10, The ribs. 11, 11, The innominata (hip-bones). 12, The humerus (arm-bone). 13, The radius. 14, The ulna (bones of the fore-arm). 15, The carpus (wrist-bones). 16, 16, The metacarpus (bones of the palm of the hand). 17, 17, The phalanges (finger-bones). 18, The femur (thigh-bone). 19, The patella (knee-pan). 20, The tibia. 21, The fibula (bones of the leg). 22, The tarsus (bones of the instep). 23, 23, The metatarsus (bones of the middle of the foot). 24, 24, The phalanges (toe-bones). 25, Ligaments of the shoulder. 26, Ligaments of the elbow. 27, Ligaments of the wrist. 28, Ligaments of the hip-joint. 29, Ligaments of the knee. 30, Interosseous membrane. 31, Ligaments of the ankle. 32, The clavicle (collar-bone). 33, The sternum (breast-bone).

B. *Bones of the Cow.*—1, The frontal bone. 2, The upper jaw (superior maxillary). 3, The lower jaw (inferior maxillary). 4, The cervical vertebræ (bones of the neck). 5, The dorsal vertebræ (bones of the back). 6, 7, The lumbar vertebræ. 11, The sacral vertebræ. 8, The caudal vertebræ. 9, The scapula. 10, 10, The ribs. 12, The humerus. 13, 14, The radius and ulna. 15, The carpus. 16, The metacarpus. 17, The phalanges. 18, The femur. 20, The tibia. 22, The tarsus. 23, The metatarsus. 24, The phalanges.

C. *Bones of the Bird.*—1, The head. 2, The superior mandible (upper jaw). 3, The inferior mandible (lower jaw). 4, The cervical vertebræ. 5, The dorsal vertebræ. 8, The lumbar and sacral vertebræ. 9, The scapula. 10, The ribs. 11, The sacrum. 12, The humerus. 13, 14, The radius and ulna. 15, The carpus. 16, The metacarpus. 17, 17, Phalanges. 18, The femur. 20, 21, The tibia and fibula. 22, 23, The tarsus and metatarsus. 24, Phalanges. 32, The coracoid bone. 33, The clavicle (furcula). 34, The sternum.

D. *Bones of the Tortoise.*—1, The head. 4, The cervical vertebræ. 5, 5, The dorsal vertebræ and marginal plates. 6, 6, The lumbar vertebræ and connecting ribs. 7, 11, The sacral bones. 8, The caudal vertebræ. 9, The scapula. 12, The humerus. 13, 14, The radius and ulna. 15, The carpus. 16, 17, Phalanges. 18, The femur. 20, 21, The tibia and fibula. 22, 23, The tarsus and metatarsus. 24, Phalanges. 32, The clavicle. 33, The coracoid bone.

E. *Bones of the Fish.*—1, The bones of the head. 2, The upper jaw. 3, The lower jaw. 4, 5, 6, The dorsal and caudal vertebræ. 8, The first dorsal fin. 9, The second dorsal fin. 10, One of the ventral fins that corresponds to the legs. 12, A pectoral fin which is analogous to arms. 18, A ventral fin.

F. *Diagram of an Annulosa.*—1, The vascular (blood-vessel) system. 2, The digestive system. 3, 3, The ganglia (nervous) system. 4, 4, A series of rings of hardened skin which forms an external skeleton.

G. *Diagram of a Mollusca.*—1, The digestive canal. 2, The heart. 3, 4, 5, Ganglia (knots of nervous matter).

H. *Diagram of a Radiata.*—The star-fish. 1, Central aperture.

CHART No. 2.

MUSCULAR SYSTEM—HUMAN AND COMPARATIVE.

A. *Muscles of Human Body.*—1, The occipito-frontalis. 2, The orbicularis palpebrarum. 3, The levator labii superioris. 4, The zygomaticus. 5, The masseter. 6, The orbicularis oris. 7, The temporal. 8, The levator anguli-oris. 9, The depressor labii inferioris. 10, 10, The deltoid. 11, 11, The pectoralis major. 13, The supinator radii longus. 14, The flexor carpi ulnaris. 15, The flexor digitorium. 16, The rectus abdominalis. 17, The sartorius. 18, The adductor longus. 19, The rectus femoris. 20, The vastus externus. 21, The vastus internus. 22, The tendon patella. 23, The gastrocnemius. 24, The tibialis anticus. 25, The extensor longus digitorium. 26, The short extensor muscles of the toes. 27, The adductor muscle of the great toe. 28, The serratus magnus anticus. 29, 29, The psoas magnus. 30, The inguinal ring. 31, 31, 31, 31, 31, 31, 31, 31, The tendons of the wrist and fingers. 32, The sterno-hyoideus. 33, The sterno-cleido-mastoideus. 34, The biceps. 35, The triceps muscle.

B. *Muscles of the Cow.*—1, The occipito-frontalis. 2, The orbicularis palpebrarum. 3, The masseter. 4, The levator labii superioris nasi. 5, The digastricus. 7, The trapezius. 10, The latissimus dorsi. 11, The pectoralis. 16, 17, The external and internal oblique muscle. 18, The opening of the mammary artery and vein (milk-veins). 19, The rectus femoris. 20, 20, 20, The gluteii muscles.

C. *Muscles of the Bird.*—1, The occipito-frontalis. 2, The orbicularis palpebrarum. 5, The masseter. 7, The temporal. 10, The deltoid. 11, The pectoralis. 13, The levator caudæ. 14, The extensor metacarpi radialis longus. 19, The rectus femoris. 20, The gluteii. 23, The gastrocnemius. 24, The extensor longus digitorium. 33, The sterno-cleido-mastoideus. 34, The biceps muscle.

D. *Muscles of the Tortoise.*—1, The digastricus. 10, 10, The deltoides. 14, The ulnaris internus. 18, The sartorius. 23, 24, The gastrocnemius. 28, The serratus magnus. 31, 32, The flexores digitorium. 34, The biceps brachialis. 35, The triceps brachialis muscle.

E. *Muscles of the Fish.*—1, 2, 3, and a, b, c, represent the zigzag arrangement of the muscles of the fish (myocomma).

F. *Diagram of an Insect.*—1, The head. 2, The first segment of the chest, with the first pair of legs. 3, The second segment, with the second pair of legs and the first pair of wings. 4, The third segment, with the third pair of legs and second pair of wings. 5, The abdomen without legs.

CHART No. 3.

NUTRITIVE SYSTEM—HUMAN AND COMPARATIVE.

A. *Organs of Man.*—1, The parotid gland. 2, The submaxillary gland. 3, The sublingual gland. 4, The œsophagus. 5, The larynx and trachea. 6, The left lung. 7, The right lung. 8, The heart. 9, The vena cava descendens. 10, The aorta. 11, The pulmonary artery. 12, The stomach. 13, The left and right lobe of the liver. 15, 15, 15, The large intestine. 16, 16, 16, 16, The small intestine. 17, The diaphragm.

B. *Internal Organs of the Sheep.*—1, The first stomach (rumen). 2, The second stomach (reticulum). 3, The third stomach (maniplies). 4, The fourth stomach (abomasum or rennet). 5, The duodenum. 6, The spleen. 7, The intestines.

C. *Organs of a winged Reptile.*—1, The ventricle of the heart. 2, 3, The auricles of the heart. 4, 5, 6, Blood-vessels. 7, The trachea. 8, The lungs. 9, 10, 11, The liver and its appendages. 12, The stomach. 13, The duodenum. 14, 15, 16, The intestines. 17, The cloaca. 18, The cæca.

D. *Diagram of the Organs of a Frog.*—1, The heart. 2, 2, Arches of the aorta. 3, 3, Pulmonary artery. 4, 4, The pulmonary veins. 6, The vena cava. 5, The digestive canal.

CHART No. 4.

DIGESTIVE SYSTEM—HUMAN AND COMPARATIVE.

A. *Digestive Organs of Man.*—1, The upper jaw. 2, The lower jaw. 3, The tongue. 4, The hard palate (roof of the mouth). 5, The parotid gland. 6, The sublingual gland. 7, The larynx. 8, 9, The œsophagus. 10, The stomach. 11, 11, The liver. 12, The gall-bladder. 13, Its duct. 14, The duodenum. 15, The pancreas. 16, The spleen (milt). 17, 17, 17, 17, The small intestine. 18, The cæcum. 19, The appendix vermiformis. 20, 20, The ascending colon. 21, The transverse colon. 22, 22, The descending colon. 23, The sigmoid flexure of the colon. 24, The rectum.

B. *Digestive Organs of a Fowl.*—9, The œsophagus. 8, The crop (ingluvies). 7. The second stomach (proventriculus). 10, The gizzard. 11, 11, The liver. 12, The gall-bladder. 13, The bile ducts. 14, 14, 14, 14, The duodenum. 15, The pancreas. 16, The cæca (pouches). 17, The large intestine. 24, The ureter and cloaca. 25, The trachea.

C. *Digestive Organs of an Ox.*—1, The œsophagus. 2, 2, The rumen (paunch). 3, The second stomach (reticulum). 4, The omasum (maniplies). 5, The fourth stomach or abomasum (rennet). 6, The duodenum (intestine).

D. *Digestive Organs of an Insect.*—8, The crop. 9, The gullet. 10, The gizzard. 14, 14, The chylific (digestive) stomach. 16, 16, Cæca (bile-tubes) 17, The intestine. 18, The renal vessels. 24, The cloaca.

E. *Digestive Organs of the Sword-Fish.*—11, 11, The liver. 13, The bile duct. 16, 16, The cæcas (pouches). 17, 17, 17, The intestine. 24, The cloaca.

F. *Digestive Organs of the Herring.*—1, 1, The air-bladder. 2, The air-duct (pneumatic). 9, The œsophagus. 10, The stomach. 16, The cæca. 17, 17, 17, The intestine.

CHART No. 5.

ABSORPTIVE SYSTEM—HUMAN AND COMPARATIVE.

A. *Absorbent Vessels in Man.*—1, 2, 3, 4, 5, 6, Lymphatic vessels and glands of the lower extremities. 8, Lymphatic vessels of the kidney. 11, 12, The thoracic duct. 10, 10, 10, The intercostal lymphatics. 13, Lymphatics of the neck. 14, 14, Carotid arteries. 15, Axillary glands. 16, 17, 18, Lymphatics of the arm and hand. 19, Lymphatics of the face. 20, The right innominata vein. 21, The junction of the thoracic duct with the left subclavian vein.

B. *Section of the Layers of the Skin.*—1, The dermis. 2, The epidermis. 3, Its cuticle. 4, Its soft layer. 5, Subcutaneous connective and adipose tissue. 6, Tactile papillæ. 7, Sweat or perspiratory glands. 8, Its duct. 9, Spiral passage of the duct through the epidermis. 10, 10, The termination of the duct on the surface of the epidermis.

C. *Section of the Papillæ and Glands of the Skin.*—1, 1, 1, 1, Ridges of the cuticle (cut perpendicularly). 2, 2, 2, 2, Furrows or wrinkles of the cuticle. 3, The epidermis. 4, Its colored layer. 5, The dermis. 6, 6, 6, The papillæ. 7, 7, Small furrows between the papillæ. 8, 8, 8, 8, Deeper furrows between each couple of the papillæ. 9, Cells filled with fat. 10, 10, 10, The adipose layer, with numerous fat vesicles. 11, 11, Cellular fibres of the adipose tissue. 12, Two hairs. 13, Sweat or perspiratory gland, with its spiral duct. 14, A perspiratory gland with a duct less spiral. 15, 15, Oil-glands, with ducts opening into the sheath of the hair.

CHART No. 6.

RESPIRATORY SYSTEM—HUMAN AND COMPARATIVE.

A. *Respiratory Organs of Man.*—1, The larynx. 2, The trachea. 3, The right bronchia. 4, The left bronchia. 5, 6, 7, Lobes of the right lung. 8, 9, Subdivisions of the bronchi or bronchial tubes. 10, 10, 10, 10, Air cells. 11, 11, The diaphragm.

B. *Diagram of the Blood-vessels in Man.*—1, The vena cava descendens. 2, The vena cava ascendens. 3, The right ventricle of the heart. 4, The left ventricle. 5, 6, The aorta. 7, The pulmonary artery. 8, 9, Divisions of the pulmonary artery.

C. *Section of a small Mammal.*—1, The œsophagus. 2, The trachea. 5, 6, The lungs. 7, The heart. 8, The stomach. 9, The liver. 10, 10, Intestines. 11, 11, The diaphragm. 12, 13, The kidney and duct. 14, The brain. 15, 15, 15, The spinal cord. 16, 16, The vertebræ. 17, Cæca.

D. *Diagram of the Lungs of a Bird.*—2, A bronchial tube. 3, 4, Divisions of the bronchi that end in sacs. 8, 8, 9, 9, Abdominal air-sacs.

E. *Lung of a Goose.*—2, A bronchus. 3, 4, The bronchial tubes laid open. 10, 10, 10, Apertures of communication with air-cells. 11, 11, Abdominal bronchial orifices.

G. *Respiratory Organs of the Water-scorpion.*—1, The head. 2, The base of the first pair of feet. 3, The first ring of the thorax. 4, The base of wings. 5, Base of the second pair of feet. 6, 6, 6, 6, Stigmata (opening at the edge of each joint). 7, 7, 7, 7, Tracheæ (air-tubes). 8, 8, Air-sacs.

F. *Diagram of the Bronchial Leaflets of the Cod.*—1, A section of a bronchial arch. 3, Bronchial leaflets or plates.

J. *Diagram of the Circulation of the Blood through the Bronchial Leaflets.*—1, A section of a bronchial arch. 2, A section of a bronchial artery. 3, 3, An arterial branch along the outer margin of the processes, giving off capillary vessels to the leaflets. 4, A vein that receives the blood from the capillaries of the inner margin of the process. 5, Bronchial vein.

H. *A Plexus of Capillary Vessels.*

K. *Diagram of the Relative Position of the Blood-vessels to the Air-cells.*—1, A bronchial tube communicating with the air-cells, 2, 2, 2. 3, A branch of the pulmonary artery containing bluish blood. 4, A branch of a pulmonary vein containing scarlet or purified blood.

CHART No. 7.

CIRCULATORY SYSTEM—HUMAN AND COMPARATIVE.

A. *Circulation in Man.*—3, 4, The heart. 5, The pulmonary artery. 6, Its branch to the left lung. 7, The vena cava descendens. 8, The vena cava ascendens. 9, The descending aorta. 10, The right femoral artery. 11, The left femoral vein. 12, The subclavian artery. 13, The subclavian vein. 14, The jugular vein. 15, The axillary vein. 16, The brachial vein and artery. 17, The kidney.

B. *Diagram of the Circulation in Reptiles.*—1, Ventricle. 2, Left auricle. 3, Right auricle. The arrows show the direction of the blood.

C. *Diagram of the Circulation in the Fish.*—1, The pericardium. 2, The ventricle that receives blood from the body. 3, The ventricle that sends blood to the gills.

D. *Diagram of the Heart of Mammals.*—1, The vena cava descendens. 2, The vena cava ascendens. 3, The right auricle. 4, The opening between the right auricle and right ventricle. 5, The right ventricle. 6, The tricuspid valves. 7, The pulmonary artery. 8, 8, Its branches. 9. The semi-lunar valves of pulmonary artery. 10, The septum between the two ventricles of the heart. 11, 11, The pulmonary veins. 12, The left auricle. 13, The opening between the left auricle and the left ventricle. 14, The left ventricle. 15, The mitral valves. 16, The aorta. 17, The semi-lunar valves of the aorta.

E. *The Heart and Arteries of a Snail.*—2, The stomach. 3, 3, The intestine. 5, The heart. 6, The aorta. 7, The pulmonary artery.

CHART No. 8.

NERVOUS SYSTEM—HUMAN AND COMPARATIVE.

A. *Section of the Human Brain and Spinal Column.*—1, The cerebrum. 2, The cerebellum. 3, The medulla oblongata. 4, 4, The medulla spinalis (spinal cord) in the canal formed by the vertebræ of the spinal column.

B. *Back view of the Brain and Nerves in Man.*—1, The cerebrum. 2, The cerebellum. 3, The spinal cord. 4, Nerves of the face. 5. Brachial plexus of nerves. 6, 7, 8, 9, Nerves of the arm. 10, Nerves that pass under the ribs. 11, The lumbar plexus of nerves. 12, The sacral plexus of nerves. 13, 14, 15, 16, Nerves of the lower extremities.

C. *The Sympathetic Nerves.*—1, The renal plexus of nerves. 2, 3, 4, The lumbar plexus. 6, The semi-lunar ganglion and solar plexus. 7, 7, 7, The thoracic ganglions. 8, 9, The right and left coronary plexus. 10, 11, 12, The cervical ganglions.

D. *Base of the Brain of a Horse.*—1, The cerebrum. 2, The optic ganglion. 3, The cerebellum. 4, The medulla oblongata and spinal cord.

E. *Brain of an Alligator.*—1, The olfactory ganglion. 2, The cerebrum. 3, The optic ganglion. 4, The cerebellum. 5, The medulla oblongata and spinal cord.

F. *Brain of a Bird.*—1, The cerebrum. 2, The optic ganglion. 3, The cerebellum. 4, The medulla oblongata.

G. *Brain of a Fish.*—1, The olfactory ganglion. 2, The cerebrum. 3, The optic ganglion. 4, The cerebellum. 5, The medulla oblongata and spinal cord.

H. *Nervous System of the Beetle.*—1, 1, 2, 2, Nervous ganglions and cords.

I. *Diagram of the Nervous System of the Centipede.*

J. *Diagram of the Star-Fish.*—The nervous matter is arranged in a ring about the mouth, which sends off branches in different directions.

CHART No. 9.

SPECIAL SENSE—HUMAN AND COMPARATIVE.

A. *The Nervous System of Man.*—1, The convolutions of the large brain (cerebrum). 2, The lesser brain (cerebellum). 3, The cervical nerves. 4, The dorsal nerves. 5, The lumbar nerves. 6, The sciatic. 7, The external popliteal nerve. 8, The tibial nerve. 9, Median and cubital nerves.

B. *Section of the Globe of the Eye.*—1, The choroid coat of the eye. 2, The sclerotic coat. 3, The retina. 4, The cornea. 5, 5, The iris. 6, The pupil. 8, 9, The chambers of the eye that contain the aqueous humor. 10, The crystalline lens. 11, 11, The vitreous humor. 12, Optic nerve. 13, The central artery of the eye.

D. *Distribution of the Trifacial (fifth pair) Nerve.*—1, The trifacial nerve. 2, A branch that passes to the eye. 3, A branch distributed to the teeth of the upper jaw. 4, The branch that passes to the tongue (5) and teeth of the lower jaw. 6, This division that passes to the tongue is the nerve of taste (gustatory). 7, The terminal branches of the upper maxillary distributed to the face.

D. *Distribution of the Olfactory Nerve.*—1, The olfactory nerve. 2, 2, The fine divisions of this nerve on the membrane of the nose. 3, A branch of the fifth pair (trifacial) nerve.

E. *View of the Ear.*—1, The auditory canal. 2, The drum of the ear (membrana tympani). The chain of bones in the ear—(3, The malleus. 4, The incus, and, 5, The stapes.) 6, The cavity of the tympanum. 7, The vestibule. 8, 9, 10, The semi-circular canals. 11, 11, 12, Channels of the cochlea. 13, The auditory nerve (nerve of hearing). 14, The opening from the middle ear to the throat (Eustachian tube).

F. *Compound Eyes of the Bee.*—Its division into facets (highly magnified).

F. Facets still more highly magnified.

F. Facets with hairs growing between them.

Questions, Diagrams and Illustrations

FOR ANALYTICAL STUDY AND RECITATION,

ALSO FOR UNIFIC AND SYNTHETIC REVIEW OF

CUTTER'S ANALYTIC ANATOMY, PHYSIOLOGY AND HYGIENE,

HUMAN AND COMPARATIVE.

PROFITABLE *reading* and *study* require the same analysis and method as clear and efficient *teaching*. *Unification* of ideas and principles is also aided by varied and frequent reviews.

To aid *pupil* and *teacher*, the following *Questions*, *Diagrams* and *Illustrations* have been prepared. The Questions in the larger type are to be used in the *analytic study* and *recitation of paragraphs;* those in the smaller type, to aid pupil and teacher to secure *unific* investigation and review of parts more or less *analogous* in *structure*, *function* or *hygiene;* while the diagrams and illustrations are to be used in *synthetical* examination and review of the sections, chapters and divisions.

I would also suggest the use of the blackboard in drawing outline figures and diagrams, and in writing the topics to be reviewed.

DIVISION I.—OUTLINE PRINCIPLES.

ANALYTIC EXAMINATION.

CHAPTER I.—General Remarks.

§ 1. *The Three Kingdoms of Nature Compared.*

1. State the Linnæan distinctions of the three kingdoms of Nature. Name the three kingdoms, and define each.
2. Of what are Organic and Inorganic bodies combinations? What is said respecting Life-force?
3. Give the distinguishing features of Organized and Unorganized matter.
4. State the distinctions between animals and plants.
5. What is said of these distinctions in the lower forms of life?

§ 2. *Definition of Terms.*

6. Define Organ, Apparatus and Function. What is Anatomy? Physiology? Hygiene?
7. Of what are organs composed? Define Histology and Chemistry.

CHAPTER II.—General Histology.

§ 3. *Cells.*

8. Where do you find Unity of Plan?
9. Define Protoplasm. What is Animal Protoplasm?
10. What is said of nucleated cells? Of the modifications of these cells?
11. Distinguish between animal and vegetable cells.
12. Of what is the simple cell the type?
13. Of what does a simple cell consist? Give an illustration.
14. To what modifications are cells subject?
15. What is the shape of the cells?
16. In what ways do cells multiply?
17. What is said of the growth and decay of cells?

§ 4. *Primary Tissues.*

18. How are the different tissues of the body formed? Upon what do their characters depend?
19. To what are the Primary Tissues reducible?

20. State the object and character of the Connective Tissues.
21. Of what is the Fibrous form composed? State its nature and forms.
22. Give the composition and forms of the White Fibrous tissue. What is Gelatin?
23 Describe the Yellow Fibrous tissue. Why called Elastic? Does it gelatinize? Where found? When found together, what proportion of White to Yellow Fibrous tissue? Observation.
24. Of what does the Areolar form consist? What is said of its cellular structure? What of its individuality? Observation.
25. Describe the Cartilaginous tissue. Mention the properties of Cartilage. Where is this tissue found? What is the relation of cartilage to bone?
26. Under what condition is Fibro-cartilage formed? State quality and adaptation.
27. What peculiarity has the Adipose tissue? Of what composed? Where found? Its use?
28. Where is the Sclerous tissue found? What is said of its composition?
29. Give the composition of Muscular tissue. Name its kinds, and describe each. What is its characteristic? What of its electrical nature?
30. Describe the Tubular tissue. What is the office of the capillary vessels? Of what are their walls composed? Where is this tissue found?
31. How is the Nervous tissue distinguished? Where found? In what respect like the Muscular tissue? Mention its elements.
32. Describe the Ganglionic Corpuscles.
33. What is said of the Gray fibres? Where found?
34. Speak of the White fibres.
35. Where are the gray and white substances found?

§ 5. *Membranes.*

36. What is the Basement membrane? What is the Epithelium? Why so called?
37. Name and describe the varieties of the Epithelium. Of what power the Cilia? Where is the Ciliated Epithelium found?
38. What is beneath the basement membrane? What are constituted by the epithelium, basement membrane, and fibro-areolar tissue?
39. Where is the Serous membrane found? Its qualities?

40. What is said of the Synovial membrane? Observation.
41. Describe the Mucous membranes.
42. Where is the Gastro-Pulmonary Mucous membrane found?
43. Where the Urinary?
44. What is continuous with the Mucous membrane? Observation.

CHAPTER III.—General Chemistry.

§ 6. *Solids and Fluids.*

45. Of what is the human body composed? What is said of the proportion of solids and fluids?
46. What are Proximate Constituents? Define Organic and Inorganic Proximate Constituents.
47. Name the Inorganic Proximate Constituents.
48. Give the classes of Organic Proximate Constituents.
49. What are contained in the Nitrogenous class? Name the most important.
50. What is the office of Albumen in the animal economy? Give the derivation of its name. Where found? What peculiarity has it?
51. Describe Albuminose.
52. What is Fibrin? Where found? What is the influence of alcohol upon it?
53. Describe Musculin. How hardened?
54. Where is Globulin and Hæmatin found?
55. Give the properties of Casein. Where does it exist?
56. Define Cartilagin. What is Osteine? Chondrigen?
57. Define and give the property of Salivin.
58. Describe Pepsin, and state its property.
59. What is Pancreatin? State its actions.
60. Describe Mucin.
61. What is Neurin?
62. Define Keratin.
63. To what is Elastin peculiar?
64. Where is Melanin found?
65. Of what use Biliverdin? Color?
66. Name the acids of the nitrogenous class.
67. Mention the non-nitrogenous groups.
68. Of what are the fats composed? From what derived? What is Glycerine?
69. Mention the different kinds of sugars. Where are starch granules found?

70. Name the ultimate chemical elements, with their percentage proportions.
71. In what condition are oxygen and hydrogen?
72. What is said of carbon? What becomes of the chemical elements in decomposition?

UNIFIC REVIEW.

[Compare 9 with 119.]

What is the relation of Protoplasm to ossification?

[Compare 11–17 with 119, 120, 152, 173, 174, 237, 240, 306, 335, 339, 340, 378, 389, 458, 464, 465, 547 and 553–556.]

Where do you find nucleated cells? Have they any influence on the plan of structure? What relation does the cellular tissue bear to the muscular? In the lining of what organs do you find epithelial cells? In the lining membranes of what organs do you find ciliated epithelium?

[Compare 20–26 with 123–126, 177, 306, 334–340, 387, 388, 463 and 464.]

Name the connective tissues. Mention some distinguishing features of each. Where do you find the white fibrous tissues? Where the yellow fibrous? Where is cartilage found?

[Compare 28 with 120–122.]

What tissue is found in the bones?

[Compare 29 with 173–176, 240–243, 306 and 337–340.]

What is the structure of muscular tissue? Where found?

[Compare 30, 341 with 459–462.]

In what blood-vessels do you find the tubular tissue? In what system?

[Compare 31–35 with 457–462.]

Tell what you can about the nervous tissue.

[Compare 36–38 with 237, 238, 240–243, 246, 335, 339, 340, 376–379, 388, 389, 463 and 464.]

Name the parts of the body where you find the Basement membrane.

[Compare 39 with 244, 246, 334, 390 and 463.]

Where is the Serous membrane found?

[Compare 40 with 125 and 177.]

What is the office of the Synovial membrane?

[Compare 41–44 with 237, 238, 240–243, 386–389, 547 and 548.]

Name the Mucous membranes. The mucous membrane lines what organs? Point out the difference between mucous and serous membranes. With what is the mucous membrane continuous?

Fig. 191.

Fig. 191 (*Leidy*). Diagram exhibiting the Relative Position of the Common Anatomical Elements of Serous and Mucous Membranes, the Glands, the Lungs and the Skin.—1, Epithelium, secreting cells or epidermis, composed of nucleated cells, and occupying the free surface of the structure mentioned 2, Basement layer, represented much thicker than natural, in comparison with the other layers. 3, Fibrous layer, in which the arteries and veins (4, 4) terminate in a capillary network. Magnified.

Fig. 192.

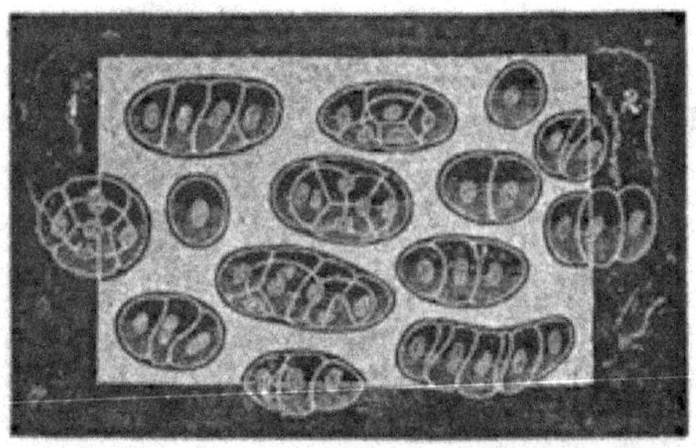

Fig. 193.

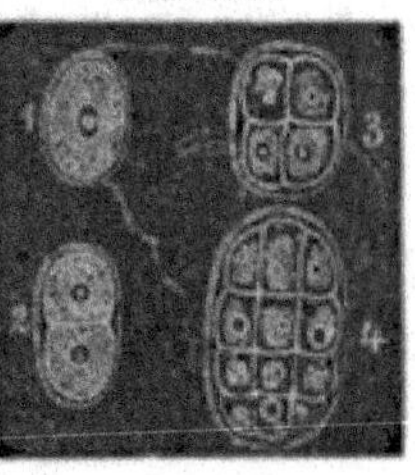

Fig. 192 (*Leidy*). Cartilage.—Section through the thickness of the oval cartilage of the nose. 1, Toward the exterior. 2, Toward the interior surface; highly magnified. It exhibits groups of cartilage cells imbedded in a homogeneous matrice.

Fig. 193 (*Leidy*). Process of Multiplication, or Cartilage Cells.—1, Simple cartilage cell from the embryo. 2, Increase of cartilage cells by division of the primary cell. 3, 4, Groups of cartilage cells, from an adult articular cartilage. Magnified.

Fig. 194.

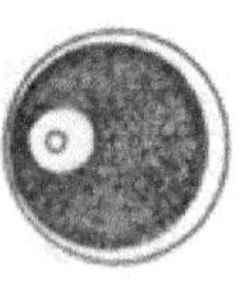
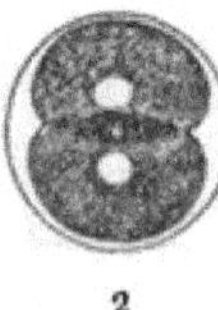
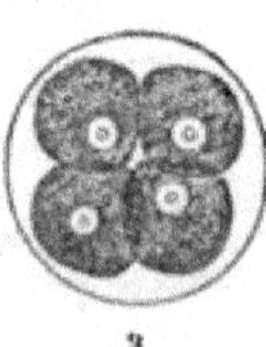
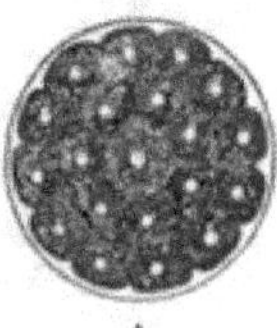

1 2 3 4

Fig. 194 (*Leidy*). An Ideal Cell.—1, Cell with its wall, protoplasm, nucleus and its nucleolus. 2, The same divided into two. 3, The same divided into four cells. 4, The same divided into many cells. The dark portion, the protoplasm; the white spot, the nucleus; the inner small circle, the nucleolus. Magnified.

SYNTHETIC REVIEW.

Topics	Section	Chapter	Division
Essential distinctions between mineral, vegetable and animal kingdoms, Nature of life-force, Vitalized and non-vitalized bodies compared, Animals and plants compared, These distinctions in higher and lower forms.	§ 1. *Three Kingdoms compared.*	CHAP. I. *General Remarks.*	Division I. *Outline Principles.*
Organ, apparatus and functions, Anatomy, Physiology and Hygiene, Structure of organs, Histology and Chemistry.	§ 2. *Definition of Terms.*		
Unity of plan in animals and plants, Protoplasm, Nucleated cell, Simple cell, Adaptation to different offices, Life and shape of cells, Modes of multiplication of cells, Growth, perfection and decay, Primary tissues, Object of the connective tissue.	§ 3. *Cells.*	CHAP. II. *General Histology.*	
Fibrous tissue, Areolar, Cartilaginous, Adipose, Sclerous, Muscular, Tubular, Nervous.	§ 4. *Primary Tissues.*		
Basement membrane, Epithelium, Serous membrane, Synovial " Mucous membranes.	§ 5. *Membranes.*		
Solids and fluids, Proximate constituents, Inorganic " Organic " Nitrogenous " Non-nitrogenous " Ultimate chemical elements.	§ 6. *Solids and Fluids.*	CHAP. III. *General Chemistry.*	

State the General Remarks, the General Histology and the General Chemistry of the human system.

A*

DIVISION II.—THE MOTORY APPARATUS.

ANALYTIC EXAMINATION.

73. Why is the Motory Apparatus so called? Name its organs.

CHAPTER IV.—The Bones.

§ 7. *Anatomy of the Bones.*

74. Of what does the Internal Framework of the body consist?
75. State the number and classes of the bones.
76. Name the divisions of bones of the Head.
77. How many bones compose the Skull? Give their names and positions.
78. What is said of the skull-bones? How are they united? Observation.
79. How many bones in the Face? Name and describe them.
80. The Ear has how many bones?
81. State the number and names of the bones of the Trunk.
82. How is the Thorax formed? What its natural form? What organs does it contain?
83. What is the situation of the Sternum?
84. Describe the Ribs. Distinguish between true and false. Why the floating ribs so called? What of their length and breadth?
85. Of what is the Spinal Column composed? What is meant by body and process of a vertebra? State their uses. What is said of the arrangement of these processes?
86. State the arrangement of the Vertebræ.
87. Describe the Cervical vertebræ.
88. What is said of the Dorsal?
89. How are the Lumbar distinguished?
90. What is found upon the Anterior and Posterior parts of the body of the vertebræ?
91. What are found between the arches of the vertebræ? How do they differ from other ligaments?
92. Speak of the Intervertebral ligaments.
93. Of what is the Pelvis composed?
94. Describe the Innominatum.
95. What is the Sacrum?
96. What changes occur in the Coccyx during life?

97. Mention the number and names of the bones of the Upper Extremities.
98. Where is the Scapula situated?
99. To what is the Clavicle attached?
100. Describe the Humerus.
101. What is the Ulna?
102. What is the position of the Radius? With what does it articulate?
103. Speak of the number and arrangement of the bones of the Carpus.
104. State the arrangement of the Metacarpal bones.
105. How many bones in the phalanges of the fingers?
106. How many in the Lower Extremities? What their names?
107. What is said of the Femur?
108. Patella?
109. Tibia?
110. Fibula?
111. Tarsus?
112. Of how many bones does the Metatarsus consist?
113. How many do the phalanges of the toes contain?
114. How are joints formed? Name their groups.
115. Mention and describe each kind of immovable joints.
116. What are the mixed joints? Give examples.
117. What is said of movable joints? How many kinds? Describe each.
118. Give special description of certain forms of movable joints.

§ 8. *Histology of the Bones.*

119. What is the character of the primitive basis of bone? State the changes prev ous to ossification.
120. Give the Intra-cartilaginous mode of ossification.
121. State the Intra-membranous mode.
122. What are the structure and texture of the long bones? Where is the Medulla found?
123. Distinguish between the Periosteum and Endosteum.
124. Of what service is Cartilage? How arranged?
125. Of what use the Synovial membrane? Name and describe its kinds.
126. What are found in connection with the Synovial membrane? Describe the several kinds of ligaments.

§ 9. *Chemistry of the Bones.*

127. Of what are the bones composed? Mention the mineral constituents. Observation.

§ 10. *Physiology of the Bones.*

128. Name the uses of the Bones.
129. What qualities found in bone?
130. What advantages result from the structure and arrangement of the skull-bones?
131. Mention the offices of the spinal column.
132. How are strength and firmness secured? How the necessary rotary movement? To what are the muscles attached? What arrangement for the spinal cord? What provision is made to prevent injury to the brain?
133. What purpose do the Ribs serve?
134. State the offices of the Pelvis.
135. What is said of the form and proportion of the Upper Extremities as relating to the hand?
136. Compare the Lower Extremities with the Upper.
137. Why are the shafts of the long bones hollow?
138. Enumerate the uses of the joints.
139. State the purposes of the different classes of joints.
140. Give the use of the Synovia.
141. What is said of Cartilage?
142. Speak of the function of the Ligaments.
143. Of what service the Periosteum?
144. What is illustrated by each bone?

§ 11. *Hygiene of the Bones.*

145. What is the influence of exercise on the health of the bones? How should it be taken?
146. To what are the lower extremities of the very young not adapted?
147. What should be avoided? Why?
148. Why should an erect position be maintained?
149. How are distortions of the body produced?
150. What statement by eminent physicians? How may slight curvatures of the spine be prevented or cured?
151. In the fracture of bones or injury of limbs, what is necessary? What is "White Swelling?" Observation.

§ 12. *Comparative Osteology*

152. Name and describe the sub-kingdoms.
153. Give the classes of the Vertebrata.
154. Compare the Vertebral Column of Mammals. What is said of it in Birds? Reptiles? Fishes?
155. What is said of the bones of the head in Mammals? Birds? Reptiles? Fishes?
156. Why not a Clavicle in the ox. Describe the clavicle of Birds. Reptiles. Fishes.
157. What of the Scapula of the lower order of animals?
158. Speak of the Sternum of Birds. Reptiles. Fishes.
159. Describe the Ribs in the different classes.
160. What is said of the Humerus?
161. What of the Radius and Ulna?
162. What of the Carpus and Metacarpus?
163. Compare Posterior and Anterior Extremities of the several classes. What suggestion by the author?

UNIFIC REVIEW.

[Compare 74 with 152.]

What constitutes the Skeleton? What is said of it in the different sub-kingdoms?

[Compare 76–80 with 155.]

Compare the Bones of the Head in man with those of the lower animals.

[Compare 81–97 with 154, 158 and 159.]

What are the bones of the Trunk? Are they all found in the lower animals? Which is the largest bone in a Bird?

[Compare 97–106 with 160–162.]

Name all the bones of the Upper Extremities in the different classes of the Vertebrata. What peculiarity in the clavicle of Birds?

[Compare 106–113 with 163.]

Describe each bone of the Lower Extremities.

[Compare 119–122 with 8–11 and 152.]

What is the earliest organic form of living things? State the process of ossification.

[Compare 123 with 21–24.]

What tissue in the Periosteum? Use of this membrane?

[Compare 126 with 21–23.]

What tissue forms the Ligaments? What does Ligament signify?

[Compare 127 with 47, 52, 56 and 70.]

Name both the organic and inorganic matter in bones.

[Compare 145 with 202, 213, 214, 281, 361 and 506.]

What is necessary to the health of the bones? What results follow a want of exercise? State the influence of exercise upon the health of the different organs.

FIG. 195.

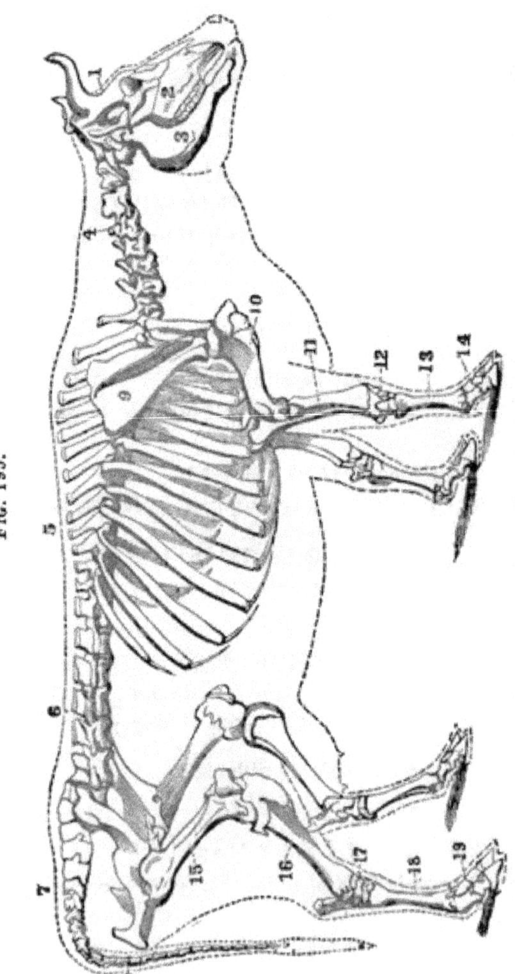

FIG. 195. SKELETON OF THE COW.—1, Frontal bone of the head. 2, Upper jaw (superior maxillary). 3, Lower jaw (inferior maxillary). 4, Cervical vertebræ. 5, Dorsal vertebræ. 6, Lumbar vertebræ. 7, Sacral vertebræ. 8, Caudal vertebræ. 9, Scapula. 10, Humerus. 11, Radius and ulna. 12, Carpus. 13, Metacarpus. 14, Phalanges (toes). 15, Femur. 16, Tibia. 17, Tarsus. 18, Metatarsus. 19, Phalanges. In this fig. the same terms are used as for the corresponding bones in man (see fig. 196). The common names vary.

Fig. 196.

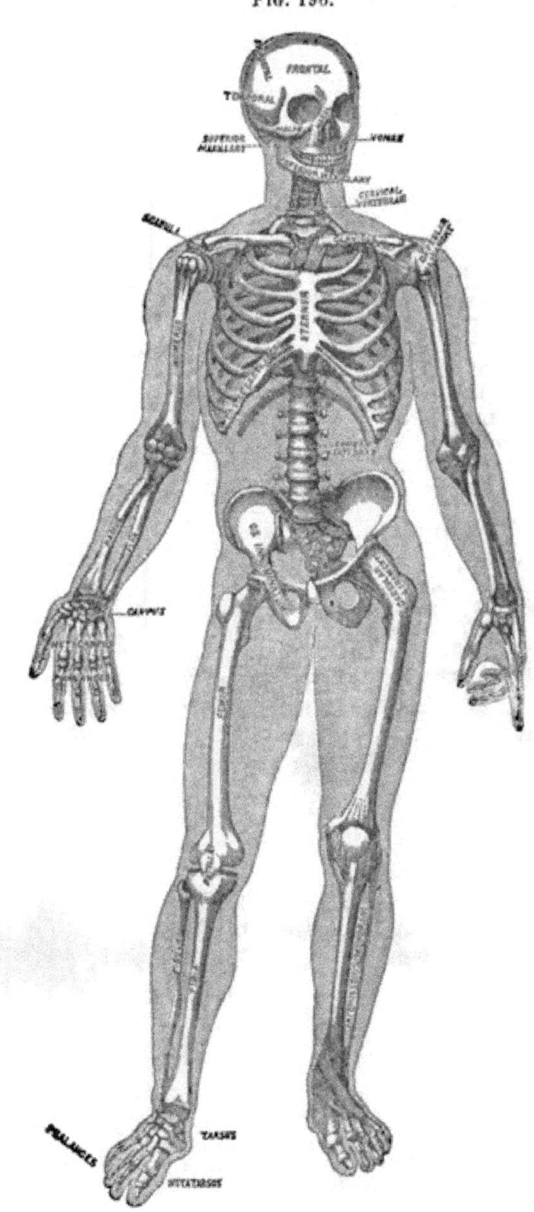
FRONTAL
VOMER
SCAPULA
CERVICAL VERTEBRAE
STERNUM
CARPUS
TARSUS
PHALANGES
METATARSUS

SYNTHETIC REVIEW.

Topic	Section	Chapter
The Skeleton and its uses,	§ 7. *Anatomy of.*	CHAP. IV. *The Bones*
Number and classes,		
Head, Trunk,		
Upper Extremities,		
Lower Extremities,		
Joints,		
Definition and classes of Joints,		
Immovable Joints,		
Mixed, Movable,		
Peculiar forms of Movable.		
Formation of Temporary Cartilage,	§ 8. *Histology of.*	
Intra-cartilaginous mode of ossification,		
Intra-membranous mode,		
Structure of the Long Bones,		
Periosteum, Endosteum,		
Cartilages of the Joints,		
Synovial membrane, Ligaments.		
Chemical Composition,	§ 9. *Chemistry of.*	
Experiment showing earthy and animal matter.		
General uses of,	§ 10. *Physiology of*	
Adaptation of their structure to their uses,		
Skill as shown in the Skull,		
" " Spinal Column,		
" " Ribs,		
" " Pelvis,		
" " Upper Extremities,		
" " Lower Extremities,		
" " Long Bones,		
The uses of the Joints,		
Classification of the Joints,		
Of Movable Joints,		
Function of the Synovia,		
" Cartilages,		
" Ligaments,		
" Periosteum,		
Perfection of this part in the animal fabric.		
Effect of exercise upon the bones of children,	§ 11. *Hygiene of.*	
" compression,		
" stooping,		
Treatment of Fractures,		
" Sprains,		
" Felons.		
Classification of Animals,	§ 12. *Comparative Osteology of.*	
" Vertebrates,		
Compare Spinal Column of Vertebrates,		
" Bones of the Head,		
" " Thorax,		
" " Extremities.		

Give the Human and Comparative Anatomy and Histology of the Bones; the Chemistry, the Physiology and the Hygiene.

ANALYTIC EXAMINATION.

CHAPTER V.—The Muscles.

§ 13. *Anatomy of the Muscles.*

164. What property do the Muscles possess? By what law governed? Give the different forms.
165. Describe the Fasciæ. Speak of the attachment of the muscles.
166. Give the number and kinds of the muscles.
167. How arranged? Define Extensors and Flexors. Examples.
168. State the office of the Occipito-Frontalis; of the Orbicularis Palpebrarum; of the Orbicularis Oris; of the Masseter and Temporal; of the Sterno-Cleido-Mastoid.
169. Of the Pectoralis Major; of the Serratus Magnus; of the Obliquus Externus and Rectus Abdominalis.
170. Of the Trapezius, Rhomboideus Major and Minor; of the Latissimus Dorsi; of the Serratus Posticus Inferior.
171. Of the Deltoid; of the Biceps; of the Triceps; of the Flexor Carpi Radialis; of the Flexor Carpi Ulnaris; of the Flexor Digitorum; of the Extensor Digitorum; of the Extensor Carpi Radialis.
172. Describe the Glutei, Sartorius, Rectus Femoris, Vastus Externus, Vastus Internus, Triceps Abductor Femoris, Biceps Femoris, Extensor Digitorum, Peroneus Longus, Gastrochnemius Externus, Tendo-Achilles.

§ 14. *Histology of the Muscles.*

173. Into what is a Muscle separable?
174. By what is each muscle invested? What is Myolemma?
175. Name and describe the classes of muscles.
176. How is the contractility of the muscles stimulated?
177. What are Tendons? In what is each tendon enveloped?
178. Where do you find the blood-vessels of the muscles?
179. What position do the Nerves occupy? What is said of the different classes of the nerves?

§ 15. *Chemistry of the Muscles.*

180. What is said of the chemical composition of the muscles? Muscle sugar is where found?
181. How does proper muscular substance differ from simple fibrous tissue?

182. Name some of the chemical changes attending muscular action. What is said of the "muscular current"? Observation.

§ 16. *Physiology of the Muscles.*

183. State the relative uses of bones and muscles.
184 Name the uses of the muscles.
185 To what are the Voluntary muscles subject? What is implied by the motion of a limb?
186. Of what aid the muscular sense? What is said of the exercise of this muscular sense?
187. What are the Involuntary muscles?
188. What involuntary muscles are somewhat under the control of the Will? Of what advantage this? Observation.
189. State the office of the Tendons. Do they possess contractility? In what respect do you see in them an exhibition of care and skill? Illustrate with the hand.
190. Define a Lever, and name its kinds.
191. Explain each kind.
192. Where are the principles of the first kind illustrated?
193. Where those of the second?
194. Of the third?
195. What is said of the oblique action of the muscles? What is important to notice in this connection? Compare the Extensors with the Flexors.
196. Where does the pulley find illustration?
197. What is said of the direction of the different layers?
198. In what is mechanical skill shown?
199. Speak of muscular force.

§ 17. *Hygiene of the Muscles.*

200. What advantage in possessing healthy muscles? Name the first essential. What is the influence of pure blood on the muscles?
201. Why should the muscles not be compressed? What is said of the pressure of dressing in case of a fractured limb? What are the results of tight dresses on health? To what is tight-lacing compared?
202. How does exercise promote the health and growth of muscles? Illustration.
203. State the relation of relaxation to contraction. Illustration.

204. Give a reason for a change of employment. Illustration.
205. How should the muscles be called into action? Observation.
206. How rested?
207. How should exercise be taken?
208. What kind of exercise? What pastimes should be chosen?
209. To what should the amount of exercise be adapted? Observation.
210. State the proper time for exercise. Observation.
211. Mention the influence of the mind on the muscles.
212 What should be taken into consideration as to the amount of exercise?
213. In what diseases are great care and discretion necessary as regards exercise?
214. What is said of the exercise of the muscles in chronic diseases of the digestive organs? What is important to secure beneficial results? Observation.
215. Why do the muscles require erect positions of the body?
216. What attention should be given to children and youth? What care in furnishing school-rooms? Observation.
217. Why relaxation of muscles necessary in walking, jumping, etc.? Observation.
218. State and illustrate the influence of education. Observation.

§ 18. *Comparative Myology.*

219. What is said of the muscles of Mammals? Of their color?
220. For what is the muscular system of Birds remarkable?
221. Speak of the muscles of Reptiles.
222. What modification of muscles in Fishes? What color?

UNIFIC REVIEW.

[Compare 164, 165, 166 with 173, 174 and 219–222.]

What is the structure of the muscles? State their relation to the bones. Compare the muscles of man with those of other mammals. What is peculiar to muscle?

[Compare 176 with 441, 450 and 469.]

What are the causes of muscular activity? State the connection between the muscular and nervous system.

[Compare 177 with 22.]

Where do you find the white fibrous and muscular tissue closely related?

[Compare 178 with 371.]

How are the muscles nourished?

[Compare 180 with 50–53.]

Of what is the muscular tissue composed?

[Compare 201 with 360 and 425.]

State the evil results of compression of the muscles.

[Compare 202 with 361 and 506.]

What is the influence of exercise on circulation and muscular power? What the effect of a want of it on the Nervous System?

[Compare 203 with 209, 210, 281 and 506.]

In taking exercise, what caution as to the age, time, amount, etc.?

Fig. 197.

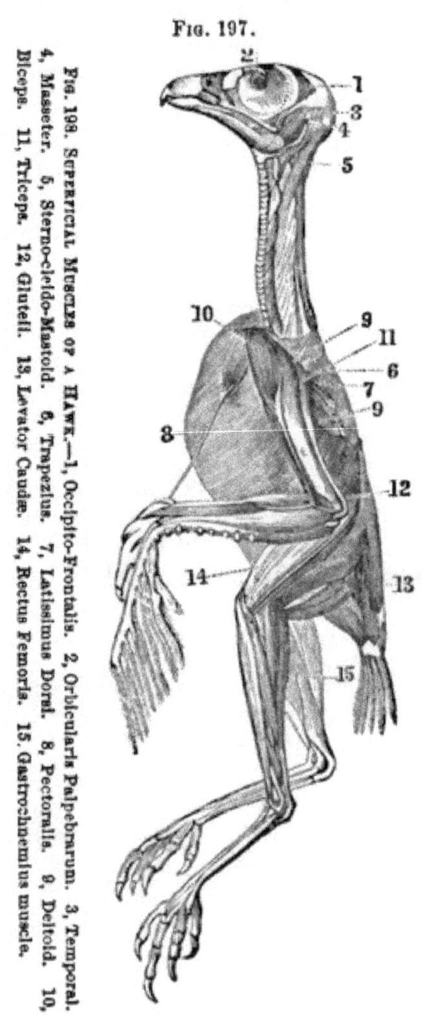

Fig. 198. Superficial Muscles of a Hawk.—1, Occipito-Frontalis. 2, Orbicularis Palpebrarum. 3, Temporal. 4, Masseter. 5, Sterno-cleido-Mastoid. 6, Trapezius. 7, Latissimus Dorsi. 8, Pectoralis. 9, Deltoid. 10, Biceps. 11, Triceps. 12, Glutei. 13, Levator Caudæ. 14, Rectus Femoris. 15. Gastrocnemius muscle.

SYNTHETIC REVIEW.

Topic	Section	Chapter
Law of muscular contraction,	§ 13. *Anatomy of.*	CHAP. V. *The Muscles.*
Consequent forms of muscles,		
Modes of attachment,		
Number and general arrangement,		
Of Head and Neck,		
" Anterior part of Trunk,		
" Posterior "		
" Upper Extremities,		
" Lower Extremities.		
Analysis,	§ 14. *Histology of.*	
Sheaths,		
Voluntary and involuntary,		
Exciting agents of contractility,		
Tendons,		
Blood-vessels,		
Nerves.		
Chemical composition,	§ 15. *Chemistry of.*	
Chemical changes attending muscular action,		
Muscular current.		
Relative uses of Bones and Muscles,	§ 16. *Physiology of.*	
Important functions,		
Relation of the Will to muscular action,		
" muscular sense " "		
The muscular sense a source of enjoyment,		
Importance of involuntary movements,		
Importance of such movements being sometimes voluntary,		
Tendons,		
Mechanical powers exhibited in muscular action,		
Lever, Pulley,		
Oblique action, etc.,		
Deep-seated,		
Minute.		
Healthy condition,	§ 17. *Hygiene of.*	
Freedom from compression,		
Exercise,		
Conditions to be observed in exercise,		
Exercise sometimes injurious,		
Effect of mental stimulus,		
Regard necessary to age and health,		
Position of the body,		
Proper tension,		
Education.		
Muscles of other mammals and man,	§ 18. *Comparative Myology of.*	
" Birds,		
" Reptiles,		
" Fishes.		

Give the Anatomy, the Histology, the Chemistry, the Physiology, the Hygiene, Human and Comparative, of the Muscles.

Fig. 198.

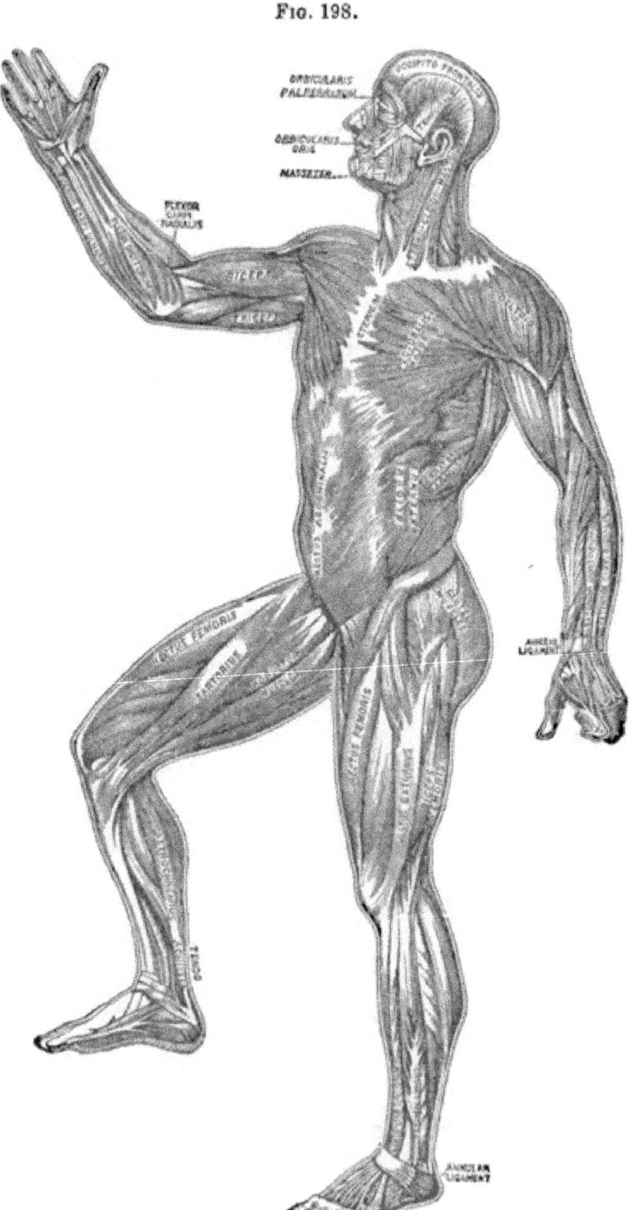

DIVISION II.—MOTORY APPARATUS.

SYNTHETIC REVIEW.

SECT. 7. Anatomy of.	CHAP. IV. *The Bones.*	Division II. *Motory Apparatus.*
" 8. Histology of.		
" 9. Chemistry of.		
" 10. Physiology of.		
" 11. Hygiene of.		
" 12. Comparative Osteology of.		
" 13. Anatomy of.	CHAP. V. *The Muscles.*	
" 14. Histology of.		
" 15. Chemistry of.		
" 16. Physiology of.		
" 17. Hygiene of.		
" 18. Comparative Myology of.		

Give the Anatomy, the Histology, the Chemistry, the Physiology, the Hygiene, Human and Comparative, of the Motory Apparatus.

DIVISION III.—THE NUTRITIVE APPARATUS.

ANALYTIC EXAMINATION.

223. In what processes are the organs of the Nutritive Apparatus used? Name the organs.

CHAPTER VI.—The Digestive Organs.

§ 19. *Anatomy of the Digestive Organs.*

224. What are included in the Digestive Organs?
225. Describe the Mouth.
226. What is said of the Teeth? How many parts has each tooth? Observation.
227. What are the temporary teeth? The permanent? Name and describe the different forms of the teeth.
228. Of how many pairs do the Salivary Glands consist? Name and describe each pair. Observation.
229. Describe the Pharynx.
230. What is the Œsophagus?
231. What is said of the Stomach?
232. Mention the divisions of the Intestines. Describe the small intestine.
233. State the length and parts of the large intestine. Describe each part.
234. Describe the Liver. By what surrounded? How many lobes? What is on the under side?
235. What is said of the Pancreas?
236. What is the Spleen? Why so named?

§ 20. *Histology of the Digestive Organs.*

237. By what is the alimentary canal lined?
238. Describe the covering of the mouth. Describe the tongue. Name and describe its muscles. Distinguish between hard and soft palate.
239. What is the relation of the teeth to the mucous membrane of the mouth? Give their composition. What is the Enamel? Describe the Cement.
240. Describe the walls of the Pharynx.

241. Name and describe the coats of the Œsophagus.
242. Describe the Stomach and its coats.
243. What is said of the coats and muscular fibres of the intestines? What are the Valvulæ Conniventes? Describe the Villi.
244. How many coats has the Liver? Describe the lobules. What is the mid-vein? What relation the hepatic system to the portal?
245. Describe the coats of the Spleen.
246. What is the Peritoneum?

§ 21. *Chemistry of the Digestive Organs.*

247. What secretions effect chemical changes during digestion?
248. What is Mucus? Its composition?
249. Describe Saliva. Its composition. What is said of it when first secreted? What salts does it contain? State its chemical effect.
250. What are the properties of Gastric Juice? Name its characteristic constituent. What saline matter? What of its solvent power? What changes does it effect?
251. Describe Bile. Its composition. What changes caused by it.
252. What is said of the Pancreatic Juice? What per cent. solid matter? Its salts? Its chemical power?
253. Speak of the Intestinal Juices.
254. State the summing up of the changes in three staminal principles of food.
255. What is the relation of acid and alkali in the digestive fluids?

§ 22. *Physiology of the Digestive Organs.*

256. What change in food is necessary? What is Primary Assimilation? What Secondary? What is Digestion?
257. To what is the alimentary canal likened? What do recent investigations show?
258. Speak of the changes of food in the stomach. Can the food return to the œsophagus? Why not? When does the food leave the stomach? What is there peculiar about the Pylorus?
259. What changes occur in the alimentary canal?
260. What is said of the absorbing surface of the intestines?

§ 23. *Hygiene of the Digestive Organs.*

261. Name the first requisite for the preservation of the Teeth. What is the effect of sudden changes of temperature? Should acids be used? What objection to the use of tobacco? Why should the teeth be frequently examined?
262. When should the temporary teeth be removed? What do the irregular permanent teeth generally require? Does toothache always indicate a necessity of extraction? Observation.
263. What is required for the health of the Digestive Organs?
264. What is said of the quantity of food?
265. What must the supply equal? When must supply exceed waste?
266. When should the quantity of food be diminished?
267. Why is more food required in winter than in summer?
268. To what should the amount be adapted?
269. What should be the quality of food?
270. What must proper aliment contain?
271. How should food be cooked? What are the best methods of preparation?
272. To what should the quality be adapted?
273. What is said of vegetable diet?
274. Who require stimulating food? Who unstimulating?
275. What is said of the manner of taking food?
276. Why should food be properly masticated?
277. Why not take drink with food?
278. Why should regard be had to the temperature of drink?
279. How and when should food be taken?
281. State the reason for not taking food just before or after exercise. What is the influence of moderate exercise? Observation.
282. Why is it not best to eat immediately before retiring to sleep?
283. What influence does the mind exert upon the digestive organs? How should indigestion arising from nervous prostration be treated?
284. After long abstinence, what kind of food should be taken?
285. What influence does the condition of the skin exert?
286. Why is pure air necessary? General Observation. Recapitulation.

§ 24. *Comparative Splanchnology.*

287. What is said of the Nutritive Apparatus of Vertebrates?
288. Compare the mouth and teeth of the Vertebrates.
289. Of Birds. 290. Of Reptiles. 291. Of Fishes.
292. How are the digestive fluids supplied?
293. Speak of the stomach and intestines of Vertebrates.
294. Give the process of digestion in Ruminants.
295. Name and describe the stomachs of Birds.
296. Compare the alimentary canal of Reptiles with that of Mammals or Birds.
297. What is said of the alimentary canal in Fishes?

UNIFIC REVIEW.

[Compare 225–227 with 287–291.]

Compare the teeth of man with those of the lower animals

[Compare 228 with 292.]

Describe the Salivary Glands in all animals.

[Compare 229–236 with 293–297.]

Contrast the Digestive Organs of Man with those of other Mammals, Birds, Reptiles and Fishes.

[Compare 237, 238 with 36–44, 289–292, 547 and 548.]

Give a full description of the lining membrane of the mouth and alimentary canal of the different classes of animals.

[Compare 239 with 288–291.]

Speak of the histological composition of the teeth in animals.

[Compare 240–243 with 293–297.]

Give the comparative Histology of the Œsophagus, Stomach and Intestines.

[Compare 244 with 292, 296 and 297.]

What is said of the Liver in the different animals?

[Compare 247–255 with 45–51, 57–60, 65 and 67–70.]

Give an outline of the Chemistry of the Digestive Organs.

[Compare 256–260 with 294 and 295.]

Compare the digestive processes in different classes of animals.

[Compare 280–286 with 209–214, 410–415 and 500–506.]

In what condition should the system be to take food without injury? State the influence of exercise upon digestion. What does the health of the human system require?

FIG. 199. FIG. 200.

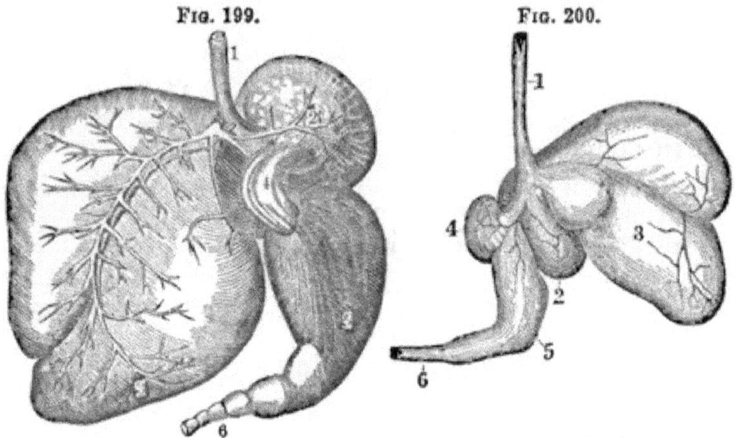

FIG. 199. STOMACH OF AN OX.—1, Œsophagus. 2, Rumen (paunch). 3, Reticulum (honeycomb). 4, Omasum (many-plies). 5, Abomasum (rennet). 6, Intestine.

FIG. 200. STOMACH OF A SHEEP.—1, Œsophagus. 2, Rumen. 3, Reticulum. 4, Omasum. 5, Abomasum, or rennet. 6, Intestine.

FIG. 201. FIG. 202.

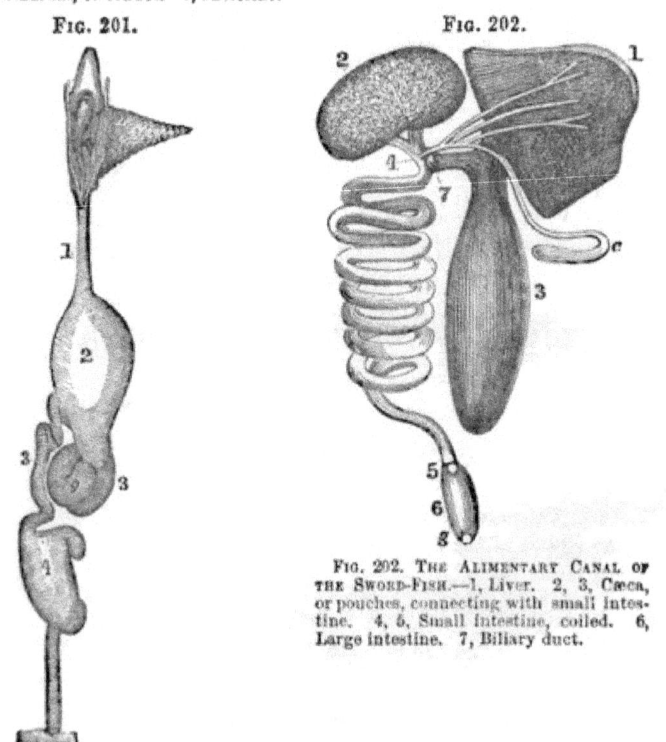

FIG. 202. THE ALIMENTARY CANAL OF THE SWORD-FISH.—1, Liver. 2, 3, Cæca, or pouches, connecting with small intestine. 4, 5, Small intestine, coiled. 6, Large intestine. 7, Biliary duct.

FIG. 201. THE ALIMENTARY CANAL OF THE FLYING LIZARD.—1, Œsophagus. 2, Stomach. 3, 3, Small intestine. 4, Large intestine.

SYNTHETIC REVIEW.

Mouth, Teeth, Salivary Glands, Pharynx, Œsophagus, Stomach, Intestines, Liver, Pancreas, Spleen.	§ 19. *Anatomy of.*	CHAP. VI. *Digestive Organs.*
Lining membrane of Alimentary Canal, " " Mouth, Composition of the Tongue, " " Teeth, Palates, Pharynx, Coats of the Œsophagus, " Stomach, " Intestines, " Liver, " Spleen, Peritoneum.	§ 20. *Histology of.*	
Secretions, Names, " Character, Mucus, Saliva, Gastric Juice, Bile, Pancreatic Juice, Intestinal Juice, Changes in Food, Acids and Alkalies.	§ 21. *Chemistry of.*	
Assimilation, Chymification, Chylifaction, Destination of Chyle,	§ 22. *Physiology of.*	
Preservation of Teeth, Removal " Quantity of Food, Quality " Manner of taking Food, Condition of the System,	§ 23. *Hygiene of.*	
Nutritive Apparatus of Vertebrates, Mouth and Teeth, Digestive Fluids, Stomach and Intestines.	§ 24. *Comparative Splanchnology of.*	

State the Anatomy, the Histology, the Chemistry, the Physiology and the Hygiene, Human and Comparative, of the Digestive Organs, figs. 199, 200, 201, 202, 203, 204.

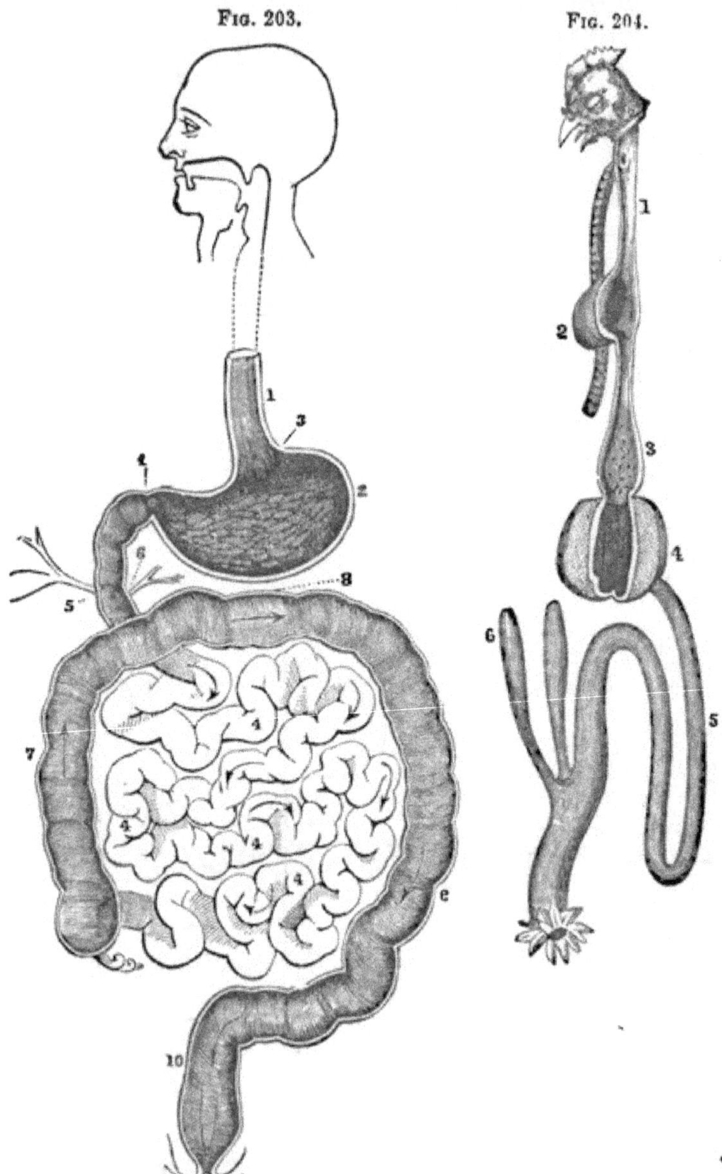

Fig. 203. The Alimentary Canal of Man.—1, Œsophagus. 2, The stomach. 3, Cardiac orifice. 11, Pylorus. 5, Biliary duct. 4, 4, 4, 4, Small intestines. 6, Pancreatic duct. 7, Ascending colon. 8, Transverse colon. 9, Descending colon. 10, Rectum.

Fig. 204. The Alimentary Canal of a Fowl.—1, The œsophagus. 2, Ingluvies (crop). 3, Proventiculus (secreting stomach). 4, Triturating stomach (gizzard). 5, Intestine 6, Two cæca.

ANALYTIC EXAMINATION.

CHAPTER VII.—ABSORPTION.

298. Define Absorption and Absorbents. State the difference between general and intrinsic absorption.

§ 25. *Anatomy of the Absorbents.*

299. Of what do the Absorbents consist? What is Lymph? Describe the Lacteals.
300. What is said of the Lymphatic Glands?
301. Where are the Lymphatic Vessels found? State the kinds of Lymphatics. How are the Thoracic and Lymphatic Ducts formed?
302. Give the course of the Thoracic Duct.
303. Describe the Lymphatic Duct.
304. Where are the Lymphatic Glands found?
305. What is the Portal Vein?

§ 26. *Histology of the Absorbents.*

306. Describe the coats of the Lymphatic Vessels. With what are the larger Lymphatic Tubes supplied?
307. What is the supposed composition of the Lymphatic Glands?
308. Give the origin of the Lymphatics and Lacteals.
309. Of what does Lymph consist?

§ 27. *Chemistry of the Absorbents.*

310. What chemical changes occur in the absorbent system?
311. Give the proportions of the chief ingredients of Chyle in the afferent Lacteals. In the efferent Lacteals. In the Thoracic Duct.
312. What changes take place in the Portal circulation?

§ 28. *Physiology of the Absorbents.*

313. What is the office of the Lymphatics?
314. What may the office of the Lymphatics include? What is said of disintegration of the tissues?
315. Speak of the absorbing power of the mucous membrane.
316. Illustrate the absorbent power of the skin.

317. When are the fluids of the serous and synovial membranes absorbed? Observations.

318. Describe Endosmosis.

§ 29. *Hygiene of the Absorbents.*

319. What should be the condition of the air? Observation.

320. What influence has moisture? Observation.

321. What is the influence of nutritious food upon absorption?

322. What care is necessary in handling poisons?

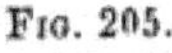

FIG. 205.

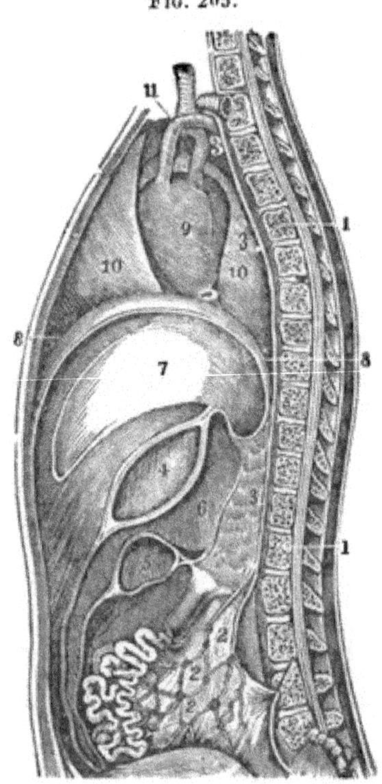

FIG. 205.—Small intestine. 2, 2, 2, Lacteals. 3, 3, 3, Thoracic duct. 4, Stomach. 5, Colon. 6, Pancreas. 7, Liver. 8, 8, Diaphragm. 9, Heart. 10, 10, Lungs. 11, Large vein into which the thoracic duct opens. 12, 12, Spinal column.

SYNTHETIC REVIEW.

Topics	Section	Chapter
Process of Absorption, Specific and General.		CHAP. VII. *The Absorbents*
Absorbents, Lymph, Lymphatic Glands, " Vessels, Thoracic Duct, Lymphatic Glands, position, Absorbent Veins.	§ 25. *Anatomy of.*	
Lymphatic Vessels, " Glands, Origin of Lymphatics, Lymph.	§ 26. *Histology of.*	
Changes in absorbent system, " portal circulation.	§ 27. *Chemistry of.*	
Office of the Lymphatics, Power of different tissues, " " membranes, Absorption in disease, Imbibition of membranes.	§ 28. *Physiology of.*	
Condition of the air, Effect of nutritious food, " removal of cuticle.	§ 29. *Hygiene of.*	

Give the **Anatomy**, the Histology, the Chemistry, the **Physiology** and the **Hygiene** of the Absorbent System of man.

Fig. 206

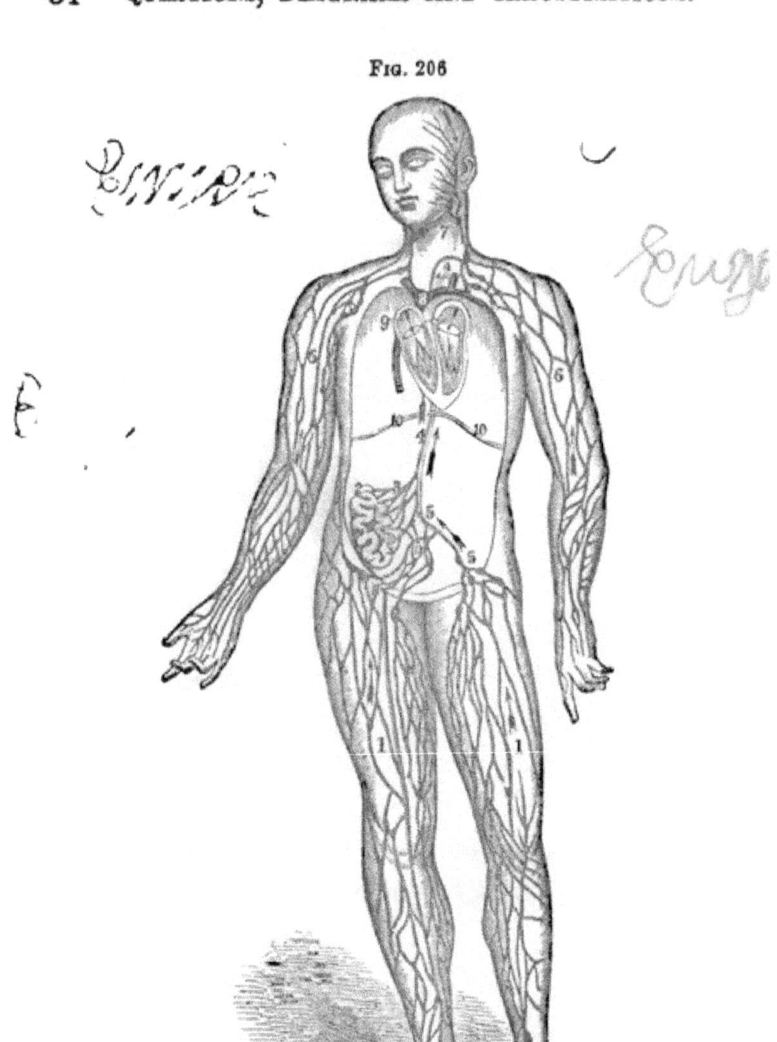

Fig. 206.—1, 1, The absorbents of the lower extremities. 2, The small intestine. 3, The lacteals. 4, 4, The thoracic duct. 5, 5, 5, Absorbent ducts. 6, 6, Absorbents of the arms. 7, Absorbents of the neck. 8, A large vein that opens into the right auricle of the heart. 9, The right auricle. 10, 10, The diaphragm.

ANALYTIC EXAMINATION.

CHAPTER VIII.—The Circulation.

§ 30. *The Blood.*

323. From what source is the blood derived? Of what does the blood consist?
324. For what purpose is the blood constantly undergoing loss? Observation.
325. Why must the blood be kept in circulation? Name the Circulatory Organs.

§ 31. *Anatomy of the Circulatory Organs.*

326. Describe the Heart.
327. What are the Arteries? To what is the Aorta likened? What are the Capillaries? Where found?
328. Give the course of the Veins. What constitutes the Systemic circulation? What the Pulmonic?
329. From what part of the heart arises the Aorta? Name its divisions. Describe the Arch.
330. State the course of the Thoracic Aorta.
331. What is said of the Abdominal Aorta, its divisions and subdivisions?
332. Give the divisions of the Carotid arteries. To what parts of the body do the subclavian arteries furnish branches? What is said of the extension of the subclavian artery?
333. How are the Veins arranged? Describe the Superior Vena Cava. Inferior Vena Cava. Portal vein. Pulmonary veins.

§ 32. *Histology of the Circulatory Organs.*

334. Of what is the Pericardium composed?
335. What can you say of the Endocardium? Where does the fibro-elastic tissue form four rings? What and where are the Semi-lunar valves?
336. Where are the Mitral valves? Where the Tricuspids?
337. Upon what is the muscular structure of the heart based? What is said of the superficial fibres? Where is the middle stratum of fibres found?
338. Of what do the muscular fibres of the auricles consist?

339. Name and describe the coats of the arteries.

340. How are the veins constructed? Describe the valves in the veins. Where found?

341. Give the structure of the Capillaries.

§ 33. *Chemistry of the Blood.*

342. State the analysis of the blood.

343. What per cent. of solid matter and water in the blood?

344. How are the mineral substances distributed in the blood? What effect has air on blood?

§ 34. *Physiology of the Circulatory Organs.*

345. Why is circulation necessary? Why a double heart?

346. Give the Systemic circulation; the Pulmonic.

347. What is said of the contraction and dilatation of the auricles and ventricles? What is the effect of such action?

348. In the construction of the circulatory system, what was necessary?

349. By what means are proper circulatory impulses given?

350. How is a backward flow from the auricles prevented? From the ventricles? From the arteries? From the Pulmonary artery?

351. How are the arteries protected against sudden action of the heart?

352. How is the current maintained?

353. Explain the capillary circulation; also the portal current.

354. How is a continuation of the flow through the veins effected?

355. How is the intermittent pressure caused by the action of the heart equalized?

356. What secures the relative amount of blood to each organ?

357. What provision is there for contingencies?

358. By the study of circulation what effect is produced upon the susceptible mind?

§ 35. *Hygiene of the Circulatory Organs.*

359. What temperature should be preserved?

360. Why should the clothing be worn loosely?

361. What is the influence of exercise on circulation?

362. What is said of the quality and quantity of the blood? Illustration.

363. In case of hæmorrhage from divided arteries, what should be done?

364. In flesh wounds, what course is to be taken? Observation. What is the treatment of wounds caused by blunt instruments? Of wounds from poisonous bites?

§ 36. *Comparative Angiology.*

365. What is said of the blood and circulatory organs of Mammals?
366. Of Birds? 367. Of Reptiles? 368. Of Fishes?

UNIFIC REVIEW.

[Compare 323 with 313-318 and 256-260.]

Give in full the change in food during primary assimilation.

[Compare 324 with 369-378.]

How does the blood contribute to the growth of the different parts of the body?

[Compare 325 with 326-333.]

Name and describe the organs by which the blood effects this contribution.

[Compare 326 with 365, 367 and 368.]

Compare the heart of man with that of other Mammals, and with those of Birds, Reptiles and Fishes.

[Compare 327-333 with 365-368.]

Describe the blood-vessels in the different classes of animals.

[Compare 359-362 with 201, 202, 211-214, 264-274, 509 and 591-607.]

What conditions favor free circulation? What can you say of the food in this connection? How is exercise essential to the health of the nervous tissue? In connection with circulation, what is said of the clothing and bathing?

SYNTHETIC REVIEW.

Topic	Section	Chapter
Blood, its circulation,	§ 30. *The Blood.*	CHAP. VIII. *The Circulatory Organs.*
" loss of,		
Circulatory Organs.		
Heart,	§ 31. *Anatomy of.*	
Arteries,		
Capillaries,		
Veins,		
Aorta, Arch,		
" Thoracic,		
" Abdominal,		
Arteries, Carotid and Subclavian.		
Veins, arrangement,		
Superior Vena Cava,		
Inferior "		
Portal Vein,		
Pulmonary Vein.		
Pericardium,	§ 32. *Histology of.*	
Endocardium,		
Valves of the heart,		
Muscular structure of the heart,		
Arteries, their coats,		
Veins, "		
Capillaries.		
Analysis of the blood,	§ 33. *Chemistry of.*	
Distribution of mineral substances.		
Necessity of double circulation,	§ 34. *Physiology of.*	
Systemic Circulation,		
Pulmonic Circulation,		
Their relation to each other,		
Necessary provisions,		
Circulatory impulse,		
Prevention of the flow,		
Current maintained,		
Flow through the capillaries,		
" " veins,		
Equalization of the current,		
Due supply to each organ,		
Provision for contingencies,		
Mechanism of the body.		
Conditions favoring free circulation,	§ 35. *Hygiene of.*	
Treatment of divided arteries.		
Blood and blood-vessels of Mammals,	§ 36. *Comparative Angiology of.*	
" " Birds,		
" " Reptiles,		
" " Fishes.		

Give the Anatomy of the several parts of the Circulatory System, Human and Comparative, the Histology, the Chemistry, the Physiology and the Hygiene.

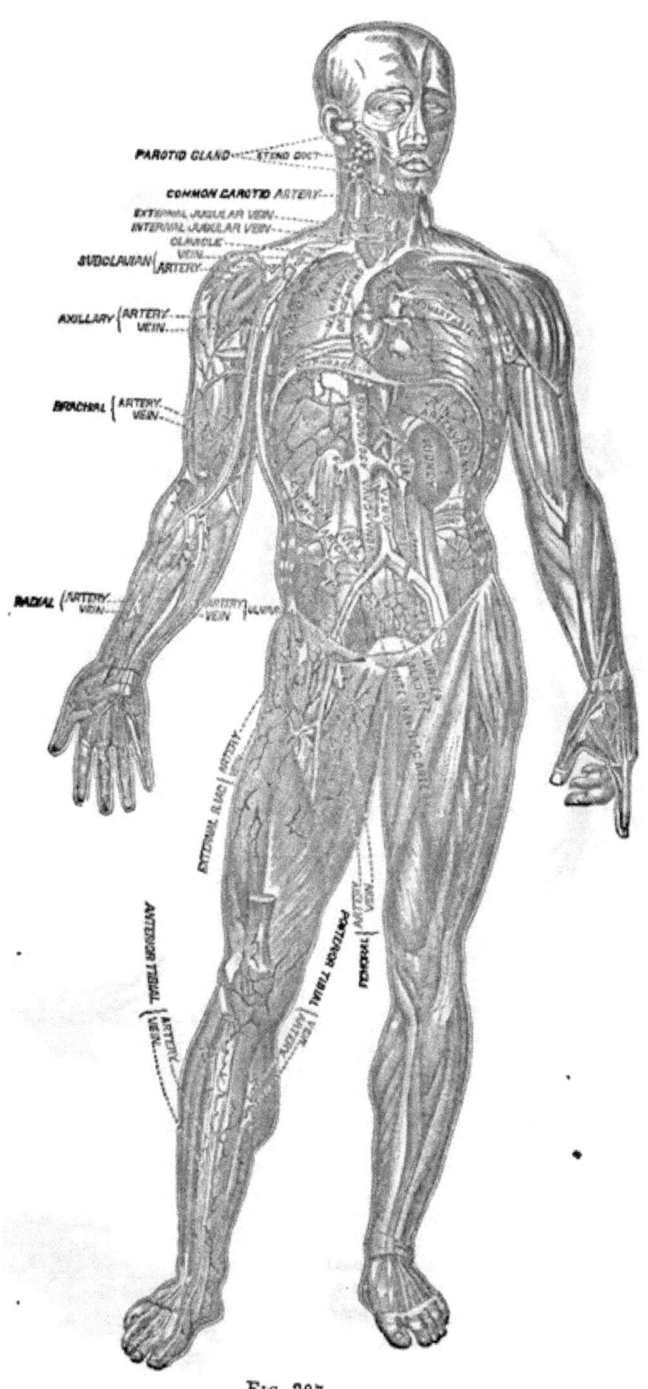
PAROTID GLAND
STENO DUCT
COMMON CAROTID ARTERY
EXTERNAL JUGULAR VEIN
INTERNAL JUGULAR VEIN
CLAVICLE
SUBCLAVIAN VEIN ARTERY
AXILLARY ARTERY VEIN
BRACHIAL ARTERY VEIN
RADIAL ARTERY VEIN
EXTERNAL ILIAC ARTERY
FEMORAL ARTERY VEIN
POSTERIOR TIBIAL ARTERY VEIN
ANTERIOR TIBIAL ARTERY VEIN

Fig. 207.

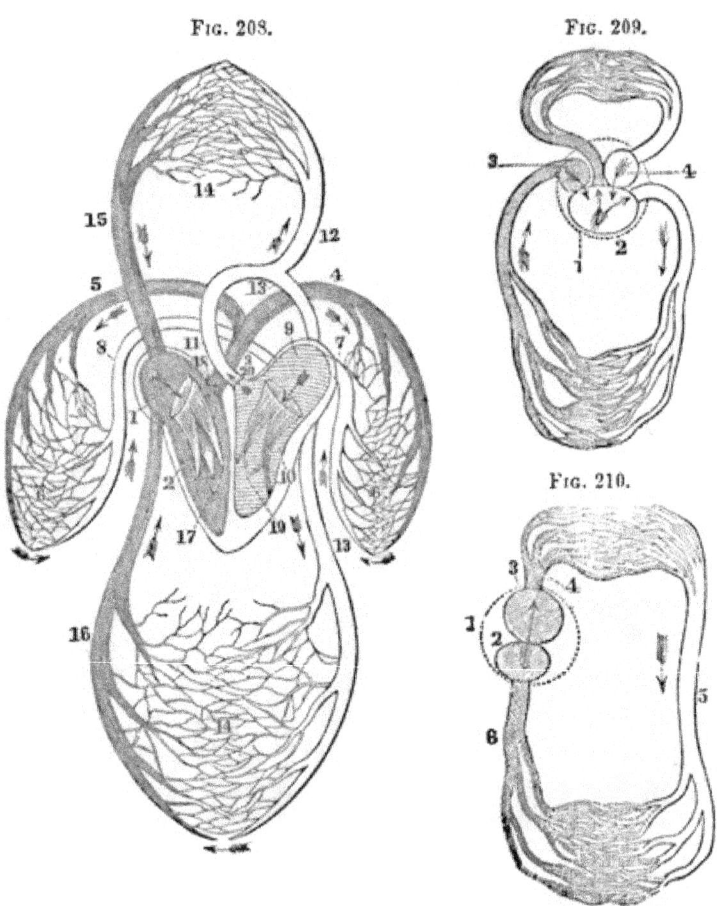

Fig. 208. A Diagram of the Circulation of Mammals.—1, Right auricle. 2, Right ventricle. 9, Left auricle. 10, Left ventricle. 4, 5, Pulmonary arteries. 7, 8, Pulmonary veins. 11, 12, 13, 13, Aorta and its branches. 6, 6, Pulmonary capillaries. 14, 14, Systemic capillaries. 17, Tricuspid valves. 19, Mitral valves. 18, 20, Semilunar valves of the pulmonary artery and the aorta.

Fig. 209. A Diagram of the Circulation of Reptiles.—1, The pericardium. 2, The ventricle. 3, The right auricle. 4, The left auricle.

Fig. 210. A Diagram of the Circulation of Fishes.—1, The pericardium. 2, The ventricle. 3, The auricle. 4, The vessel that conveys the blood to the branchia (gills). 5, The vessel that conveys the blood from the gills to the body of the fish. 6, The vessel that conveys the blood from the body of the fish to the heart.

In these three diagrams the arrows indicate the direction of the blood.

ANALYTIC EXAMINATION.

CHAPTER IX.—ASSIMILATION.

§ 37. *Assimilation, General and Special.*

369. How is life maintained? Distinguish between General and Special Assimilation.
370. What is said of the corpuscles of the blood? What of the blood-plasma?
371. State the first stage in the nutrition of the organs and tissues. What is the second? The third? The fourth? The fifth?
372. How are new cell-elements reproduced? When does this process occur?
373. What is Special Assimilation?
374. Name the secreting glands and membranes. What is said of substances not found in the blood?
375. How is excretion effected? Name the excretory organs. How are the substances which are eliminated from the blood in excretion produced?
376. Speak of the secretory and excretory processes.
378. Describe the kidneys. Observation.

UNIFIC REVIEW.

[Compare 369 with 3.]

In studying assimilation, with what distinctions between organized and unorganized bodies do you become acquainted?

[Compare 370 with 256-260.]

Give the successive stages in Primary Assimilation.

[Compare 371, 372 with 13-17, 45, 46, 119-121, 173, 178, 180, 181 and 460.]

Speak of the structure of cells, and tell how their growth is promoted.

[Compare 373, 374 with 247-255 and 36-44.]

Name the secretory organs, and state the changes caused by their secretions.

[Compare 375-379 with 13, 14, 247, 251, 253, 391-395 and 554.]

Distinguish between Excretion and Secretion. In what processes do the epithelial cells become ruptured? Of what advantage is excretion?

SYNTHETIC REVIEW.

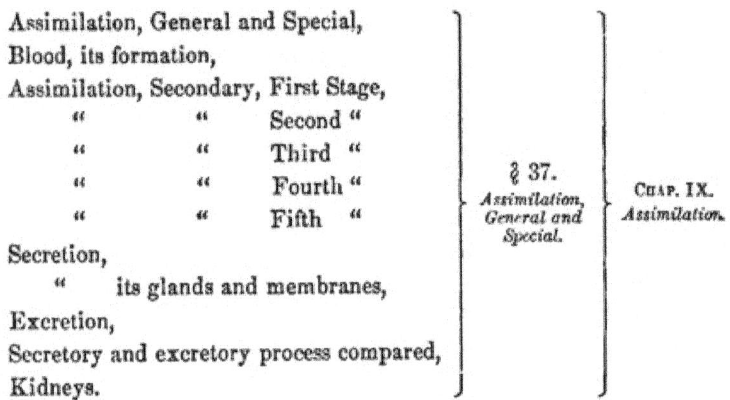

Topics	Section	Chapter
Assimilation, General and Special, Blood, its formation, Assimilation, Secondary, First Stage, " " Second " " " Third " " " Fourth " " " Fifth " Secretion, " its glands and membranes, Excretion, Secretory and excretory process compared, Kidneys.	§ 37. *Assimilation, General and Special.*	Chap. IX. *Assimilation.*

State what you know of Assimilation, general and special, Secretion and Excretion.

Fig. 211.

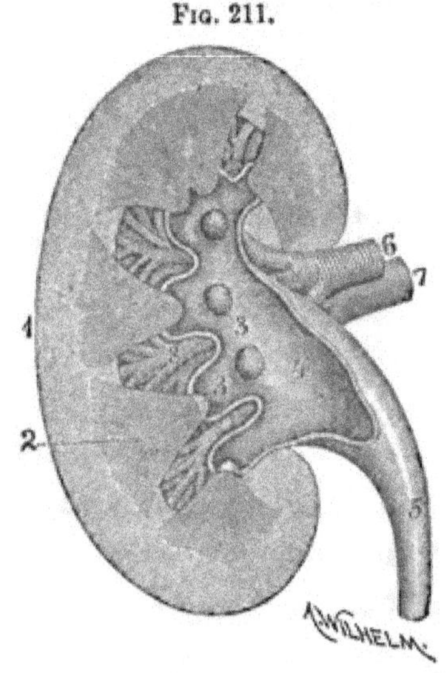

Fig. 211 (*Leidy*). Longitudinal Section of a Kidney.—1, Cortical substance. 2, Renal pyramid. 3, Renal papillæ. 4, Pelvis. 5, Ureter. 6, Renal artery. 7, Renal vein. 8, Branches of the latter vessels in the sinus of the kidney.

ANALYTIC EXAMINATION.

CHAPTER X.—The Respiratory and Vocal Organs.

§ 38. *Anatomy of the Respiratory and Vocal Organs.*

380. Of what do the Respiratory and Vocal organs consist?
381. Describe the Larynx. Of what is it composed? What is said of the Thyroid cartilage? Of the Cricoid? Of the Arytenoid? Of the Epiglottis?
382. What is the Trachea?
383. Give the divisions and subdivisions of the Bronchi.
384. Of how many divisions do the Lungs consist? Of what form are they? What is the Pleura? Compare the Lungs.

§ 39. *Histology of the Respiratory and Vocal Organs.*

385. What is said of the structure of the Larynx?
386. Describe the Vocal cords.
387. Of what is the Trachea made up? Speak of each part.
388. Distinguish between the Bronchi and Trachea.
389. How are the Lungs constructed? In what way are the air-cells connected together?
390. Describe the Pleura.

§ 40. *Chemistry of the Respiratory and Vocal Organs.*

391. Of what does Respiration consist?

392, 393. State the sources of carbonic acid.

394. Give the proportions of oxygen and carbonic acid in the arterial and venous blood.
395. State the physical process by which an exchange of oxygen and carbonic acid in the capillaries is effected, also the chemical process.
396. In what respect does expired air differ from that inspired?
397. What is the source of animal heat? Of what temperature the tissues? Of what the blood?

§ 41. *Physiology of the Respiratory and Vocal Organs.*

398. What are the objects of Respiration? What are the results of the chemical changes?
399. Of what acts does respiration consist? How is inspiration effected? Give the motion of the ribs and diaphragm.

400. What is said of the movements in expiration? What muscles are called into action?
401. Define abdominal and pectoral respiration.
402. How is the air in the air-cells renovated?
403. Is the amount of air taken in and given out in respiration always the same?
404. Speak of the frequency of respiration.
405. What are the actions of sighing, yawning, sobbing, laughing, coughing and sneezing?
406. What is the office of the Larynx in respiration? Of what is the Larynx the special organ?
407. What laws govern the vibrations of the vocal cords?
408. What modify the tones? How further modified? Upon what does the general strength of the voice depend?

§ 42. *Hygiene of the Respiratory and Vocal Organs.*

409. Why is proper respiration important?
410. Why must there be a constant and sufficient supply of pure air?
411. What is the influence of carbonic acid?
412. Mention its sources.
413. What regard should be had for the surroundings of our dwelling-houses?
414. Where is the chief danger?
415. What remarks as to the necessity of ventilation of school-rooms? Of churches?
416. Of concert-halls?
417. State the influence of habit in accustoming ourselves to foul air.
418. What is said of the ventilation of sleeping-rooms? Observations.
419. What attention should be paid to the sick-room?
420. Speak of the means of ventilation in summer.
421. What means in winter?
422. What is the healthiest known means for ventilating a small room?
423. What is said of the use of stoves?
424. Give the quotation on the use of steam for warming rooms.
425. What besides purity of air is required for proper respiration? What objectionable fashion is noticed?
426. Compare the custom of the Chinese women with that of the American.

427. What effect has compression of the mother's chest on her offspring?
428. How can the chest made small by compression be enlarged? Observation.
429. By what is respiration much influenced?
430. State the process of resuscitating persons asphyxiated from drowning, strangulation, electricity, or breathing poisonous gases. Observation.

§ 43. *Comparative Pneumonology.*

431. How does the Respiratory apparatus in all the mammalia compare with that in man?
432. Describe the Lungs of Birds.
433. What is said of the Ultimate Pulmonary Capillaries?
434. What marked modification of respiration in Birds?
435. Speak of respiration in Reptiles. 436. In Fishes.
437. Describe the Gills.
438. What remarkable feature in the organization of some fish?

UNIFIC REVIEW.

[Compare 380–385 with 431–438.]

Compare each respiratory organ in man with that of the lower classes of animals.

[Compare 385–388 with 21, 22, 23 and 25.]

Name the tissues found in the organs of respiration. How disposed?

[Compare 389 with 26, 36, 37 and 341.]

What tissue in the Lungs? Describe the variety of Epithelium in the organs of respiration, and name those organs. Describe the capillaries.

[Compare 390 with 39.]

What membrane forms the Pleura? What is said of it and its secretion?

[Compare 391–396 with 45, 46, 50 and 70–72.]

Give the chemical changes which occur during respiration.

[Compare 397, 398 with 182, 186 and 187.]

What chemical actions produce heat? State the influence of respiration on motion.

[Compare 425–428 with 206.]

Of what advantage is exercise of the lungs? What is necessary after exercise?

[Compare 429 with 211, 215, 509, 514 and 515.

What connection is there between respiration and mental energy? What caution is given?

SYNTHETIC REVIEW.

Topics	Section	Chapter
Larynx, " its parts, Trachea, Bronchi, Lungs.	§ 38. *Anatomy of.*	Chap. X. *The Respiratory and Vocal Organs.*
Larynx, Vocal Cords, Trachea, Bronchi, Lungs, Pleura.	§ 39. *Histology of.*	
Respiration, Carbonic Acid, Exchange of Oxygen and Carbonic Acid, Expired and inspired air, Animal heat.	§ 40. *Chemistry of.*	
Object of respiration, Modes " Renovation of air in air-cells, Amount of air in respiration, Number of respirations, Modifications of respiratory movements, Double function of the Larynx, Special " " Vibration of the Vocal Cords, Conditions affecting tones, " " strength of voice.	§ 41. *Physiology of.*	
Importance of proper respiration, Pure blood, how obtained, Carbonic Acid, its influence, " its sources, Dwelling-houses, location, " impure air in, Public Buildings, ventilation, Sleeping-rooms, " Sick-rooms, " Pure air and warmth, how obtained, Importance of moisture, Compression of respiratory organs, Enlargement of the chest, Influence of nervous system, Resuscitation of asphyxiated persons.	§ 42. *Hygiene of.*	
Mammalia, Respiratory Organs of, Birds, " " Reptiles, " " Fishes, " "	§ 43. *Comparative Pneumonology of.*	

Give the Anatomy, the Histology, the Chemistry, the Physiology and the Hygiene, Human and Comparative, of the Organs of Respiration, figs. 212, 213, 214, 215, 216.

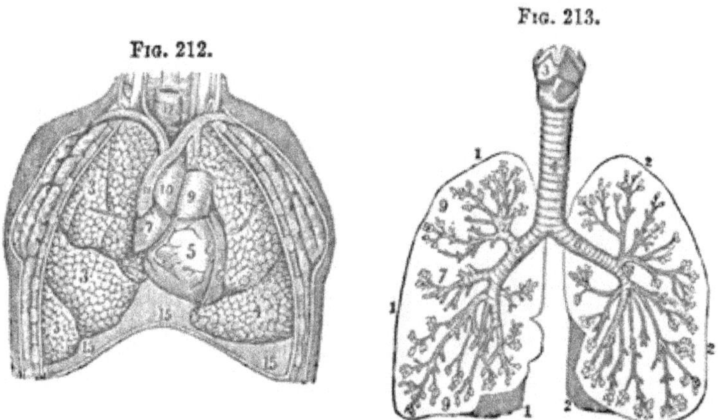

Fig. 212. 3, 3, 3, The lobes of the right lung. 4, 4, The lobes of the left lung. 5, 6, 7, The heart. 9, 10, 11, The large blood-vessels. 12, The trachea. 15, 15, 15, The diaphragm.

Fig. 213. 1, Outline of right lung. 2, Outline of left lung. 3, 4, Larynx and trachea. 5, 6, 7, 8, Bronchial tubes. 9, 9, Air-cells.

Fig. 214.

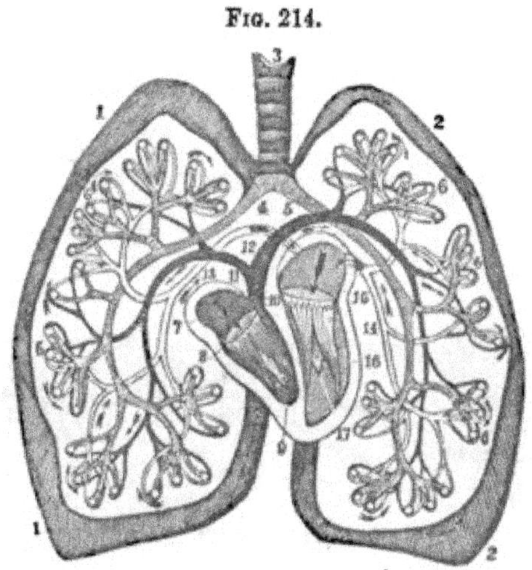

Fig. 214. An Ideal View of the Pulmonic Circulation.—1, 1, The right lung. 2, 2, The left lung. 3, The trachea. 4, The right bronchial tube. 5, The left bronchial tube. 6, 6, 6, 6, Air-cells. 7, The right auricle. 8, The right ventricle. 9, The tricuspid valves. 10, The pulmonic artery. 11, The branch to the right lung. 12, The branch to the left lung. 13, The right pulmonic vein. 14, The left pulmonic vein. 15, The left auricle. 16, The left ventricle. 17, The mitral valves.

Fig. 215.

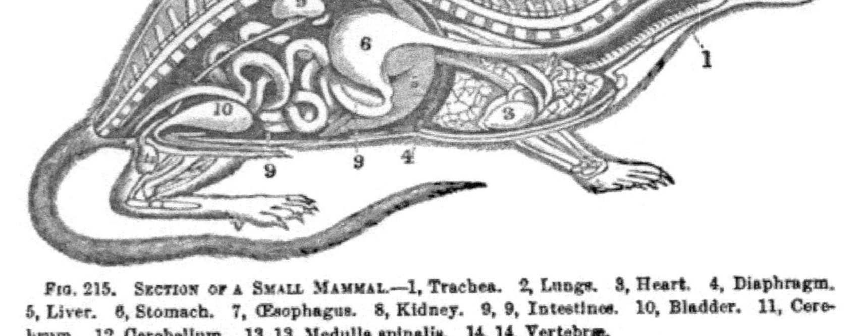

Fig. 215. Section of a Small Mammal.—1, Trachea. 2, Lungs. 3, Heart. 4, Diaphragm. 5, Liver. 6, Stomach. 7, Œsophagus. 8, Kidney. 9, 9, Intestines. 10, Bladder. 11, Cerebrum. 12. Cerebellum. 13, 13, Medulla spinalis. 14, 14, Vertebræ.

Fig. 216.

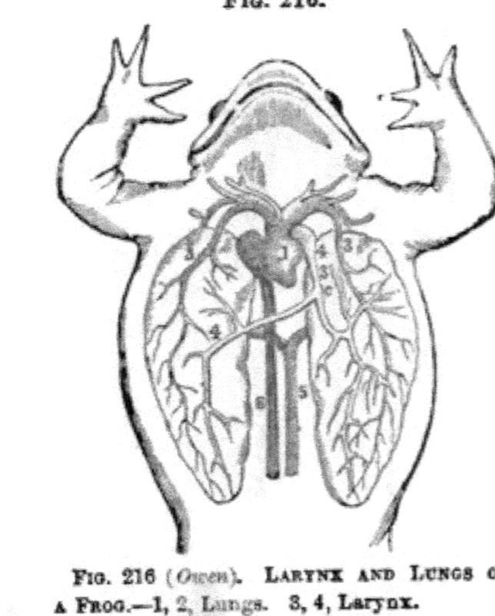

Fig. 216 (*Owen*). Larynx and Lungs of a Frog.—1, 2, Lungs. 3, 4, Larynx.

DIVISION III.—THE NUTRITIVE APPARATUS.

SYNTHETIC REVIEW.

Sections	Chapter	Division
SECT. 19. Anatomy of. " 20. Histology of. " 21. Chemistry of. " 22. Physiology of. " 23. Hygiene of. " 24. Comparative Splanchnology of.	CHAP. VI. *The Digestive Organs.*	Division III. *Nutritive Apparatus.*
" 25. Anatomy of. " 26. Histology of. " 27. Chemistry of. " 28. Physiology of. " 29. Hygiene of.	CHAP. VII. *The Absorbents.*	
" 30. The Blood. " 31. Anatomy of. " 32. Histology of. " 33. Chemistry of. " 34. Physiology of. " 35. Hygiene of. " 36. Comparative Angiology of.	CHAP. VIII. *The Circulation.*	
" 37. Assimilation, General and Special.	CHAP. IX. *Assimilation.*	
" 38. Anatomy of. " 39. Histology of. " 40. Chemistry of. " 41. Physiology of. " 42. Hygiene of. " 43. Comparative Pneumonology of.	CHAP. X. *Respiratory Organs.*	

Give succinctly the Anatomy, the Chemistry, the Physiology and the Hygiene, Human and Comparative, of the Nutritive Apparatus.

DIVISION IV.—SENSORIAL APPARATUS.

ANALYTIC EXAMINATION.

CHAPTER XI.—Nervous System.

§ 44. *Anatomy of the Nervous System.*

439. What two formal characters does Nervous Tissue present? Give the arrangement and names of each.

440. How are the Ganglia, Nerves and Commissures arranged? What is included in each system?

441. Describe the Spinal Cord. What is the Medulla Oblongata? To what is this enlargement due? What may be seen in each of the lateral halves of the Medulla Oblongata? What forms the Decussation of the Anterior Pyramids? How is the Fourth Ventricle formed?

442. Where is the Cerebellum? How is the Pons Varolii formed? Describe the Inferior Peduncles of the Cerebellum. What are the Peduncles of the Cerebrum, and why so called? Give the course of these bundles. How are these ganglia connected with the Spinal Cord? Of what does the Quadrigeminal Body consist?

443. What is said of the connections of all the above-mentioned ganglia?

444. How are the hemispheres of the Cerebrum united? How are the ventricles formed?

445. Are the above-mentioned all the ganglia, membranes and galleries which exist in the brain?

446. What is the relation of the Cerebrum to the other parts? How many lobes has each hemisphere? How does the surface appear?

447. How do the convolutions in the two hemispheres compare? What is a remarkable fact respecting these convolutions?

448. What is said of the Cerebellum?

449. What do the brain and spinal cord constitute?

450. Into what classes are the nerves divided? How are the motor and sensory tracts formed?

451. Distinguish between cranial and spinal nerves.

452. Give the grouping and arrangement of the cranial nerves.

453. How many pairs of spinal nerves? How do they differ from the cranial as to their origin?
454. What are the divisions of the spinal nerves? What are plexuses? Name them, and give their formation.
455. Describe the Sympathetic System.
456. What is a peculiarity of the sympathetic nerves?

§ 45. *Histology of the Nervous System.*

457. Name the elements of nervous tissue.
458. Describe the nerve-cells. Where found?
459. Of what do the White Fibres consist?
460. Where are the nerve-filaments distributed? What is said of their individuality? How are they arranged? What their mode of termination?
461. Where are the Tubular Fibres found? What of their size?
462. What are the Gray Fibres?
463. Name the membranes of the Cerebro-spinal System. Describe the Dura Mater, Pia Mater and Arachnoid Membrane.
464. Give a further description of the Dura Mater.
465. What is the Ependyma?

§ 46. *Physiology of the Nervous System.*

466. What opinions have men in different ages held respecting the relation of soul and body?
467. How is the Nervous System related to the compound nature of man?
468. What influence has this system on the different organs?
469. Speak of the connection between the Nervous Centres and the motor and sensitive fibres.
470. Classify the Nervous Centres.
471. Give a full description of the relations existing between the different Centres.
472. What is the function of the Sympathetic Centres?
473. What is said of their connections?
474. Name and illustrate the different kinds of reflex action.
475. Give a marked peculiarity of the Sympathetic System. Illustrate it by the iris of the eye.
476. What is the office of the white substance of the spinal cord? What that of the gray?
477. How is reflex action acquired? State the theory of acquired reflex action as respects repetition.

478. Mention the influence of association. Why is such an arrangement wise?
479. Describe the Sensational Centres. Show that these centres have an independent reflex action. Can they acquire reflex action?
480. What theory is applicable to these centres?
481. How are these centres excited to action?
482. What power have the Ideational Centres?
483. Upon what depends the character of ideas?
484. What is the first way in which the independent reflex action in these centres is manifested? What the second? Third? Fourth?
485. Of what are these centres the seat?
486. What relation is there between the centre of idea and volition?
487. What is the highest energy of which these centres are capable?
488. Upon what does the power of the Will depend?
489. What relations to the Emotions does the Will sustain?
490. What does a free action of the Will require?
491. What influence has the body over the thoughts, emotions and volitions? How does the theory already given find application here?
492. Where does the character of a man leave visible tracings?

§ 47. *Hygiene of the Nervous System.*

493. Why is a knowledge of the laws of the hygiene of this system important?
494. What agencies affect the health of this system? Name the requirements of its health and vigor.
495. What in addition to the features of parents do children inherit? May acquired habits be transmitted?
496. What history is given by M. Morel?
497. What is said of the evil effects of tobacco?
498. What is the effect of all vices in parents?
499. What results spring from nervous diseases in parents? How can such natural constitutions be improved?
500. State the second requirement of health and vigor.
501. Speak of the evil of breathing impure air.
502. What are the results of improper diet?
503. Speak of the effects of alcohol, opium, etc.

504. How does the use of opium compare with that of intoxicating drinks?
505. What is said of the use of tobacco, tea and coffee?
506. What will a want of physical exercise produce?
507. Speak of the benefits of sleep, and the amount needed.
508. Name the third requirement of health.
509. Why is mental exercise essential?
510. Give the remarks of Dr. Ray. What is said of steady employment?
511. Where are seen the saddest effects of an absence of stated employment? What remarks as to the little accomplishments of needlework?
512. To what should the amount of exercise be adapted? What differences are there in the quality of different brains?
513. What is the present tendency in education?
514. State the effect of intense activity.
515. Give the influence of recreation and amusement. Observation.
516. What is essential to the highest mental vigor? What is said of the use of the imagination?
517. What attention is it important to pay to the æsthetic faculty?
518. What is the moral faculty? Upon what depend the happiness and destiny of man?
519. Give Dr. Ray's remarks concerning the hygienic influence of a Harmonious Development of the Mental Powers.

§ 48. *Comparative Neurology.*

520. In what respects does the Nervous System of man differ from that of the lower orders of animals?
521. Compare the brain of other Mammals with that of man.
522. Compare that of Birds.
523. Of Reptiles.
524. What is said of the relative size of the Cerebrum? Of the Cerebellum, Medulla Oblongata and some of the organs of Special Sense?
525. Speak of the spinal cord and nerves.
526. Describe the brain of the Fish.
527. Describe the Torpedo.
528. Describe the Electric Eel. What is said of the structure and nervous system of the Articulata? What of them in the Centipede?

529. Speak of the nervous system in Mollusks.

530. Describe the nervous system in Radiata.

531. How is stimulus received in the lowest forms of animals? How is it perceived? As we ascend in the animal kingdom, what tissue appears first? What is the simplest type? Of what do the relations of the animal kingdom afford an evidence?

UNIFIC REVIEW.

[Compare 439–456 with 520–530 and 471, 472.]

Compare the Nervous System in man with that in the lower orders of animals.

[Compare 457, 458 with 10, 31, 32 and 36–38.]

Give the composition of Nervous Tissue. Describe its first element.

[Compare 459–462 with 33, 34 and 35.]

Describe the White and Gray Fibres. Where are they found?

[Compare 463–465 with 21, 22, 36, 37, 38 and 39.]

What membranes belong to the Cerebro-spinal System? What names do they assume there?

[Compare 469–474, 479 and 482 with 441, 442, 446, 455 and 456.]

Name the Nervous Centres. Give their functions. What do they comprise? Speak of the Sympathetic System.

[Compare 500–502 with 264–279 and 409–412.]

What is essential to the health of the nervous system? What is said of food and air in this connection?

[Compare 506 with 200–215.]

What can you say of the influence of physical exercise on the health of the nervous system?

SYNTHETIC REVIEW.

Nervous Tissue. Forms, " " Arrangement, Ganglia, Nerves and Commissures, Spinal Cord, Medulla Oblongata, Cerebellum, Peduncles, Cerebrum, " Corpora Striata, Optici Thalami, Corpora Quadrigesima, Corpus Callosum, Ventricles, Cerebrum Hemispheres, Convolutions, Cerebro-Spinal Nerves, Cranial Nerves, Spinal Nerves, Sympathetic System.	§ 44. *Anatomy of.*	Chap. XI. *Nervous System.*
Nervous Tissue, Composition, Nerve-Cells, Nerve-Fibres, Membranes,	§ 45. *Histology of.*	
Man's compound nature. Nervous System. Its relation to this nature, " " Its rank, Nervous Centres. Function, " " Classes, " " Arrangement, Organic Centres. Function, " " Connection, " " Modes of reflex action, " " Marked peculiarity, Reflex Centres. Function, " " Acquired action, " " Importance of acquired action, Sensational Centres. Character and action, " " How excited to activity, Ideational Centres. Function, Different persons have different ideas, Ideational Centres. Independent reflex action, " " Emotional character, " " Volitional, Relation of the Emotions to the Will, Free action of the Will, Influence of the body for good or evil, Language of the muscles.	§ 46. *Physiology of.*	
Agencies affecting the health, Natural heritage, Impure Air, influence of, Improper Diet, Poisons, Physical Exercise, want of, Sleep, Mental Exercise, Employment, Amount of exercise, Intense Activity, Recreation, Each faculty to be educated, The Æsthetic faculty, The Moral "	§ 47. *Hygiene of.*	
Mammals, Nervous System, Birds, " Reptiles, " Fishes, " Mollusks, " Radiata, " Lower forms of Life, "	§ 48. *Comparative Neurology of.*	

Fig. 217. Fig. 218.

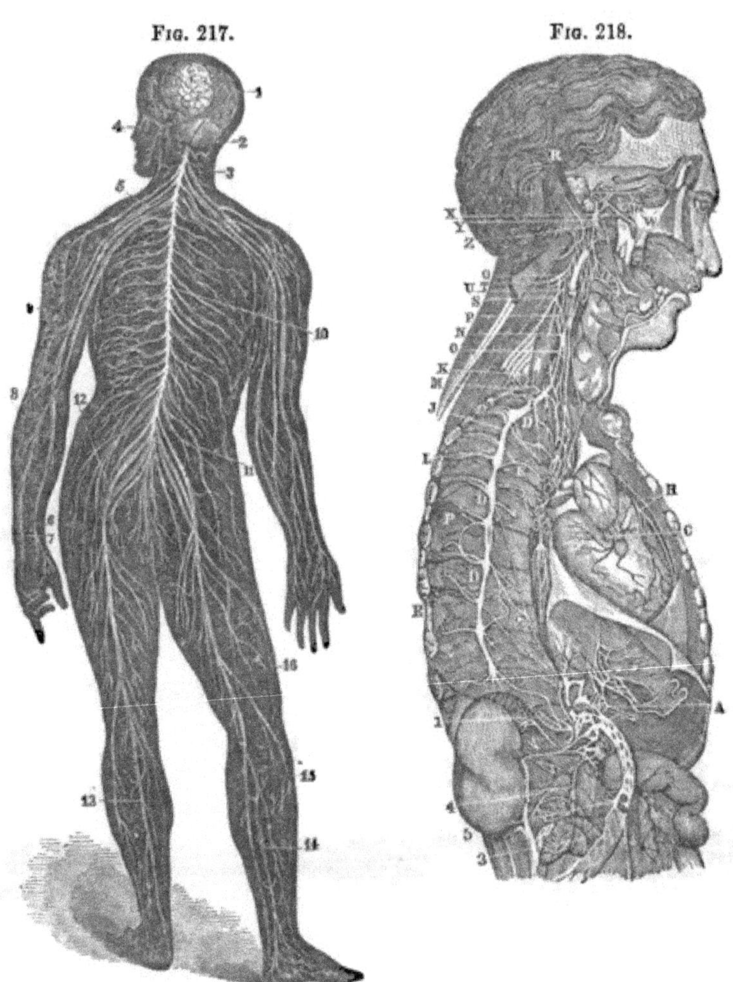

Fig. 217. A Back View of the Brain and Spinal Cord.—1, The cerebrum. 2, The cerebellum. 3, The spinal cord. 4, Nerves of the face. 5, The brachial plexus of nerves. 6, 7, 8, 9, Nerves of the arm. 10, Nerves that pass under the ribs. 11, The lumbar plexus of nerves. 12, The sacral plexus of nerves. 13, 14, 15, 16, Nerves of the lower limbs.

Fig. 218 represents the Sympathetic Ganglia, and their Connection with other Nerves, from the grand engraving of Manec, reduced in size. A, A, A, The semilunar ganglion and solar plexus, situated below the diaphragm and behind the stomach. This ganglion is situated in the region (pit of the stomach) where a blow gives severe suffering. D, D, D, The thoracic (chest) ganglia, ten or eleven in number. E, E, The external and internal branches of the thoracic ganglia. G, H, The right and left coronary plexus, situated upon the heart. I, N, Q, The inferior, middle and superior cervical (neck) ganglia. 1, The renal plexus of nerves that surrounds the kidneys. 2, The lumbar (loin) ganglion. 3, Their internal branches. 4, Their external branches. 5, The aortic plexus of nerves that lies upon the aorta. The other letters and figures represent nerves that connect important organs and nerves with the sympathetic ganglia.

FIG. 219.

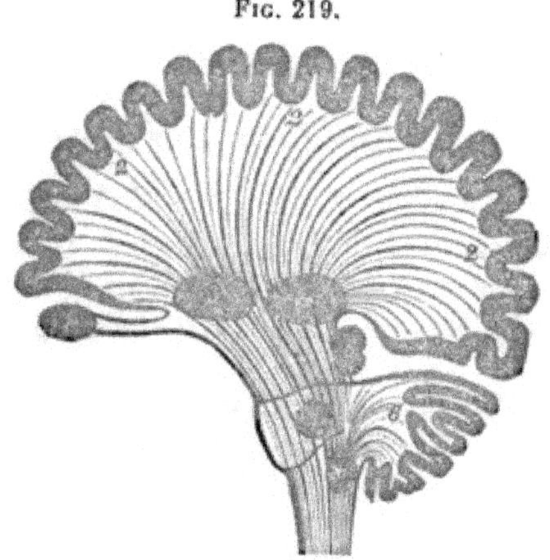

FIG. 219. DIAGRAM OF HUMAN BRAIN, IN VERTICAL SECTION, showing the situation of the different ganglia and the course of the fibres. 1, Olfactory ganglion. 2, Hemisphere. 3, Corpus striatum. 4, Optic thalamus. 5, Tubercula quadrigemina. 6, Cerebellum. 7, Ganglion of tuber annulare. 8, Ganglion of medulla oblongata.

FIG. 220. FIG. 221. FIG. 222.

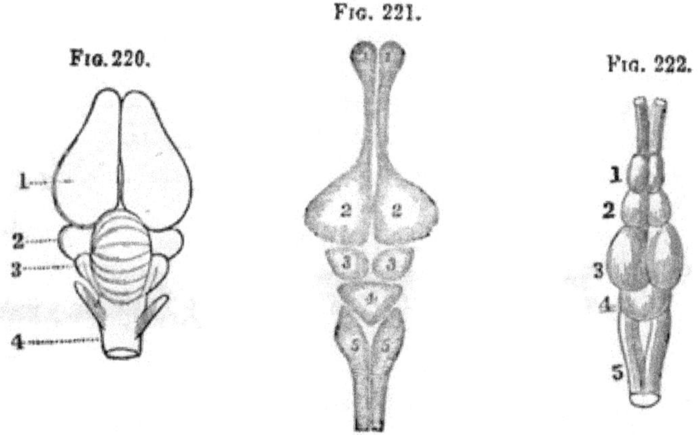

FIG. 220. BRAIN OF A BIRD.—1, Cerebrum. 2, Optic ganglion. 3, Cerebellum. 4, Medulla oblongata.

FIG. 221. BRAIN OF AN ALLIGATOR.—1, Olfactory ganglia. 2, Cerebrum. 3, Optic ganglia. 4, Cerebellum. 5, Medulla oblongata and spinal cord.

FIG. 222. BRAIN OF A FISH.—1, Olfactory ganglia. 2, Cerebrum. 3, Optic ganglia. 4, Cerebellum. 5, Medulla oblongata and spinal cord.

ANALYTIC EXAMINATION.

CHAPTER XII.—The Organs of Special Sense.

§ 49. *The Anatomy of the Organs of Special Sense.*

532. What is the organ of the sense of Taste? Give a description of the Tongue. From what nerves are filaments received?
533. Describe the organ of the sense of Smell. Mention the nerves.
534. What is the Eye? Name its parts. Of what service is the Sclerotica? Describe the Choroidea. What is its composition? Of what do the ciliary processes consist? What is said of the Iris? What is the Retina?
535. Describe the Aqueous Humor. Crystalline lens. Observation.
536. What is the Vitreous Humor? Distinguish between it and Aqueous Humor.
537. Speak of the muscles of the Eye. Observation.
538. What are the Orbits? Eyebrows? Eyelids? Give the Observation. Of what does the Lachrymal Apparatus consist? Where is the Lachrymal Gland situated? Describe the Lachrymal Canals. Nasal Duct.
539. What is said of the sense of Hearing?
540. Why the Labyrinth so called? Give its divisions.
541. Describe the Vestibule.
542. Describe the Semicircular Canals.
543. Speak of the Cochlea. Of the Fenestra Ovalis.
544. What is the Tympanum? Why called the Drum? Where is the Eustachian Tube? What are found in the tympanic cavity?
545. Describe the External Ear.
546. What is concerned in the Sense of Touch? Give its layers.
547. What is said of the Skin and its connection with the mucous membrane?
548. Give the relation of the Epidermis to the Dermis. What change does the Epidermis experience? What is the seat of color?
549. What is the Cuticle?
550. What is said of the Dermis? What are found with the fibrous and elastic tissues?

551. Describe the Papillæ.

552. Speak of the blood-vessels, nerves and lymphatics of the Cutis Vera.

553. Where are the Hair-Follicles? Describe the different parts of a hair. What results from the contraction of the unstriated muscular fibres?

554. Describe the Oil-Glands.

555. Where are the Sweat-Glands? What are "pores"? What is "insensible perspiration"?

556. Speak of the Nails. Of what is the horny part composed? How do they grow?

§ 50. *Physiology of the Organs of Special Sense.*

557. State the primary use of the sense of Taste. What is said of this sense in man? What is the effect of cultivation?

558. Is the sense of Smell one of great importance? Why not?

559. When light passes through different media, to what changes are its rays subject? What effect has convex or concave surfaces? Illustrate and apply the above principles.

560. Give the shape of those parts of the eye which act as media. State the use of so many lenses.

561. In what case will a more convex and in what a less convex lens be required? How is the eye able to change the convexity of its lenses and vary its focal distances?

562. What is the cause of short-sightedness and long-sightedness? What suggestion in the selection of glasses?

563. What is the function of the Sclerotic coat? What that of the pigment of the Choroid coat? How may the functions of some parts of the eye be beautifully shown?

564. Speak of the accessory parts of the eye. What enables the eye to move without friction? How are the eyelids drawn together? Give the functions of the Eyelashes and Eyebrows.

565. What is Hearing?

566. What is the function of the External Ear?

567. What that of the Auditory Canal? State the design of the Eustachian Tube. Give the uses of the Vestibule, Cochlea and Semicircular Canals.

568. What are distinguished by this sense? How does this apparatus compare with that of vision?

569. Speak of the special organ of the sense of Touch.

570. State the threefold functions of the skin.
571. Give the uses of the Epidermis.
572. What is said of the Cuticle?
573. Of what service are the cutaneous Papillæ?
574. Where does vitality reside? Why there?
575. What power does the surface of the skin possess?
576. What are the uses of the oil derived from the oil-glands?
577. State the uses of Perspiration. By what is the quantity influenced?
578. What is the influence of the condition of the atmosphere?
579. Give the functions of the Hair and Nails.

§ 51. *Hygiene of the Organs of Special Sense.*

580. What perverts the sense of Taste? By what is this sense varied?
581. By what does the sense of Smell become impaired?
582. What care is necessary in using the eye?
583. What is the effect of sudden transitions of light?
584. What should be avoided?
585. How should the eye of the child be trained?
586. What is beneficial? Observation.
587. Can the sense of Hearing be improved?
588. How may this sense be impaired? Observation.
589. What parts are absolutely essential, and what not?
590. To what must attention be given?
591. What is said of the use of clothing?
592. Of what material should it be? Compare furs, woolen cloth silk, cotton and linen.
593. Why should the clothing be porous and loosely fitted?
594. To what must it be suited? Observation.
595. Who require the more clothing?
596. What is said of clothing when a vital organ is diseased?
597. What persons need less clothing?
598. What is said of cleanliness of the clothing?
599. What of damp clothing?
600. What is indispensable to health?
601. What effect has bathing on the internal organs?
602. State the simplest mode of bathing.
603. Speak of the shallow bath.
604. Upon what must depend the frequency of bathing?
505. What should the time be?

606. In what diseases is bathing of *great* importance?
607. State the rules to be observed.
608. State the influence of pure air.
609. What influence does light exercise?
610. What is a blister? What care should be taken? How is vesication prevented?
611. What are Corns? From what comes the pain?
612. What is said of Frost-bite? How is Chilblain caused?

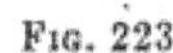
Fig. 223.

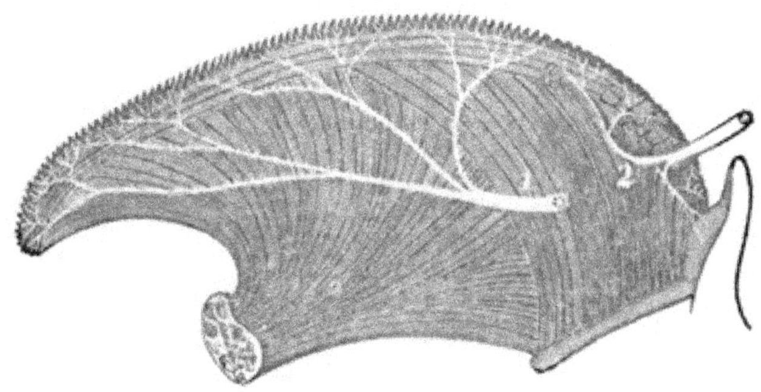

Fig. 223 (*Dalton*). Diagram of the Tongue, with its sensitive nerves and papillæ. 1, Lingual branch of fifth pair. 2, Glosso-pharyngeal nerve.

Fig. 224.

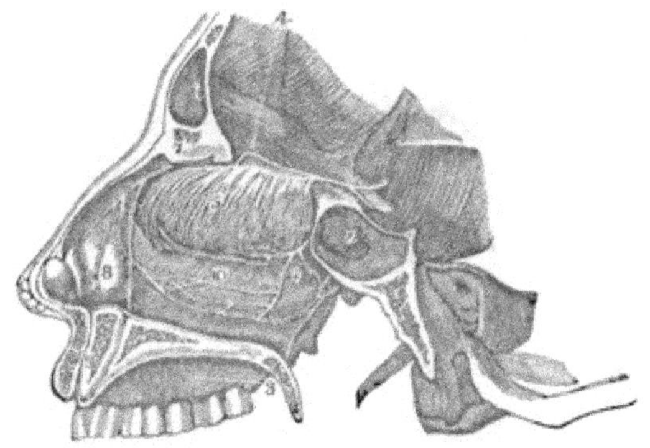

Fig. 224. A Side View of the Passage of the Nostrils, and the Distribution of the Olfactory Nerve.—4, The olfactory nerve. 5, The fine divisions of this nerve on the membrane of the nose. 6, A branch of the fifth pair of nerves.

SYNTHETIC REVIEW.

Topics	Section	Chapter
Organs of Taste, Smell and Sight, Sclerotica, Choroidea, Ciliary Processes, Iris, Retina, Aqueous, Crystalline and Vitreous Humors, Muscles of the Eye, Orbits, Eyebrows, Eyelids, Lachrymal Glands and Canals, Nasal Duct, Organs of Hearing, Labyrinth, Vestibule, Semicircular Canals, Cochlea, Tympanum, External Ear, Organs of Touch, Two layers of skin—Epidermis and Dermis, Hairs, Sebaceous and Respiratory Glands, Nails.	§ 49. *Anatomy of.*	Chap. XII. *The Organs of Special Sense.*
Sense of Taste, Primary use, " Smell, " Laws of Light, " Adaptation of the eye, Short-sightedness, Cause, Long-sightedness, " Defect remedied, Coats, Function, Accessory parts of the eye, Hearing, External Ear, Function, Auditory Canal, " Eustachian Tube, " Cochlea and Semicircular Canals, Function, Hearing, " Organ of Touch, Skin, Function, Epidermis and Cuticle, Function. Cutaneous Papillæ, Corium, Vessels, Oil-Glands, Function, Perspiration. Use, " Quantity, " External condition, Hair and Nails.	§ 50. *Physiology of.*	
Sense of Taste. Perversion, " Smell, " Eye, how to be used, Amaurosis, Oblique positions, long-continued, Viewing objects at different distances, Bathing the eye, Dust, removal, Defective Hearing, Cause, Hearing, parts essential, Clothing, Material, Class of persons needing more clothing, Clothing, Cleanliness, Bathing. Modes, " Time, " General Rules. Water a curative agent, Skin. Air beneficial, " Effect of light, Burns and Scalds, Treatment, Corns, Frost-Bite.	§ 51. *Hygiene of.*	

State the Anatomy, the Physiology and the Hygiene of the Organs of Special Sense, the Care of the Sick, of Poisoned Persons and of persons injured in any way.

Fig. 225.

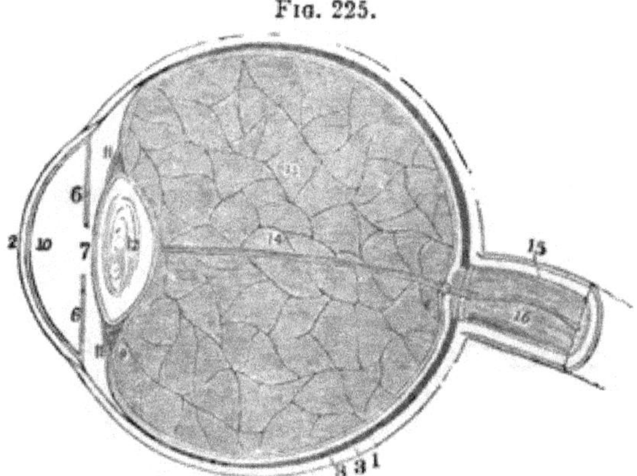

Fig. 225. A Section of the Globe of the Eye.—1, The sclerotic coat. 2, The cornea. (This connects with the sclerotic coat by a beveled edge.) 3, The choroid coat. 6, 6, The iris. 7, The pupil. 8, The retina. 10, 11, 11, Chambers of the eye that contain the aqueous humor. 12, The crystalline lens. 13, The vitreous humor. 15, The optic nerve. 16, The central artery of the eye.

Fig. 226.

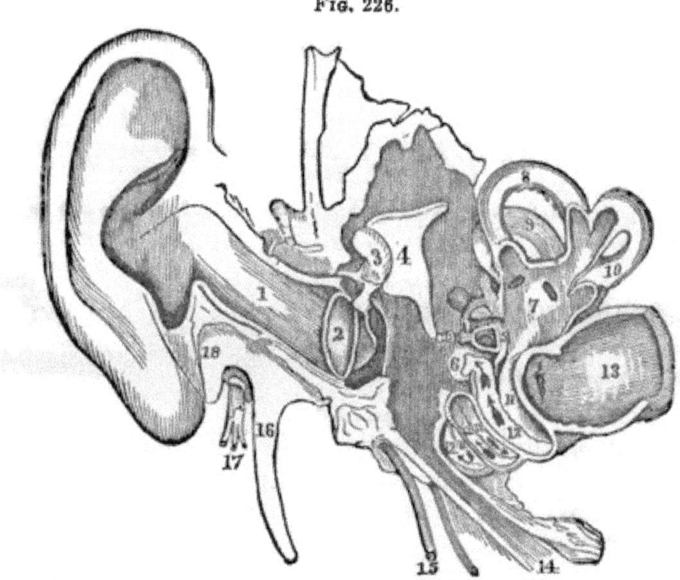

Fig. 226. A View of all the Parts of the Ear.—1, The canal that leads to the internal ear. 2, The membrana tympani. 3, 4, 5, The bones of the ear. 7, The central part of the labyrinth (vestibule). 8, 9, 10, The semicircular canals. 11, 12, The channels of the cochlea. 13, The auditory nerve. 14, The opening from the middle ear, or tympanum, to the throat (Eustachian tube).

APPENDIX.

CHAPTER XIII.—Care of the Sick.

§ 1. *The Nurse.*

[Compare 591–599.]

Cleanliness.—What regard should be had for cleanliness?

[Compare 601–608.]

Bathing.—Mention what is said respecting bathing.

[Compare 264–286.]

Food and Drink.—What is said of the food and drink of the sick? Name the means of nourishment, and tell how they are prepared.

Temperature.—Speak of the temperature of the sick-room.

Light.—What suggestions are made as to light?

Quiet.—How may quiet be had? Mention other duties of the nurse.

§ 2. *The Watcher.*

Give the duties of the Watcher.

§ 3. *Poisons and their Antidotes.*

When poisons have been taken, what is to be done? Name the most common poisons, and their antidotes.

[Compare 363.]

How can hæmorrhage be arrested?

[Compare 364.]

Give the manner of dressing wounds.

[Compare 430.]

How may asphyxiated persons be recovered?

[Compare 610–612.]

Speak of Burns, Scalds and Frost-Bite, and their treatment.

DIVISION IV.—SENSORIAL APPARATUS.

SYNTHETIC REVIEW.

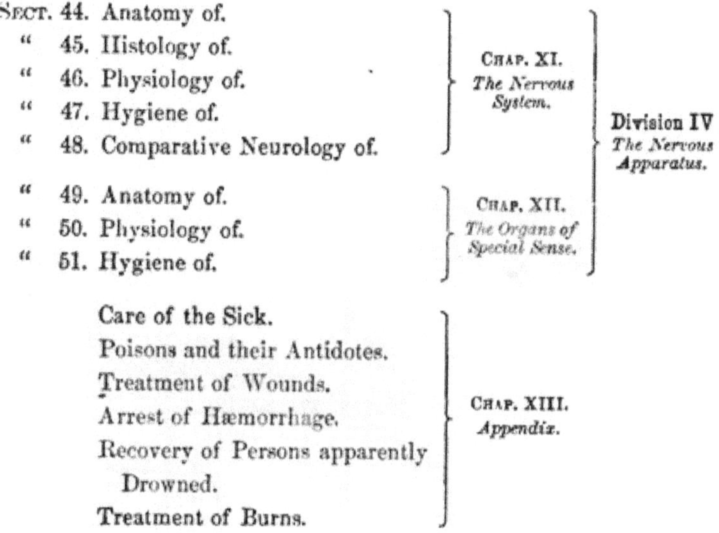

SECT. 44. Anatomy of.	CHAP. XI. *The Nervous System.*	Division IV *The Nervous Apparatus.*
" 45. Histology of.		
" 46. Physiology of.		
" 47. Hygiene of.		
" 48. Comparative Neurology of.		
" 49. Anatomy of.	CHAP. XII. *The Organs of Special Sense.*	
" 50. Physiology of.		
" 51. Hygiene of.		
Care of the Sick.	CHAP. XIII. *Appendix.*	
Poisons and their Antidotes.		
Treatment of Wounds.		
Arrest of Hæmorrhage.		
Recovery of Persons apparently Drowned.		
Treatment of Burns.		

State the Anatomy, the Histology, the Physiology and the Hygiene, Human and Comparative, of the Nervous Apparatus, and the Care of the Sick, Poisons and their Antidotes, Treatment of Wounds, Hæmorrhage, Burns, and persons apparently drowned.

SUMMARY.—SYNTHETIC REVIEW.

Sect. 1. The Three Kingdoms of Nature Compared. " 2. Definitions.	Chap. I. *General Remarks.*	Division I. *Outline Principles.*	Mammals.
" 3. Cells. " 4. Tissues. " 5. Membranes.	Chap. II. *General Histology.*		
" 6. Solids and Fluids.	Chap. III. *General Chemistry.*		
" 7. Anatomy of. " 8. Histology of. " 9. Chemistry of. " 10. Physiology of. " 11. Hygiene of. " 12. Comparative Osteology.	Chap. IV. *The Bones.*	Division II. *Motory Apparatus.*	
" 13. Anatomy of. " 14. Histology of. " 15. Chemistry of. " 16. Physiology of. " 17. Hygiene of. " 18. Comparative Myology.	Chap. V. *The Muscles.*		
" 19. Anatomy of. " 20. Histology of. " 21. Chemistry of. " 22. Physiology of. " 23. Hygiene of. " 24. Comparative Splanchnology.	Chap. VI. *The Digestive Organs.*	Division III. *Nutritive Apparatus.*	
" 25. Anatomy of. " 26. Histology of. " 27. Chemistry of. " 28. Physiology of. " 29. Hygiene of.	Chap. VII. *The Absorbents.*		
" 30. The Blood. " 31. Anatomy of. " 32. Histology of. " 33. Chemistry of. " 34. Physiology of. " 35. Hygiene of. " 36. Comparative Angiology.	Chap. VIII. *The Circulation.*		
" 37. Assimilation, General and Specific.	Chap. IX. *Assimilation.*		
" 38. Anatomy of. " 39. Histology of. " 40. Chemistry of. " 41. Physiology of. " 42. Hygiene of. " 43. Comparative Pneumonology.	Chap. X. *The Organs of Respiration.*		
" 44. Anatomy of. " 45. Histology of. " 46. Physiology of. " 47. Hygiene of. " 48. Comparative Neurology.	Chap. XI. *The Nervous System.*	Division IV. *Nervous Apparatus.*	
" 49. Anatomy of. " 50. Physiology of. " 51. Hygiene of.	Chap. XII. *The Organs of Special Sense.*		
Care of the Sick. Poisons and Antidotes. Treatment of Wounds, Hæmorrhage, of apparently Drowned Persons and of Burns.	Chap. XIII *Appendix.*		

State succinctly the Anatomy, the Histology, the Chemistry, the Physiology and the Hygiene of Mammals.

www.ingramcontent.com/pod-product-compliance
Lightning Source LLC
Chambersburg PA
CBHW021353210326
41599CB00011B/861
9783337365349